C
P
L

ELASTODYNAMICS

VOLUME I

Finite Motions

ELASTODYNAMICS

VOLUME I

Finite Motions

A. Cemal Eringen

School of Engineering and Applied Science
Princeton University
Princeton, New Jersey

Erdoğan S. Şuhubi

Faculty of Science
Technical University of Istanbul
Istanbul, Turkey

ACADEMIC PRESS New York and London 1974
A Subsidiary of Harcourt Brace Jovanovich, Publishers

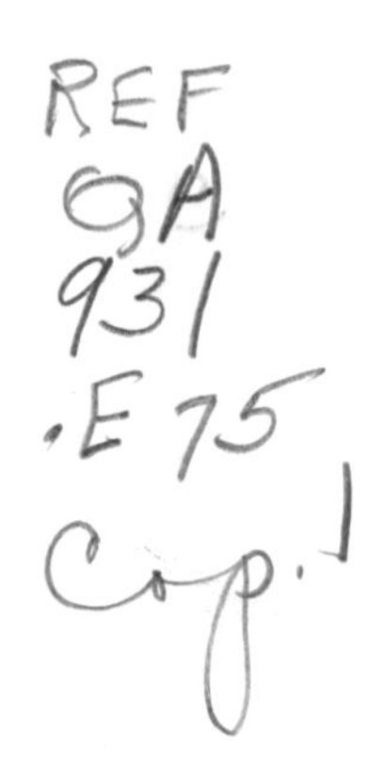

ACADEMIC PRESS, INC.
111 Fifth Avenue, New York, New York 10003

United Kingdom Edition published by
ACADEMIC PRESS, INC. (LONDON) LTD.
24/28 Oval Road, London NW1

Library of Congress Cataloging in Publication Data

Eringen, A. Cemal
Elastodynamics.

Bibliography: v. 1, p.
CONTENTS: v. 1. Finite motions.
1. Elasticity. 2. Elastic solids. 3. Dynamics.
I. Suhubi, E. S., 1934- joint author. II. Title.
QA931.E75 531'.3825 73-18987
ISBN 0-12-240601-X

PRINTED IN THE UNITED STATES OF AMERICA

To Jean and Birsen

Contents

Chapter II. **Propagation of Singular Surfaces**

Chapter III. **Finite Motions of Elastic Bodies**

Chapter IV. **Small Motions Superimposed on Large Static Deformations**

Appendix A. **Tensor Analysis**

Preface

From the inception of the theory of elasticity with Navier (1821) and Cauchy (1828), the dynamical problems of elasticity and the subject of wave propagations in elastic solids have been under intense study by a large number of workers. In fact the literature is so extensive that any desire to write accounts on the subject immediately induces discouragement on the part of a prospective author. For it is impossible to do justice to all aspects of this wide field in any one- or two-volume treatise. Perhaps, partially, it is this concern that kept this important field barren of books for nearly two centuries.

During the last two decades the research in dynamical problems has acquired even stronger emphasis. The demands for economical uses of materials in space technology, introductions of composite materials for industrial application, underground explosions, and important problems in seismology and the oil industry have stimulated intense research in finite deformation theories, propagation of shock waves in solids, diffraction theory, dynamic stress concentrations, and wave propagations in inhomogenous and anisotropic materials. Nearly all graduate studies in engineering science, mechanics, applied mathematics, and geophysics include one or more graduate courses on the subject. Yet, excluding a few special accounts (mostly out of print), at the completion of this work there was no textbook or treatise available to teach and to guide research.

The origin of the present work goes back to 1954, when one of the authors (A.C.E.) was engaged in teaching a graduate course in linear elastodynamics. Over the years mimeographed notes were written, endlessly revised, and distributed to students and interested colleagues. A book was also planned on the subject, in 1960; however, it proved to be too diverse in some sense and naturally not inclusive of the major developments to come. The present two volumes have been reorganized afresh and rewritten in their entirety to include some of the major developments of the last decade and other earlier ones. These books are intended hopefully to remedy the need for a rigorous study of the mathematical theory of elastodynamics.

Excluding Chapter I, which constitutes the development of the fundamental equations, each volume can be studied independently. Those who had a first course in continuum mechanics may omit Chapter I and study the theory of singular surfaces, shock waves, finite amplitude oscillations, and propagations of waves in deformed and stressed elastic solids in Volume I or the linear theory in Volume II. One- or two-semester courses can be given on either volume depending on the need. We believe that these volumes may be suitable for use in any one of the current engineering departments, and in the departments of mechanics, geophysics, geology, applied mathematics, applied physics, and engineering science where a rigorous program exists for the cultivation of mathematical and physical foundation of the wave propagation theory.

Volume I consists of four chapters and two appendices. Chapter I deals with the development of the fundamental equations of the dynamical theory of thermoelasticity. The concepts relevant to finite deformations, strains, rotations, their invariants, length, area, volume changes, compatibility conditions of strain, kinematics of continua, global and local balance laws, thermodynamics of continua, and constitutive theory are discussed concisely. The field equations are obtained for the exact nonlinear theory and the linear and quadratic theories. Chapter II deals with the propagation of singular surfaces. Geometrical, kinematical, and dynamical conditions are obtained and used for the study of shock waves and acceleration waves. In Chapter III we present the solution of various problems for the finite motions of elastic bodies: radial oscillations of cylinders and spheres and a discussion of the simple wave theory. Chapter IV is devoted to a study of small motions superimposed on large static deformations, a topic which is fundamental to stability theory. Appendices A and B on tensor analysis and the theory of characteristics provide the necessary background for study of Volume I.

Volume II presents a thorough study of the mathematical problems relevant to the linear theory. Various fundamental theorems (uniqueness, reciprocal, integral representations, variational principles, etc.) are discussed in Chapter V. In Chapters VI–VIII there are presented a plethora of exact solutions of significant problems in one, two, and three space dimensions. In Chapter IX we give a concise account of diffraction theory.

The present volumes omit many important topics, such as anisotropic solids, mixtures, and thermal effects, even though, for future purposes, the fundamentals of thermoelasticity are fully developed. Only exact solutions are presented, thus omitting a large number of approximate and numerical treatments. We believe, however, that those who are equipped with the fundamental theory and the mathematical methods for which these volumes are intended can go on dealing with approximations and numerical techniques. The theory being well founded and verified over a century and a half, we did not include any discussion on experimental work.

Acknowledgments

During the writing of these volumes, many students and colleagues have provided help. With great pleasure we mention Drs. J. W. Dunkin and N. F. Jordan, who read and checked various parts of the early versions of Chapters VI and VII. Miss G. Segol, Mr. S. Dost, and Mr. M. Teymur read and checked part of the first draft. Drs. H. Demiray, W. D. Claus, Jr., J. D. Lee, P. O'Leary, and others helped in making Xeroxes of many references.

The authors would like to express their gratitude and thanks to the Department of Aerospace and Mechanical Sciences of Princeton University, the Faculty of Civil Engineering of the Technical University of Istanbul, and to the Office of Naval Research for providing an opportunity for one of the authors (E.S.Ş.) to spend two years at Princeton as a visiting professor.

Valuable typing and reproduction services were provided by Princeton University and Esso Production Research Company. The burden of this work fell upon Mrs. S. Frazier and Miss P. Halliday.

CHAPTER 1

Basic Theory

1.1 SCOPE OF THE CHAPTER

This chapter is devoted to the derivation of basic equations of thermoelastic solids. To this end we study the geometry of deformation, kinematics, and balance laws of bodies, irrespective of their shapes and constitutions, and construct a set of constitutive equations for thermoelastic solids through thermodynamical and other constitutive axioms. Later, the field equations are obtained for linear and nonlinear solids.

In Sections 1.2 and 1.3 we introduce coordinates and discuss the motion in general terms. In Section 1.4, deformation gradients and deformation and strain measures, basic to all deformable bodies, are introduced. Through a discussion of length and angle changes, in Section 1.5 we give the geometrical significance of strain tensors. Various strain invariants are presented and related to each other in Section 1.6. In Section 1.7 we calculate the area and volume changes in arbitrary deformations of a body, and in Section 1.8 we derive the compatibility conditions of strain. The kinematics of continua and the rate measures are presented in Section 1.9. The global balance laws, conservation of mass, balance of momenta, conservation of energy, and the axiom of entropy are the subject of Section 1.10. Their local forms including the jump conditions over a moving discontinuity surface are obtained in Section 1.11. In Section 1.12 we obtain the expressions of local balance laws in the reference frame, expressions which are often more convenient in the study of nonlinear problems. Thus far no reference has been made to the constitution of bodies, so all derived results are valid for all bodies. With Section 1.13 we begin the discussion of the constitutive equations of thermoelastic solids. The exact forms of these equations restricted by thermodynamics, objectivity, and material symmetry are derived. In Section 1.14 various forms of the constitutive equations are given for isotropic solids. In Sections 1.15 and 1.16 we obtain the linear and quadratic approximations.

The field equations of thermoelastic solids are the union of balance laws and the constitutive equations. The combined results for the nonlinear and linear theories are found in Section 1.17. Typical boundary–initial value problems are posed. In Section 1.18 we discuss the restrictions, arising from physical considerations, that may be placed on the material moduli. In Section 1.19 we give a summary of the crucial equations necessary for the solutions of problems of elastodynamics. The final section of this chapter (Section 1.20) gives a method by which passage can be made to curvilinear coordinates. Specific forms of the balance laws and the field equations are presented for general orthogonal curvilinear coordinates and particular cylindrical and spherical coordinates.

1.2 COORDINATES

The material points of a body at a fixed time $t = 0$ occupy a region B consisting of the material volume V and its surface S. The position of a material point P in B may be denoted by a rectangular coordinate system X_1, X_2, X_3 (or simply by X_K, $K = 1, 2, 3$). Alternatively a vector $\mathbf{P}$ (or equivalently $\mathbf{X}$) that extends from an origin 0 to the point P may be used (see Fig. 1.2.1). After deformation takes place at time t, the material points of $V + S$ occupy a *spatial region* $\mathscr{B}$ which consists of a volume $\mathscr{V}$ having surface $\mathscr{S}$. In this deformed state a material point may be located by its rectangular coordinates x_k, $k = 1, 2, 3$. Sometimes, it is advantageous to select these two coordinate systems nonidentical. The choice of two different coordinate systems, one for the undeformed body and one for the deformed body, is particularly suitable in the case when curvilinear coordinates are used. For example, when a rectangular block is deformed into a circular cylinder, the use of rectangular coordinates for the undeformed block and cylindrical coordinates for the deformed body may prove particularly advantageous, especially in expressing the boundary conditions. The use of two sets of coordinates has other advantages in making many subtle kinematical problems clear.

When explicit components of equations are desired, we write $X_1 = X$, $X_2 = Y$, $X_3 = Z$, and $x_1 = x$, $x_2 = y$, $x_3 = z$. The coordinates X_K are called

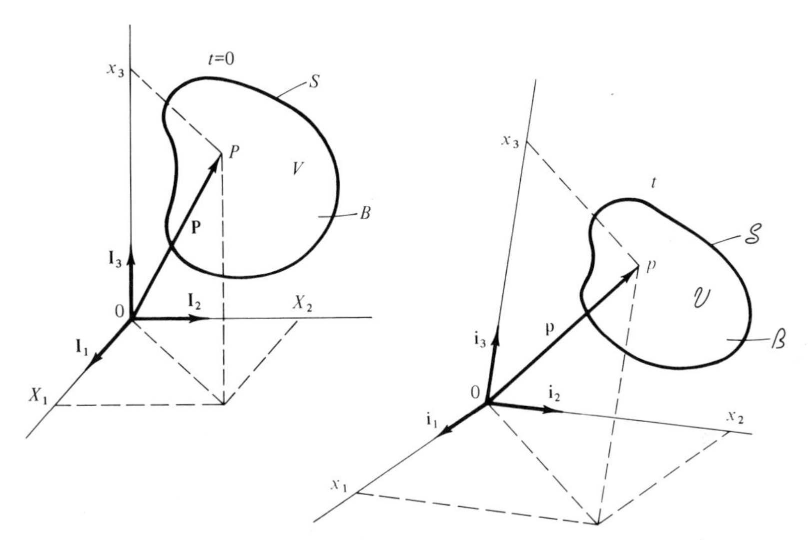

Fig. 1.2.1 Coordinate system for an undeformed body B and a deformed body $\mathscr{B}$.

the *material* or *Lagrangian* coordinates and x_k the *spatial* or *Eulerian* coordinates.

The rectangular *base vectors* $\mathbf{I}_K$ and $\mathbf{i}_k$ are unit vectors lying along the coordinates X_K and x_k, respectively. The position vectors $\mathbf{P}$ and $\mathbf{p}$ of the points P and p are given by

$$\mathbf{P} = X_K \mathbf{I}_K, \qquad \mathbf{p} = x_k \mathbf{i}_k \tag{1.2.1}$$

where, and henceforth, the usual summation convention is understood over the repeated indices, e.g.,

$$\mathbf{P} = X_K \mathbf{I}_K = X_1 \mathbf{I}_1 + X_2 \mathbf{I}_2 + X_3 \mathbf{I}_3$$

The infinitesimal differential vectors $d\mathbf{P}$ at P and $d\mathbf{p}$ at p are given by

$$d\mathbf{P} = dX_K \mathbf{I}_K, \qquad d\mathbf{p} = dx_k \mathbf{i}_k \tag{1.2.2}$$

Thus the squares of the arc length in B and $\mathscr{B}$ are, respectively,

$$\begin{aligned} dS^2 &= d\mathbf{P} \cdot d\mathbf{P} = \delta_{KL}\, dX_K\, dX_L = dX_K\, dX_K \\ ds^2 &= d\mathbf{p} \cdot d\mathbf{p} = \delta_{kl}\, dx_k\, dx_l = dx_k\, dx_k \end{aligned} \tag{1.2.3}$$

where

$$\delta_{KL} = \mathbf{I}_K \cdot \mathbf{I}_L, \qquad \delta_{kl} = \mathbf{i}_k \cdot \mathbf{i}_l$$

are the *Kronecker symbols*, which are equal to unity when the two indices are equal and zero otherwise.

Sometimes a vector associated with $\mathscr{B}$ is projected on the rectangular base vectors of B and vice versa. Thus, for example,

$$p_K \equiv \mathbf{p} \cdot \mathbf{I}_K = x_k \mathbf{i}_k \cdot \mathbf{I}_K = x_k \, \delta_{kK}$$

is the parallel projection of $\mathbf{p}$ on $\mathbf{I}_K$, where

$$\delta_{kK} = \delta_{Kk} \equiv \mathbf{i}_k \cdot \mathbf{I}_K \tag{1.2.4}$$

is *not* a Kronecker symbol except when $\mathbf{I}_K$ and $\mathbf{i}_k$ are coincident. Quantities defined by (1.2.4) are called *shifters*. They play an important role in expressing a vector in one frame in terms of its projections in another frame. It is clear that (1.2.4) is none other than the cosine directors of the two frames of reference x_k and X_K.

Similarly, for any vector we may write

$$\mathbf{v} = v_k \mathbf{i}_k = V_K \mathbf{I}_K \tag{1.2.5}$$

By taking scalar products of this by $\mathbf{I}_L$ and $\mathbf{i}_l$ we find for the components V_K of $\mathbf{v}$ in X_K and the components v_k of $\mathbf{v}$ in x_k

$$V_K = \delta_{Kk} v_k, \qquad v_k = \delta_{kK} V_K \tag{1.2.6}$$

By carrying one of these into the other, we find that

$$\delta_{Kk} \, \delta_{kL} = \delta_{KL}, \qquad \delta_{kK} \, \delta_{Kl} = \delta_{kl} \tag{1.2.7}$$

Note that for the components of $\mathbf{v}$ in material coordinates we employ capital letters with capital indices, and in spatial coordinates lower case letters with lower case indices. This convention is used throughout the book.

1.3 DEFORMATION, MOTION

Under the influence of the external loads, the body B moves and deforms. The motion carries the material points of B to new spatial positions. This is expressed by a one-parameter family of mappings

$$x_k = x_k(X_1, X_2, X_3, t) \quad \text{or} \quad x_k = x_k(X_K, t), \qquad k = 1, 2, 3 \tag{1.3.1}$$

or, alternatively,

$$X_K = X_K(x_1, x_2, x_3, t) \quad \text{or} \quad X_K = X_K(x_k, t), \qquad K = 1, 2, 3 \tag{1.3.2}$$

For brevity, we may also write these as

$$\mathbf{x} = \mathbf{x}(\mathbf{X}, t), \qquad \mathbf{X} = \mathbf{X}(\mathbf{x}, t) \tag{1.3.3}$$

Equation (1.3.1) states that at time t a material point X_K of B occupies the spatial position x_k in $\mathscr{B}$. The inverse motion (1.3.2) states the inverse phenomenon, namely, that at time t the material point occupying the spatial position x_k can be traced back to its original position X_K.

We assume that (1.3.2) is the unique inverse of (1.3.1) in a neighborhood of X_K, and conversely that (1.3.1) is a unique inverse of (1.3.2) in a neighborhood of x_k corresponding to this. This is secured by the well-known implicit function theorem (cf. Widder [1947, p. 47]).

Theorem *(Implicit Function). If, for a fixed time t, the functions $x_k(X_K, t)$ are continuous and possess continuous first-order partial derivatives, with respect to X_K in a neighborhood $|X_K^1 - X_K| < \Delta$ of the point P, and if the Jacobian*

$$J \equiv \det\left(\frac{\partial x_k}{\partial X_K}\right) = \begin{vmatrix} \partial x_1/\partial X_1 & \partial x_1/\partial X_2 & \partial x_1/\partial X_3 \\ \partial x_2/\partial X_1 & \partial x_2/\partial X_2 & \partial x_2/\partial X_3 \\ \partial x_3/\partial X_1 & \partial x_3/\partial X_2 & \partial x_3/\partial X_3 \end{vmatrix} \tag{1.3.4}$$

does not vanish there, then a unique inverse of the form (1.3.2) *exists in a neighborhood of $|x_k^1 - x_k| < \delta$ of a point p.*

Henceforth we assume that these conditions are satisfied for (1.3.1) and (1.3.2) to be unique inverses of each other throughout the entire regions occupied by the body at all times, except possibly at some singular surfaces, lines, and points. In fact, we assume that (1.3.3) are single valued, continuous, and possess continuous partial derivatives with respect to their arguments for whatever order is desired. Whenever some singular submanifold exists in which this assumption is violated, a separate discussion will be provided.

The previous assumption is called the *axiom of continuity*. It expresses the notion that matter is *indestructible*. No region of a positive, finite volume of matter goes into zero or infinite volume. Moreover, matter is also *impenetrable*, that is, motion carries every region into a region, every surface into a surface, and every curve into a curve. No portion of matter can penetrate into another. In practice, there are examples in which this axiom is violated. For example, the materials may fracture or transmit shock waves or other types of discontinuities. Special attention must be given to these cases.

Given the initial shape of the body, if we determine the function $x_k(\mathbf{X}, t)$ we can construct the position of all material points of the body at time t. This enables one to calculate the length change between any two points and angle change between any two directions. The ultimate goal of the theory is to relate these changes to external loads (external forces, moments, thermal changes, etc.). With this knowledge one can then design machines and buildings or analyze existing structures so that not only will failure be avoided but also efficient use will be made of materials. Thus the ultimate goal of the theory is to determine (1.3.1) when the external effects and initial and boundary conditions of the body are known.

1.4 DEFORMATION GRADIENTS, DEFORMATION, AND STRAIN TENSORS

From (1.3.1) and (1.3.2), for fixed time, we have

$$dx_k = x_{k,K}\, dX_K, \qquad dX_K = X_{K,k}\, dx_k \tag{1.4.1}$$

where indices following a comma represent partial differentiation with respect to X_K when they are magiscules, and with respect to x_k when they are miniscules, i.e.,

$$x_{k,K} \equiv \partial x_k / \partial X_K, \qquad X_{K,k} \equiv \partial X_K / \partial x_k \tag{1.4.2}$$

The set of quantities defined by (1.4.2) is called *deformation gradients.* By the chain rule of differentiation, we have

$$x_{k,K}\, X_{K,l} = \delta_{kl}, \qquad X_{K,k}\, x_{k,L} = \delta_{KL} \tag{1.4.3}$$

Each of these two sets consists of nine linear equations for the nine unknowns $x_{k,K}$ or $X_{K,k}$. Since the Jacobian is assumed not to vanish, a unique solution exists and, according to Cramer's rule of determinants, is given by

$$X_{K,k} = \frac{1}{j} \operatorname{cofactor}(x_{k,K}) = \frac{1}{2j} e_{KLM}\, e_{klm}\, x_{l,L}\, x_{m,M} \tag{1.4.4}$$

where e_{KLM} and e_{klm} are the permutation symbols, and

$$j \equiv \det(x_{k,K}) = \frac{1}{3!} e_{KLM}\, e_{klm}\, x_{k,K}\, x_{l,L}\, x_{m,M} \tag{1.4.5}$$

By differentiating (1.4.4) and (1.4.5) we get two important identities:

$$(j X_{K,k})_{,K} = 0, \qquad \partial j / \partial x_{k,K} = \text{cofactor } x_{k,K} = j X_{K,k} \tag{1.4.6}$$

of which the latter is attributed to Jacobi.

Substituting (1.4.1) into (1.2.3) we have

$$dS^2 = c_{kl}\, dx_k\, dx_l, \qquad ds^2 = C_{KL}\, dX_K\, dX_L \tag{1.4.7}$$

where

$$c_{kl}(\mathbf{x}, t) \equiv \delta_{KL}\, X_{K,k}\, X_{L,l}, \qquad C_{KL}(\mathbf{X}, t) = \delta_{kl}\, x_{k,K}\, x_{l,L} \tag{1.4.8}$$

are, respectively, *Cauchy's deformation tensor* and *Green's deformation tensor.* Two other equally important tensors are the reciprocal tensors $\overset{-1}{c}_{kl}$ and $\overset{-1}{C}_{KL}$ (known as the *Finger* and *Piola deformation tensors*, respectively) defined by

$$\overset{-1}{c}_{kl}(\mathbf{x}, t) \equiv \delta_{KL}\, x_{k,K}\, x_{l,L}, \qquad \overset{-1}{C}_{KL}(\mathbf{X}, t) \equiv \delta_{kl}\, X_{K,k}\, X_{L,l} \tag{1.4.9}$$

which can be shown to satisfy

(1.4.10) $$\overset{-1}{c}_{kl}\, c_{lm} = \delta_{km}, \qquad \overset{-1}{C}_{KL}\, C_{LM} = \delta_{KM}$$

by mere substitution of (1.4.8) and (1.4.9).

The *Lagrangian* and *Eulerian* strain tensors are defined, respectively, by

(1.4.11) $$E_{KL} \equiv \tfrac{1}{2}(C_{KL} - \delta_{KL}), \qquad e_{kl} \equiv \tfrac{1}{2}(\delta_{kl} - c_{kl})$$

Carrying C_{KL} and c_{kl} from these into (1.4.8), we obtain

(1.4.12) $$ds^2 - dS^2 = 2E_{KL}\, dX_K\, dX_L = 2e_{kl}\, dx_k\, dx_l$$

From this it also follows that

(1.4.13) $$E_{KL} = e_{kl}\, x_{k,K}\, x_{l,L}, \qquad e_{kl} = E_{KL}\, X_{K,k}\, X_{L,l}$$

which exhibit the fact that both E_{KL} and e_{kl} are second-order absolute tensors.

When the body undergoes only a *rigid displacement* there will be no change in the differential length, in which case the difference $ds^2 - dS^2$ given by (1.4.12) vanishes. If this is true for all directions dX_K and dx_k, then E_{KL} and e_{kl} vanish. Therefore, these tensors represent a measure of deformation of the body.

We may express the strain tensors in terms of the *displacement vector* **u** that extends from a material point P on the undeformed body to its spatial location p at time t (Fig. 1.4.1):

(1.4.14) $$\mathbf{u} \equiv \mathbf{p} - \mathbf{P} + \mathbf{b} = x_l \mathbf{i}_l - X_L \mathbf{I}_L + \mathbf{b}$$

The displacement vector may be expressed in terms of its Lagrangian and Eulerian components U_K and u_k as

$$\mathbf{u} = U_L \mathbf{I}_L = u_l \mathbf{i}_l$$

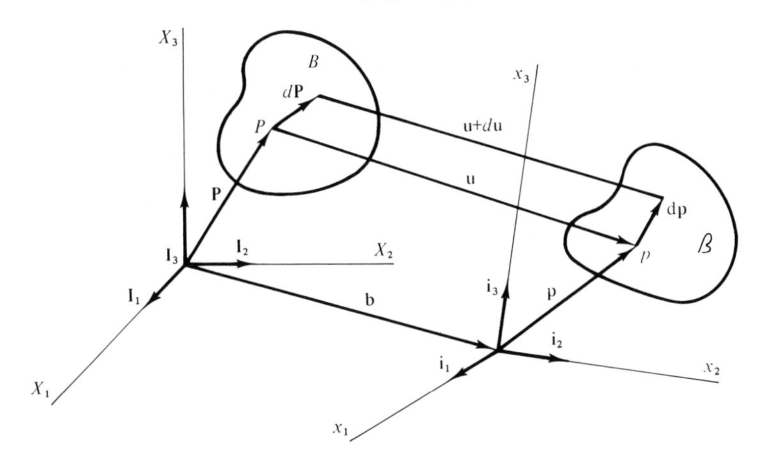

Fig. 1.4.1 Displacement vector.

Taking the inner product of both sides of (1.4.14) by $\mathbf{i}_k$ and $\mathbf{I}_K$ we obtain

$$u_k = x_k - \delta_{kL} X_L + b_k, \qquad U_K = \delta_{Kl} x_l - X_K + B_K \tag{1.4.15}$$

where $b_k \equiv \mathbf{b} \cdot \mathbf{i}_k$ and $B_K \equiv \mathbf{b} \cdot \mathbf{I}_K$ are the components of $\mathbf{b}$. From (1.4.15) we have

$$dx_k = (\delta_{MK} + U_{M,K})\, \delta_{Mk}\, dX_K, \qquad dX_K = (\delta_{mk} - u_{m,k})\, \delta_{mK}\, dx_k \tag{1.4.16}$$

Upon substituting $x_{k,K}$ and $X_{K,k}$, obtained from these, into (1.4.8) we find

$$\begin{aligned} c_{kl} &= \delta_{kl} - 2e_{kl} = \delta_{kl} - u_{k,l} - u_{l,k} + u_{m,k} u_{m,l} \\ C_{KL} &= \delta_{KL} + 2E_{KL} = \delta_{KL} + U_{K,L} + U_{L,K} + U_{M,K} U_{M,L} \end{aligned} \tag{1.4.17}$$

which give the strain tensors in terms of the displacement gradients.

The strain tensors, as well as deformation tensors, are symmetric tensors, i.e.,

$$\begin{aligned} e_{kl} &= e_{lk}, \qquad E_{KL} = E_{LK} \\ c_{kl} &= c_{lk}, \qquad C_{KL} = C_{LK} \end{aligned} \tag{1.4.18}$$

These tensors may be arranged in matrix form, e.g.,

$$\|E_{KL}\| = \begin{bmatrix} E_{11} & E_{12} & E_{13} \\ E_{21} & E_{22} & E_{23} \\ E_{31} & E_{32} & E_{33} \end{bmatrix} \tag{1.4.19}$$

This matrix is symmetric with respect to its main diagonal. The entries E_{11}, E_{22}, and E_{33} placed on the main diagonal of (1.4.19) are called the *normal strains*, and other entries are called the *shear strains.*

Finally, we write down explicit expressions of all components of e_{kl} to see the advantage of the present notation:

$$\begin{aligned} 2e_{xx} &= 1 - c_{xx} = 2\frac{\partial u}{\partial x} - \left(\frac{\partial u}{\partial x}\right)^2 - \left(\frac{\partial v}{\partial x}\right)^2 - \left(\frac{\partial w}{\partial x}\right)^2 \\ 2e_{yy} &= 1 - c_{yy} = 2\frac{\partial v}{\partial y} - \left(\frac{\partial u}{\partial y}\right)^2 - \left(\frac{\partial v}{\partial y}\right)^2 - \left(\frac{\partial w}{\partial y}\right)^2 \\ 2e_{zz} &= 1 - c_{zz} = 2\frac{\partial w}{\partial z} - \left(\frac{\partial u}{\partial z}\right)^2 - \left(\frac{\partial v}{\partial z}\right)^2 - \left(\frac{\partial w}{\partial z}\right)^2 \\ 2e_{xy} &= -c_{xy} = \frac{\partial u}{\partial y} + \frac{\partial v}{\partial x} - \frac{\partial u}{\partial x}\frac{\partial u}{\partial y} - \frac{\partial v}{\partial x}\frac{\partial v}{\partial y} - \frac{\partial w}{\partial x}\frac{\partial w}{\partial y} \\ 2e_{yz} &= -c_{yz} = \frac{\partial v}{\partial z} + \frac{\partial w}{\partial y} - \frac{\partial u}{\partial y}\frac{\partial u}{\partial z} - \frac{\partial v}{\partial y}\frac{\partial v}{\partial z} - \frac{\partial w}{\partial y}\frac{\partial w}{\partial z} \\ 2e_{zx} &= -c_{zx} = \frac{\partial w}{\partial x} + \frac{\partial u}{\partial z} - \frac{\partial u}{\partial x}\frac{\partial u}{\partial z} - \frac{\partial v}{\partial x}\frac{\partial v}{\partial z} - \frac{\partial w}{\partial x}\frac{\partial w}{\partial z} \end{aligned} \tag{1.4.20}$$

where we wrote $x_1 = x$, $x_2 = y$, $x_3 = z$, $u_1 = u$, $u_2 = v$, $u_3 = w$, and e_{xx}, e_{xy}, ... for e_{11}, e_{12}, ... and c_{xx}, c_{xy}, ... for c_{11}, c_{12} ....

Often employed in linear theories of continua are the *infinitesimal strain tensors* $\tilde{E}_{KL}$, $\tilde{e}_{kl}$ and *infinitesimal rotation tensors* $\tilde{R}_{KL}$, $\tilde{r}_{kl}$ defined by

$$\begin{aligned} \tilde{E}_{KL} &\equiv \tfrac{1}{2}(U_{K,L} + U_{L,K}) \equiv U_{(K,L)}, \qquad \tilde{e}_{kl} \equiv \tfrac{1}{2}(u_{k,l} + u_{l,k}) \equiv u_{(k,l)} \\ \tilde{R}_{KL} &\equiv \tfrac{1}{2}(U_{K,L} - U_{L,K}) \equiv U_{[K,L]}, \qquad \tilde{r}_{kl} = \tfrac{1}{2}(u_{k,l} - u_{l,k}) \equiv u_{[k,l]} \end{aligned} \tag{1.4.21}$$

where parentheses enclosing indices indicate the symmetric part and brackets the antisymmetric part of the quantities. From (1.4.21) we have

$$U_{K,L} = \tilde{E}_{KL} + \tilde{R}_{KL}, \qquad u_{k,l} = \tilde{e}_{kl} + \tilde{r}_{kl} \tag{1.4.22}$$

Carrying these into (1.4.17) we obtain

$$\begin{aligned} c_{kl} &= \delta_{kl} - 2e_{kl} = \delta_{kl} - 2\tilde{e}_{kl} + (\tilde{e}_{mk} + \tilde{r}_{mk})(\tilde{e}_{ml} + \tilde{r}_{ml}) \\ C_{KL} &= \delta_{KL} + 2E_{KL} = \delta_{KL} + 2\tilde{E}_{KL} + (\tilde{E}_{MK} + \tilde{R}_{MK})(\tilde{E}_{ML} + \tilde{R}_{ML}) \end{aligned} \tag{1.4.23}$$

from which, in various physical situations, approximate expressions are obtained by dropping various combinations of products. For example, when $\tilde{E}_{KL} \ll 1$ we can drop $\tilde{E}_{KM}\tilde{E}_{ML}$ as compared to other terms. When $\tilde{R}_{KL}$ is also small as compared to $\tilde{E}_{KL}$, then we obtain $E_{KL} = \tilde{E}_{KL}$ and $e_{kl} = \tilde{e}_{kl}$. In this case (1.4.13) and (1.4.16) also give upon linearization

$$\tilde{E}_{KL} = \tilde{e}_{kl}\,\delta_{kK}\,\delta_{lL}, \qquad \tilde{e}_{kl} = \tilde{E}_{KL}\,\delta_{Kk}\,\delta_{Ll} \tag{1.4.24}$$

Where $\delta_{Kk} = \delta_{kK}$ is now the Kronecker delta. Thus, in the *linear theory* (*the infinitesimal deformation theory*) *the distinction between the Lagrangian and Eulerian strain tensors disappears.*

The infinitesimal rotation tensor $\tilde{r}_{kl}$ is skew symmetric. Thus, in three-dimensional space axial vectors $\tilde{r}_k$ and $\tilde{R}_K$ exist such that

$$\begin{aligned} 2\tilde{r}_k &= e_{klm}\,\tilde{r}_{ml}, \qquad \tilde{r}_{kl} = -e_{klm}\,\tilde{r}_m \\ 2\tilde{R}_K &= e_{KLM}\,\tilde{R}_{ML}, \qquad \tilde{R}_{KL} = -e_{KLM}\,\tilde{R}_M \end{aligned} \tag{1.4.25}$$

where e_{klm} and e_{KLM} are the permutation symbols.

1.5 LENGTH AND ANGLE CHANGES

An infinitesimal rectangular parallelepiped with edge vectors $\mathbf{I}_1\,dX_1$, $\mathbf{I}_2\,dX_2$, and $\mathbf{I}_3\,dX_3$ at $\mathbf{X}$ after deformation becomes a rectilinear parallelepiped at $\mathbf{x}$ with corresponding edge vectors $\mathbf{C}_1\,dX_1$, $\mathbf{C}_2\,dX_2$, and $\mathbf{C}_3\,dX_3$ (Fig. 1.5.1):

$$d\mathbf{X} = \mathbf{I}_K\,dX_K, \qquad d\mathbf{x} = \mathbf{C}_K\,dX_K \tag{1.5.1}$$

where

$$\mathbf{C}_K \equiv \mathbf{x}_{,K} \tag{1.5.2}$$

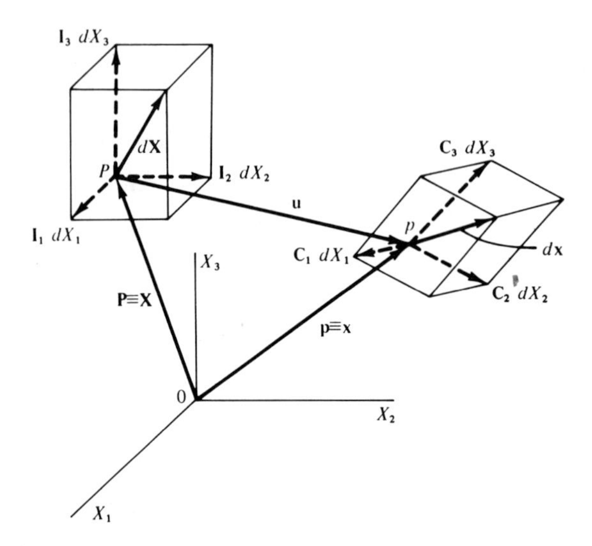

Fig. 1.5.1 Deformation of an infinitesimal rectangular parallelepiped.

If $\mathbf{N}$ and $\mathbf{n}$ are, respectively, unit vectors along $d\mathbf{X}$ and $d\mathbf{x}$, we have

$$N_K \equiv dX_K/|d\mathbf{X}| = dX_K/dS, \qquad n_k \equiv dx_k/|d\mathbf{x}| = dx_k/ds \tag{1.5.3}$$

where dS and ds are the lengths of $d\mathbf{X}$ and $d\mathbf{x}$, respectively. The *stretch* $\Lambda_{(\mathbf{N})} = \lambda_{(\mathbf{n})}$ is the ratio of ds/dS. When it is considered as a function of $\mathbf{N}$ we write $\Lambda_{(\mathbf{N})}$, and when it is considered as a function of $\mathbf{n}$ we write $\lambda_{(\mathbf{n})}$:

$$\Lambda_{(\mathbf{N})} = ds/dS = (C_{KL} N_K N_L)^{1/2}, \qquad \lambda_{(\mathbf{n})} = ds/dS = 1/(c_{kl} n_k n_l)^{1/2} \tag{1.5.4}$$

From (1.5.4) it is clear that the *normal components* of $\mathbf{C}$ *and* $\mathbf{c}$ *in the direction of* $\mathbf{N}$ *and* $\mathbf{n}$ *are, respectively, the square and the inverse square of stretches in these directions.* Thus, for example, when $\mathbf{N}$ is taken along the X_1-axis we have $N_1 = 1$, $N_2 = N_3 = 0$, and $(1.5.4)_1$ gives

$$\Lambda_{(1)} = (C_{11})^{1/2} = (1 + 2E_{11})^{1/2} \tag{1.5.5}$$

The *extension* $E_{(\mathbf{N})} = e_{(\mathbf{n})}$ is defined by

$$E_{(\mathbf{N})} = e_{(\mathbf{n})} = \Lambda_{(\mathbf{N})} - 1 = (ds - dS)/dS \tag{1.5.6}$$

When $\mathbf{N}$ is taken along the X_1-axis, this gives

$$2E_{11} = (1 + E_{(1)})^2 - 1 = \Lambda_{(1)}^2 - 1 \tag{1.5.7}$$

Thus, the normal component of the Lagrangian strain is one-half the square of stretch minus one. When extension is small, $E_{(1)} \ll 1$, then expanding (1.5.7) we get

$$E_{11} \simeq E_{(1)} \simeq \tilde{E}_{11} \tag{1.5.8}$$

Similar results are valid for E_{22} and E_{33}.

The angle $\Theta_{(12)}$ between $\mathbf{C}_1\, dX_1$ and $\mathbf{C}_2\, dX_2$ is calculated by

$$\cos \Theta_{(12)} = \frac{\mathbf{C}_1\, dX_1}{|\mathbf{C}_1\, dX_1|} \cdot \frac{\mathbf{C}_2\, dX_2}{|\mathbf{C}_2\, dX_2|}$$

or

$$\cos \Theta_{(12)} = \frac{C_{12}}{(C_{11}C_{22})^{1/2}} = \frac{2E_{12}}{(1 + 2E_{11})^{1/2}(1 + 2E_{22})^{1/2}} \tag{1.5.9}$$

This provides a geometrical significance for E_{12}. In fact,

$$2E_{12} = (1 + E_{(1)})(1 + E_{(2)}) \cos \Theta_{(12)} \tag{1.5.10}$$

When extensions are small as compared to unity, we have approximately

$$2E_{12} \simeq 2\tilde{E}_{12} \simeq \cos \Theta_{(12)} \tag{1.5.11}$$

For an infinitesimal deformation the angle change is given by

$$(\pi/2) - \Theta_{(12)} \equiv \Gamma_{(12)} \simeq 2\tilde{E}_{12}$$

Therefore, *infinitesimal shear strain* (the off-diagonal components of the strain tensor) *is approximately one-half the angle change between the coordinate axes for small deformations*. The infinitesimal rotation tensors can be shown to represent the mean rotation at coordinate planes (cf. Eringen [1967], Section 1.7).

1.6 STRAIN INVARIANTS

It is of interest to determine, at a given point $\mathbf{X}$, the directions for which the stretch takes extremum values. For this we must differentiate

$$\Lambda^2_{(\mathbf{N})} = C_{KL} N_K N_L = 2E_{KL} N_K N_L + 1 \tag{1.6.1}$$

with respect to N_M, where $N_M \equiv dX_M/dS$ is subject to the condition

$$\delta_{KL} N_K N_L = 1 \tag{1.6.2}$$

Using Lagrange's method of multipliers, we set

$$\frac{\partial}{\partial N_M} \Lambda^2_{(\mathbf{N})} = \frac{\partial}{\partial N_M} [C_{KL} N_K N_L - C(\delta_{KL} N_K N_L - 1)] = 0$$

where C is the unknown Lagrange multiplier. This gives

$$(C_{KL} - C\, \delta_{KL}) N_L = 0 \tag{1.6.3}$$

Alternatively, if we use $C_{KL} = 2E_{KL} + \delta_{KL}$, we may also write this as

$$(E_{KL} - E\,\delta_{KL})N_L = 0 \tag{1.6.4}$$

where

$$2E \equiv C - 1 \tag{1.6.5}$$

We may solve either set of equations (1.6.3) or (1.6.4) for N_L. Using the relations $C_{KL} = 2E_{KL} + \delta_{KL}$ and (1.6.5) one can always convert C-language to E-language.

A nontrivial solution of (1.6.4) exists if the coefficient determinant vanishes, i.e.,

$$\det(E_{KL} - E\,\delta_{KL}) = \begin{vmatrix} E_{11} - E & E_{12} & E_{13} \\ E_{21} & E_{22} - E & E_{23} \\ E_{31} & E_{32} & E_{33} - E \end{vmatrix} = 0 \tag{1.6.6}$$

Upon expanding this determinant we obtain a cubic equation, known as the *characteristic equation* of the strain tensor:

$$-E^3 + I_E E^2 - II_E E + III_E = 0 \tag{1.6.7}$$

where

$$\begin{aligned} I_E &\equiv E_{KK} = E_{11} + E_{22} + E_{33} \\ II_E &\equiv E_{22}E_{33} + E_{33}E_{11} + E_{11}E_{22} - E_{23}^2 - E_{31}^2 - E_{12}^2 \\ III_E &\equiv \det E_{KL} \end{aligned} \tag{1.6.8}$$

The quantities I_E, II_E, and III_E are known as the *principal invariants* of the strain. These quantities remain invariant upon the transformation of coordinates at $\mathbf{X}$. A second-order tensor E_{KL} in three dimensions possesses only three independent invariants, that is, all other invariants of E_{KL} can be shown to be functions of the above three invariants. For example, three other invariants are

(1.6.9)

$$I_1 \equiv \operatorname{tr} \mathbf{E} \equiv E_{KK}, \qquad I_2 \equiv \operatorname{tr} \mathbf{E}^2 \equiv E_{KL}E_{LK}, \qquad I_3 \equiv \operatorname{tr} \mathbf{E}^3 \equiv E_{KM}E_{ML}E_{LK}$$

where tr abbreviates trace. The relations of these to I_E, II_E, and III_E are

$$\begin{aligned} I_E &= I_1, & I_1 &= I_E \\ II_E &= \tfrac{1}{2}(I_1{}^2 - I_2), & I_2 &= I_E{}^2 - 2II_E \\ III_E &= \tfrac{1}{3}(I_3 - \tfrac{3}{2}I_1 I_2 + \tfrac{1}{2}I_1{}^3), & I_3 &= I_E{}^3 - 3I_E II_E + 3III_E \end{aligned} \tag{1.6.10}$$

The characteristic equation (1.6.7) possesses three roots E_α ($\alpha = 1, 2, 3$) called *principal strains*. The coefficients I_E, II_E, and III_E of the characteristic equation are the sums of products of these roots taken one, two, and three at a time, i.e.,

$$I_E = E_1 + E_2 + E_3, \quad II_E = E_2 E_3 + E_3 E_1 + E_1 E_2, \quad III_E = E_1 E_2 E_3 \tag{1.6.11}$$

The three linear equations (1.6.4) determine a direction $\mathbf{N}_\alpha$ ($\alpha = 1, 2, 3$) corresponding to each principal strain E_α. If the principal strains are *real* and *distinct*, then the directions N_{1K}, N_{2K}, and N_{3K} are real and uniquely determined. By use of (1.6.4) and the symmetry of the strain tensor it is not difficult to prove the following lemmas (cf. Eringen [1967], Section 1.10).

Lemma 1. *All principal strains are real.*

Lemma 2. *The principal directions corresponding to two distinct principal strains are orthogonal.*

When the characteristic equation (1.6.7) possesses multiple roots, then the associated directions become indeterminate. For a double root there will be an infinite number of directions in a plane satisfying (1.6.4). Any two orthogonal directions in this plane can be chosen as proper directions, the third being perpendicular to the plane. If all three roots are identical, then any three mutually orthogonal directions at $\mathbf{X}$ constitute a principal triad. Thus we see that it is always possible to find at a point $\mathbf{X}$, at least three mutually orthogonal directions for which the stretch takes the stationary values.

The state of strain takes a particularly simple form when the reference frame is selected to coincide with the principal directions. In this case $N_{LK} = 0$ whenever $L \neq K$. We may write $N_{LK} = \delta_{LK}$, and from (1.6.4) it follows that

$$E_{KM} = E_{\underline{M}}\, \delta_{\underline{M}K} \tag{1.6.12}$$

where underscored indices are not summed. In matrix notation (1.6.12) reads

$$\begin{bmatrix} E_{11} & E_{12} & E_{13} \\ E_{21} & E_{22} & E_{23} \\ E_{31} & E_{32} & E_{33} \end{bmatrix} = \begin{bmatrix} E_1 & 0 & 0 \\ 0 & E_2 & 0 \\ 0 & 0 & E_3 \end{bmatrix} \tag{1.6.13}$$

Hence, the *determination of principal directions and principal strains of a tensor E_{KL} is equivalent to finding a rectangular frame of reference in which the matrix $\|E_{KL}\|$ takes the diagonal form.* The following lemma is also clear from (1.6.13).

Lemma 3. *The principal strains are the normal components of the strain tensor, and the shear components of the strain tensor in the principal frame of reference vanish.*

In the principal triad the square of the arc length is given by

(1.6.14) $$ds^2 = C_{KL}\, dX_K\, dX_L = \sum_\alpha C_\alpha\, (dX_\alpha)^2$$

where according to (1.6.5)

(1.6.15) $$C_\alpha = 1 + 2E_\alpha$$

are the *proper numbers*. For $ds^2 = k^2 =$ fixed, Eq. (1.6.14) represents an ellipsoid called the *strain ellipsoid* of Cauchy. The stretches $\Lambda_{(\alpha)} = \lambda_{(\alpha)} = ds/dS$ along the principal axes of this ellipsoid are given by

(1.6.16) $$\Lambda_{(\alpha)} = k/dX_\alpha = C_\alpha^{1/2} = k/a_\alpha$$

where a_α are the lengths of the semiaxes (Fig. 1.6.1).

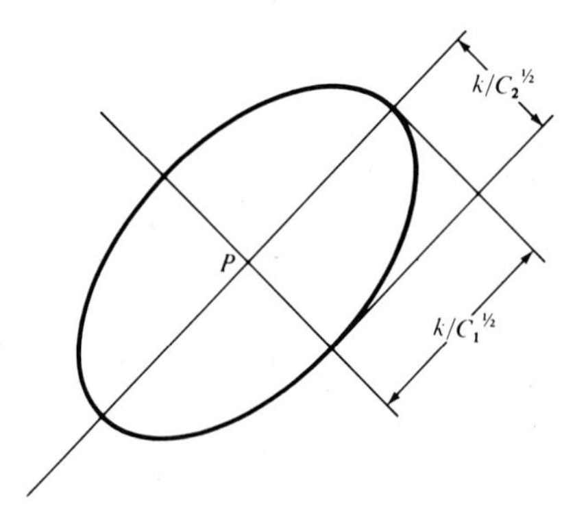

Fig. 1.6.1 Strain ellipsoid of Cauchy.

Another Cauchy ellipsoid is the *material ellipsoid* given by

(1.6.17) $$dS^2 = c_{kl}\, dx_k\, dx_l = K^2$$

Referred to the principal axes of **c**, this may be expressed as

(1.6.18) $$dS^2 = \sum_\alpha c_\alpha\, (dx_\alpha)^2 = K^2$$

The stretches $\lambda_{(\alpha)} \equiv \Lambda_{(\alpha)} = ds/dS$ for the principal axes of the material ellipsoid are given by

$$\lambda_{(\alpha)} = dx_\alpha/K = 1/c_\alpha^{1/2} = a_\alpha/K$$

Since $\lambda_{(\alpha)} = \Lambda_{(\alpha)}$, we find that

(1.6.19) $$C_\alpha = 1/c_\alpha = \lambda_{(\alpha)}^2$$

hence the following theorem.

Theorem *(Cauchy). The lengths of semiaxes of the strain ellipsoids of Cauchy are reciprocals of each other. The proper numbers C_α are equal to the squares of stretches along the principal directions of the strain ellipsoid at* **X**. *Along the principal direction the stretches have extremal values.*

Invariants may be formed from strain tensors $\tilde{E}_{KL}$, e_{kl}, $\tilde{e}_{kl}$, C_{KL}, c_{kl}, $\overset{-1}{C}_{KL}$, $\overset{-1}{c}_{kl}$, etc. These are obtained by replacement of **E** by the foregoing quantities and using

$$2E_{KL} = C_{KL} - \delta_{KL}, \qquad 2e_{kl} = \delta_{kl} - c_{kl}$$

Thus we find the relations

$$I_C = 3 + 2I_E, \quad II_C = 3 + 4I_E + 4II_E, \quad III_C = 1 + 2I_E + 4II_E + 8III_E \tag{1.6.20}$$

$$I_c = 3 - 2I_e, \quad II_c = 3 - 4I_e + 4II_e, \quad III_c = 1 - 2I_e + 4II_e - 8III_e \tag{1.6.21}$$

In terms of stretches $\lambda_{(\alpha)}$ we also have

$$\begin{aligned} I_C &= I_{\overset{-1}{c}} = \lambda_{(1)}^2 + \lambda_{(2)}^2 + \lambda_{(3)}^2 \\ II_C &= II_{\overset{-1}{c}} = \lambda_{(1)}^2 \lambda_{(2)}^2 + \lambda_{(2)}^2 \lambda_{(3)}^2 + \lambda_{(3)}^2 \lambda_{(1)}^2 \\ III_C &= III_{\overset{-1}{c}} = \lambda_{(1)}^2 \lambda_{(2)}^2 \lambda_{(3)}^2 \end{aligned} \tag{1.6.22}$$

$$\begin{aligned} I_c &= I_{\overset{-1}{C}} = \lambda_{(1)}^{-2} + \lambda_{(2)}^{-2} + \lambda_{(3)}^{-2} \\ II_c &= II_{\overset{-1}{C}} = \lambda_{(1)}^{-2} \lambda_{(2)}^{-2} + \lambda_{(2)}^{-2} \lambda_{(3)}^{-2} + \lambda_{(3)}^{-2} \lambda_{(1)}^{-2} \\ III_c &= III_{\overset{-1}{C}} = \lambda_{(1)}^{-2} \lambda_{(2)}^{-2} \lambda_{(3)}^{-2} \end{aligned} \tag{1.6.23}$$

From these we obtain the following fundamental identities:

$$I_c = II_C / III_C, \qquad II_c = I_C / III_C, \qquad III_c = 1/III_C \tag{1.6.24}$$

Since $0 < \lambda_{(\alpha)} < \infty$, we see that

$$0 < I_C, II_C, III_C < \infty, \qquad 0 < I_c, II_c, III_c < \infty \tag{1.6.25}$$

For a rigid deformation $\lambda_{(1)} = \lambda_{(2)} = \lambda_{(3)} = 1$. Hence,

$$I_C = II_C = 3, \qquad III_C = 1 \tag{1.6.26}$$

which is necessary and sufficient for a local rigid deformation.

1.7 AREA AND VOLUME CHANGES

The element of area built on the edge vectors $\mathbf{I}_1\, dX_1$ and $\mathbf{I}_2\, dX_2$ after deformation becomes the area with the edge vectors $\mathbf{C}_1\, dX_1$ and $\mathbf{C}_2\, dX_2$ (Fig. 1.7.1). Thus the deformed area is given by

$$\begin{aligned} d\mathbf{a}_3 &= \mathbf{C}_1\, dX_1 \times \mathbf{C}_2\, dX_2 = x_{k,1} x_{l,2}\, \mathbf{i}_k \times \mathbf{i}_l\, dX_1\, dX_2 \\ &= x_{k,1} x_{l,2}\, e_{klm}\, \mathbf{i}_m\, dA_3 \end{aligned}$$

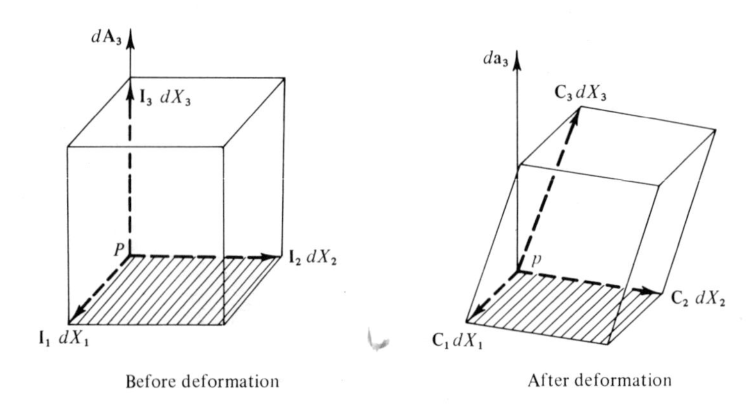

Fig. 1.7.1 Deformation of an infinitesimal rectangular parallelepiped.

where we wrote $dA_3 = dX_1\, dX_2$. But from (1.4.4) we have

$$jX_{3,m} = e_{klm}\, x_{k,1} x_{l,2}$$

so that

$$d\mathbf{a}_3 = jX_{3,m}\, dA_3\, \mathbf{i}_m$$

Similar expressions are valid for $d\mathbf{a}_1$ and $d\mathbf{a}_2$. Thus

$$d\mathbf{a} = d\mathbf{a}_1 + d\mathbf{a}_2 + d\mathbf{a}_3 = jX_{K,k}\, dA_K\, \mathbf{i}_k \tag{1.7.1}$$

whose kth component is

$$da_k = jX_{K,k}\, dA_K \tag{1.7.2}$$

To calculate the deformed volume element, we take the scalar product of $d\mathbf{a}_3$ with $\mathbf{C}_3\, dX_3$:

$$\begin{aligned} dv = d\mathbf{a}_3 \cdot \mathbf{C}_3\, dX_3 &= jX_{3,k}\, \mathbf{i}_k \cdot (x_{m,3}\, \mathbf{i}_m)\, dA_3\, dX_3 \\ &= jX_{3,k}\, x_{m,3}\, \delta_{km}\, dV \end{aligned}$$

Hence

$$dv = j\,dV \tag{1.7.3}$$

which can also be obtained from the transformation law of volume elements. Recalling

$$C_{KL} = x_{k,K}\,x_{k,L}$$

we have

$$\det C_{KL} = (\det x_{k,K})(\det x_{l,L}) = (\det x_{k,K})^2 = j^2$$

Thus (1.7.3) is also equivalent to

$$dv = j\,dV = (III_C)^{1/2}\,dV \tag{1.7.4}$$

From (1.6.20), (1.6.21), and (1.6.24) we also have the alternative expressions

$$\begin{aligned} dv/dV = j = (III_C)^{1/2} &= 1/(III_c)^{1/2} = (1 + 2I_E + 4II_E + 8III_E)^{1/2} \\ &= (1 - 2I_e + 4II_e - 8III_e)^{-1/2} \end{aligned} \tag{1.7.5}$$

Upon binomial expansion from (1.7.5) we obtain

$$(dv - dV)/dV \simeq I_{\tilde{E}} \simeq I_{\tilde{e}} \tag{1.7.6}$$

which is valid only for the infinitesimal deformations (the linear theory).

1.8 COMPATIBILITY CONDITIONS

In three-dimensional space the deformation tensor C_{KL} and the strain tensor E_{KL} each possess six components, but they are expressible in terms of three components U_K of the displacement vectors, i.e.,

$$C_{KL} = \delta_{KL} + 2E_{KL} = \delta_{KL} + U_{K,L} + U_{L,K} + U_{M,K}U_{M,L} \tag{1.8.1}$$

Thus given three U_K we can calculate six C_{KL}. If, on the other hand, we are given six C_{KL} or E_{KL} can we find a *single-valued displacement field* U_K corresponding to these strains? It is clear that this requires the integration of six partial differential equations (1.8.1) for the *three* unknowns U_K. Unless certain integrability conditions, known as the *compatibility conditions*, are satisfied, this may not be possible.

An obvious way of finding the compatibility conditions is through the elimination of U_K from (1.8.1) by partial differentiation, but this is extremely awkward. An alternative method is to use a theorem of Riemann. Both undeformed and deformed bodies are located in Euclidean space. The arc lengths in undeformed and deformed bodies are given by

$$dS^2 = \delta_{KL}\,dX_K\,dX_L = c_{kl}\,dx_k\,dx_l, \quad ds^2 = C_{KL}\,dX_K\,dX_L = \delta_{kl}\,dx_k\,dx_l$$

If we look at the motion

$$x_k = x_k(X_1, X_2, X_3, t), \qquad k = 1, 2, 3$$

as the coordinate transformation from rectangular coordinates X_K to curvilinear coordinates x_k at a fixed time, then c_{kl} would play the role of a metric tensor in the curvilinear coordinates x_k, and the same is valid for C_{KL} for the inverse motion. In curvilinear coordinates any six quantities cannot be a metric tensor unless they satisfy the theorem of Riemann: *For a symmetric tensor a_{kl} to be a metric tensor for a Euclidean space, it is necessary and sufficient that a_{kl} be a nonsingular positive definite tensor and the Riemann–Christoffel tensor $R^{(\mathbf{a})}_{klmn}$ formed from it vanish identically.* By definition (Equation A.6.3) we have

$$\begin{aligned} R^{(\mathbf{a})}_{klmn} \equiv {} & \tfrac{1}{2}(a_{kn,lm} + a_{lm,kn} - a_{km,ln} - a_{ln,km}) \\ & + \overset{-1}{a}_{rs}([lm,s][kn,r] - [ln,s][km,r]) \end{aligned} \tag{1.8.2}$$

where

$$[kl,m] \equiv \tfrac{1}{2}(a_{km,l} + a_{lm,k} - a_{kl,m}), \qquad \overset{-1}{a}_{ns}\, a_{sl} = a_{ns}\, \overset{-1}{a}_{sl} = \delta_{nl} \tag{1.8.3}$$

Both C_{KL} and c_{kl} are nonsingular, symmetric, positive definite tensors of the Euclidean three-dimensional space. Therefore we must have

$$R^{(\mathbf{C})}_{KLMN} = 0, \qquad R^{(\mathbf{c})}_{klmn} = 0 \tag{1.8.4}$$

In the former, partial differentiation is understood with respect to X_K and in the latter with respect to x_k. In three-dimensional space, of the eighty-one components of R_{klmn} only six are independent and nonvanishing. Thus $(1.8.4)_1$ and $(1.8.4)_2$ give six partial differential equations for C_{KL} and c_{kl}, respectively. By substituting

$$C_{KL} = \delta_{KL} + 2E_{KL}, \qquad c_{kl} = \delta_{kl} - 2e_{kl}$$

we obtain the compatibility conditions for E_{KL} and e_{kl}. The following is one of these two sets:

$$\begin{aligned} & e_{kn,lm} + e_{lm,kn} - e_{km,ln} - e_{ln,km} \\ & \quad - \overset{-1}{c}_{rs}[(e_{kr,n} + e_{nr,k} - e_{kn,r})(e_{ls,m} + e_{ms,l} - e_{lm,s}) \\ & \quad - (e_{kr,m} + e_{mr,k} - e_{km,r})(e_{ls,n} + e_{ns,l} - e_{ln,s})] = 0 \end{aligned} \tag{1.8.5}$$

When the strains are small we drop their products to obtain the compatibility conditions for the infinitesimal strain tensor:

$$\tilde{e}_{kn,lm} + \tilde{e}_{lm,kn} - \tilde{e}_{km,ln} - \tilde{e}_{ln,km} = 0 \tag{1.8.6}$$

Alternative forms of (1.8.6) are

$$e_{pkl}\, e_{qmn}\, \tilde{e}_{km,ln} = 0 \tag{1.8.7}$$
$$\tilde{e}_{kk,ll}\,\delta_{pq} + \tilde{e}_{pk,kq} + \tilde{e}_{qk,kp} = \tilde{e}_{kl,kl}\,\delta_{pq} + \tilde{e}_{pq,kk} + \tilde{e}_{kk,pq}$$

When these conditions are satisfied, the single-valued integral of

$$\tilde{e}_{kl} = \tfrac{1}{2}(u_{k,l} + u_{l,k}) \tag{1.8.8}$$

exists and is given by

$$u_k = u_k{}^0 + \tilde{R}_{lk}\, x_l + b_k \tag{1.8.9}$$

where $u_k{}^0$ is any solution of (1.8.8), $\tilde{R}_{lk}$ a skew-symmetric tensor and b_k a vector, both independent of x_k. Physically the solution (1.8.9) determines a single-valued displacement field corresponding to a given $\tilde{e}_{kl}$, uniquely within a rigid body motion.

For the finite deformation theory the exact compatibility conditions are, of course, given by (1.8.5).

1.9 KINEMATICS, TIME RATES OF VECTORS AND TENSORS

Definition 1. *Material time rate of a vector* $\mathbf{f}$ *(or tensor) is defined by*

$$\frac{d\mathbf{f}}{dt} = \left.\frac{\partial \mathbf{f}}{\partial t}\right|_{\mathbf{X}} \tag{1.9.1}$$

where subscript $\mathbf{X}$ accompanying a vertical bar indicates that $\mathbf{X}$ is held constant in the differentiation of $\mathbf{f}$. If $\mathbf{f}$ is a function of material coordinates, e.g.,

$$\mathbf{f} = \mathbf{f}(\mathbf{X}, t) = F_K(\mathbf{X}, t)\mathbf{I}_K$$

then clearly

$$d\mathbf{f}/dt = (\partial F_K/\partial t)\mathbf{I}_K$$

If $\mathbf{f}$ is a function of spatial coordinates, e.g.,

$$\mathbf{f} = \mathbf{f}(\mathbf{x}, t) = f_k(\mathbf{x}, t)\mathbf{i}_k$$

then

$$\frac{d\mathbf{f}}{dt} = \left(\left.\frac{\partial f_k}{\partial t}\right|_{\mathbf{x}} + \frac{\partial f_k}{\partial x_l}\frac{\partial x_l}{\partial t}\right)\mathbf{i}_k$$

since through the motion we have

$$\mathbf{x} = \mathbf{x}(\mathbf{X}, t), \qquad \mathbf{X} = \mathbf{X}(\mathbf{x}, t) \tag{1.9.2}$$

The foregoing expression may also be written as

(1.9.3) $$d\mathbf{f}/dt = \dot{\mathbf{f}} = (Df_k/Dt)\mathbf{i}_k = \dot{f}_k \mathbf{i}_k$$

where

(1.9.4) $$Df_k/Dt = \dot{f}_k = (\partial f_k/\partial t) + f_{k,l}\,\partial x_l/\partial t$$

is called the *material derivative* of f_k. The first term on the right of (1.9.4) is called the *local* or *nonstationary rate*, and the second term is the *convective time rate*. Clearly, expressions (1.9.3) and (1.9.4) apply equally well to material vectors and tensors with no ambiguity, since

$$DF_K(\mathbf{X}, t)/Dt = \dot{F}_K = \partial F_K/\partial t$$

The material derivative obeys the ordinary rules of the partial differentiation involving sums and products, i.e.,

$$\frac{D}{Dt}(f_k + g_k) = \frac{Df_k}{Dt} + \frac{Dg_k}{Dt}, \qquad \frac{D(f_k g_l)}{Dt} = \frac{Df_k}{Dt} g_l + f_k \frac{Dg_l}{Dt}$$

Definition 2. *The velocity vector* $\mathbf{v}$ *is the time rate of change of the position vector*

(1.9.5) $$\mathbf{v} = d\mathbf{p}/dt = (\partial x_k/\partial t)\mathbf{i}_k, \qquad v_k \equiv \partial x_k/\partial t$$

If we substitute $(1.9.2)_2$ for $\mathbf{X}$ in $\mathbf{V}(\mathbf{X}, t)$, we have

(1.9.6) $$\mathbf{V} = \mathbf{V}(\mathbf{X}(\mathbf{x}, t), t) = \mathbf{v}(\mathbf{x}, t) = v_k(\mathbf{x}, t)\mathbf{i}_k$$

which gives the velocity field at each spatial point $\mathbf{x}$ at time t with no indication of its relation to the material point $\mathbf{X}$. This is the Eulerian point of view.

Definition 3. *The acceleration vector* $\mathbf{a}$ *is the time rate of change of the velocity vector*

(1.9.7) $$\mathbf{a} \equiv d\mathbf{v}/dt$$

According to (1.9.3) and (1.9.4), we have

(1.9.8) $$\mathbf{a}(\mathbf{x}, t) = (Dv_k/Dt)\mathbf{i}_k, \qquad Dv_k/Dt = (\partial v_k/\partial t) + v_{k,l} v_l$$

In the *Lagrangian viewpoint* the material point is known, so that

(1.9.9) $$\mathbf{a}(\mathbf{X}, t) = \left.\frac{\partial V_K(\mathbf{X}, t)}{\partial t}\right|_{\mathbf{X}} \mathbf{I}_K$$

In the kinematics of continua the material derivative of displacement gradients play a fundamental role.

Fundamental Lemma. *The material derivative of the displacement gradients is given by*

$$\frac{D}{Dt}(x_{k,K}) = v_{k,l}\,x_{l,K}, \qquad \frac{D}{Dt}(dx_k) = v_{k,l}\,dx_l \tag{1.9.10}$$

The proof is immediate, for

$$\frac{D}{Dt}\left(\frac{\partial x_k}{\partial X_K}\right) = \frac{\partial}{\partial X_K}\left(\frac{Dx_k}{Dt}\right) = v_{k,K} = v_{k,l}\,x_{l,K}$$

since in the operation D/Dt we have X_K fixed so that D/Dt and $\partial/\partial X_K$ commute. Equation $(1.9.10)_2$ follows from this by multiplying $(1.9.10)_1$ by dX_K.

A corollary to this lemma is

$$\frac{D}{Dt}(X_{K,k}) = -v_{k,l}\,X_{K,l} \tag{1.9.11}$$

which follows from differentiation of $x_{k,K}\,X_{K,l} = \delta_{kl}$ and using $(1.9.10)_1$ and this expression.

The following two theorems are important.

Theorem 1. *The material derivative of the square of the arc length is given by*

$$\frac{D}{Dt}(ds^2) = 2\,d_{kl}\,dx_k\,dx_l \tag{1.9.12}$$

where

$$d_{kl} \equiv v_{(k,l)} \equiv \tfrac{1}{2}(v_{k,l} + v_{l,k}) \tag{1.9.13}$$

is called the *deformation rate tensor*. The proof follows by differentiating $ds^2 = dx_k\,dx_k$ and using $(1.9.10)_2$.

In the Lagrangian description (1.9.12) is expressed as

$$\frac{D}{Dt}(ds^2) = 2\,d_{kl}\,x_{k,K}x_{l,L}\,dX_K\,dX_L = 2\dot{E}_{KL}\,dX_K\,dX_L \tag{1.9.14}$$

where

$$\dot{E}_{KL} \equiv \frac{D}{Dt}(E_{KL}) = \frac{1}{2}\,\dot{C}_{KL} = d_{kl}\,x_{k,K}\,x_{l,L} \tag{1.9.15}$$

is the material time rate of the Lagrangian strain tensor. From (1.9.12) it follows that for arbitrary dx_k, $D(ds^2)/Dt = 0$ if and only if $d_{kl} = 0$, which leads to the next theorem.

Theorem 2 (*Killing*). *The necessary and sufficient condition for the motion of a body to be locally rigid is* $d_{kl} = 0$.

Lemma. *The material derivative of the Jacobian is given by*

$$Dj/Dt = jv_{k,k} \tag{1.9.16}$$

Proof. We have

$$\frac{Dj}{Dt} = \frac{D}{Dt}(\det x_{k,K}) = \frac{\partial j}{\partial x_{k,K}} \frac{Dx_{k,K}}{Dt} = \frac{\partial j}{\partial x_{k,K}} v_{k,l} x_{l,K}$$

Using $(1.4.6)_2$, this gives (1.9.16).

Since $dv = j\,dV$, it follows that

$$\frac{D}{Dt}(dv) = \frac{Dj}{Dt}(dV) = v_{k,k}\,dV \tag{1.9.17}$$

which finds important uses in the differentiation of integrals taken over the material volume.

In the kinematics of continua the *spin tensor* w_{kl} and the *vorticity vector* w_k defined by

$$w_{kl} \equiv v_{[k,l]} = \tfrac{1}{2}(v_{k,l} - v_{l,k}) \tag{1.9 18}$$

$$w_k = e_{klm} w_{ml} = e_{klm} v_{m,l} \qquad \text{or} \qquad \mathbf{w} = \operatorname{curl} \mathbf{v} \tag{1.9.19}$$

find important applications. For the physical significance of these quantities and other related accounts the reader is referred to Eringen [1962, 1967].

1.10 GLOBAL BALANCE LAWS

In continuum mechanics the following five laws are postulated irrespective of material constitution and geometry. They are valid for all bodies subject to thermomechanical effects. The domain of applicability of these laws is restricted by the relativistic speeds (special relativity) and dimensions (general relativity) and microscopic and quantum-mechanical phenomena.

Fundamental Axiom 1 (*Conservation of Mass*). *The total mass of a body is unchanged with motion.*

In continuum mechanics the existence of a continuous mass measure (the mass density) ρ is postulated. The total mass M is given by

$$M = \int_{\mathscr{V}} \rho\,dv, \qquad 0 \le \rho < \infty$$

where the integral is taken over the material volume $\mathscr{V}$ of the body.

The law of conservation of mass states that the initial total mass of the body is the same as the total mass of the body at any other time, i.e.,

$$\int_V \rho_0 \, dV = \int_{\mathscr{V}} \rho \, dv \tag{1.10.1}$$

Using the transformation law $dv = j\,dV$ we may write this as

$$\int_V (\rho_0 - \rho j) \, dV = 0 \tag{1.10.2}$$

Alternatively, we may take the material derivative of (1.10.1). Thus,

$$\frac{d}{dt} \int_{\mathscr{V}} \rho \, dv = 0 \tag{1.10.3}$$

Either expression (1.10.2) or (1.10.3) expresses the law of conservation of mass.

Fundamental Axiom 2 (*Balance of Momentum*). *The time rate of change of momentum is equal to the resultant force* $\mathscr{F}$ *acting on the body.*

Mathematically,

$$\frac{d}{dt} \int_{\mathscr{V}} \rho \mathbf{v} \, dv = \mathscr{F} \tag{1.10.4}$$

where the left-hand side is the time rate of change of the total momentum of the body. The forces acting on a body consist of body forces such as gravity and the surface forces arising from the contact of the body with other bodies. Thus we may write

$$\mathscr{F} = \oint_{\mathscr{S}} \mathbf{t}_{(\mathbf{n})} \, da + \int_{\mathscr{V}} \rho \mathbf{f} \, dv$$

where $\mathbf{t}_{(\mathbf{n})}$ is the surface contact force per unit area of the surface $\mathscr{S}$ of the body having exterior unit normal $\mathbf{n}$. This force depends on the normal vector $\mathbf{n}$ as indicated by a subscript ($\mathbf{n}$). The body force $\mathbf{f}$ is the force per unit mass of the body resulting from long-range effects such as gravity. Thus the balance of momentum reads

$$\frac{d}{dt} \int_{\mathscr{V}} \rho \mathbf{v} \, dv = \oint_{\mathscr{S}} \mathbf{t}_{(\mathbf{n})} \, da + \int_{\mathscr{V}} \rho \mathbf{f} \, dv \tag{1.10.5}$$

Fundamental Axiom 3 (*Balance of Moment of Momentum*). *The time rate of change of the moment of momentum is equal to the resultant moment of all forces and couples acting on the body.*

Mathematically,

$$\frac{d}{dt}\int_{\mathscr{V}} \rho \mathbf{p} \times \mathbf{v}\, dv = \oint_{\mathscr{S}} \mathbf{p} \times \mathbf{t}_{(\mathbf{n})}\, da + \int_{\mathscr{V}} \rho \mathbf{p} \times \mathbf{f}\, dv \tag{1.10.6}$$

where the left-hand side is the time rate of the total moment of momentum about the origin. On the right-hand side the surface integral is the moment of the surface tractions about the origin, and the volume integral is the total moment of body forces about the origin.

Fundamental Axiom 4 (*Conservation of Energy*). *The time rate of the sum of kinetic energy $\mathscr{K}$ and internal energy $\mathscr{E}$ is equal to the sum of the rate of work of all forces and couples $\mathscr{W}$ and all other energies $\mathscr{U}_\alpha$ that enter and leave the body per unit time.*

Mathematically,

$$\frac{d}{dt}(\mathscr{K} + \mathscr{E}) = \mathscr{W} + \sum_\alpha \mathscr{U}_\alpha \tag{1.10.7}$$

In continuum mechanics the existence of the internal energy density ε is postulated:

$$\mathscr{E} = \int_{\mathscr{V}} \rho\varepsilon\, dv \tag{1.10.8}$$

The total kinetic energy of the body is given by

$$\mathscr{K} = \frac{1}{2}\int_{\mathscr{V}} \rho \mathbf{v} \cdot \mathbf{v}\, dv \tag{1.10.9}$$

The power of the surface traction $\mathbf{t}_{(\mathbf{n})}$ and body force $\mathbf{f}$ is given by

$$\mathscr{W} = \oint_{\mathscr{S}} \mathbf{t}_{(\mathbf{n})} \cdot \mathbf{v}\, da + \int_{\mathscr{V}} \rho \mathbf{f} \cdot \mathbf{v}\, dv \tag{1.10.10}$$

Other energies $\mathscr{U}_\alpha$ $(\alpha = 1, 2, \ldots, n)$ that enter and leave the body may be of thermal, electromagnetic, chemical, or some other origin. Here we consider only the thermal energy. The heat energy consists of the *heat* $\mathbf{q}$ that enters or leaves through the surface of the body and the *heat source* h per unit mass that may be present in the body. Thus we set $\mathscr{U}_\alpha = 0$ except for

$$\mathscr{U}_1 \equiv \mathscr{Q} = \oint_{\mathscr{S}} \mathbf{q} \cdot d\mathbf{a} + \int_{\mathscr{V}} \rho h\, dv \tag{1.10.11}$$

where $\mathbf{q}$ is directed outward from the surface of the body. Thus (1.10.7) reads

$$\frac{d}{dt}\int_{\mathscr{V}} (\rho\varepsilon + \tfrac{1}{2}\rho \mathbf{v} \cdot \mathbf{v})\, dv = \oint_{\mathscr{S}} (\mathbf{t}_{(\mathbf{n})} \cdot \mathbf{v} + \mathbf{q} \cdot \mathbf{n})\, da + \int_{\mathscr{V}} (\rho \mathbf{f} \cdot \mathbf{v} + \rho h)\, dv \tag{1.10.12}$$

where we wrote $d\mathbf{a} = \mathbf{n}\, da$.

Fundamental Axiom 5 (*Entropy*). *The time rate of change of the total entropy H is never less than the sum of the influx of entropy* **s** *through the surface of the body and the entropy B supplied by the body sources. This law is postulated to hold for all independent processes.*

Mathematically,

$$\Gamma \equiv \frac{dH}{dt} - B - \oint_{\mathscr{S}} \mathbf{s} \cdot d\mathbf{a} \geq 0 \tag{1.10.13}$$

where Γ so defined is the *total entropy production.* In classical continuum mechanics the entropy density η and entropy source b, per unit mass, are postulated to exist such that

$$H = \int_{\mathscr{V}} \rho\eta \, dv, \qquad B = \int_{\mathscr{V}} \rho b \, dv$$

Moreover, the entropy influx **s** and entropy source b are taken as[1]

$$\mathbf{s} = \mathbf{q}/\theta, \qquad b = h/\theta$$

Thus the entropy inequality (1.10.13) reads

$$\frac{D}{Dt}\int_{\mathscr{V}} \rho\eta \, dv - \oint_{\mathscr{S}} \frac{1}{\theta} \mathbf{q} \cdot \mathbf{n} \, da - \int_{\mathscr{V}} \frac{\rho h}{\theta} \, dv \geq 0 \tag{1.10.14}$$

The scalar θ so introduced is called the *absolute temperature.* It is subject to

$$\theta > 0, \qquad \inf \theta = 0 \tag{1.10.15}$$

The foregoing five laws are postulated to hold for all bodies irrespective of their geometries and constitutions. To obtain local equations, further restrictions are necessary, which are made in the next section.

1.11 LOCAL BALANCE LAWS

The following two integral theorems are fundamental in the derivation of local laws:

Consider a material volume $\mathscr{V}$ intersected by a discontinuity surface $\sigma(t)$ moving with velocity **v** (Fig. 1.11.1). The material derivative of the volume integral of a tensor field ϕ over $\mathscr{V} - \sigma$ is given by

$$\frac{d}{dt}\int_{\mathscr{V}-\sigma} \phi \, dv = \int_{\mathscr{V}-\sigma} \left[\frac{\partial \phi}{\partial t} + \operatorname{div}(\phi \mathbf{v})\right] dv + \int_{\sigma} [\phi(\mathbf{v} - \boldsymbol{\nu})] \cdot d\mathbf{a} \tag{1.11.1}$$

[1] In Section 1.13 we shall show that, under certain conditions, this expression for **s** follows from the axiom of admissibility.

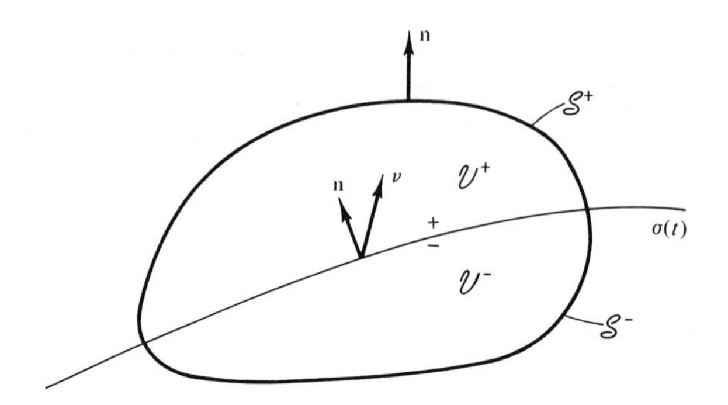

Fig. 1.11.1 Discontinuity surface ($\mathscr{V} - \sigma \equiv \mathscr{V}^+ + \mathscr{V}^-$, $\mathscr{S} - \sigma \equiv \mathscr{S}^+ + \mathscr{S}^-$).

The generalized Green–Gauss theorem for a tensor field τ_k over the surface $\mathscr{S} - \sigma$ of the body $\mathscr{V} - \sigma$ states that

$$\oint_{\mathscr{S}-\sigma} \tau_k \, da_k = \int_{\mathscr{V}-\sigma} \tau_{k,k} \, dv + \int_\sigma [\tau_k] n_k \, da \tag{1.11.2}$$

The volume integral over $\mathscr{V} - \sigma$ means the volume $\mathscr{V}$ of the body excluding the material points located on the discontinuity surface σ. Similarly, the integral over the surface $\mathscr{S} - \sigma$ excludes the line of intersection of σ with $\mathscr{S}$, i.e.,

$$\mathscr{V} - \sigma \equiv \mathscr{V}^+ + \mathscr{V}^-, \qquad \mathscr{S} - \sigma \equiv \mathscr{S}^+ + \mathscr{S}^-$$

The boldface brackets indicate the jump of their enclosures (the difference of the enclosure at the surface σ approached from the positive and negative sides of its positive normal), e.g.,

$$[f] \equiv f^+ - f^-$$

(For proof of these two theorems, see Eringen [1967, Section 2.5].)

We now apply these theorems to balance laws postulated in the previous section. If we take $\phi \equiv \rho$ in (1.11.1) we have the equation of conservation of mass (1.10.3):

$$\int_{\mathscr{V}-\sigma} \left[\frac{\partial \rho}{\partial t} + \operatorname{div}(\rho \mathbf{v})\right] dv + \int_\sigma [\rho(\mathbf{v} - \boldsymbol{\nu})] \cdot \mathbf{n} \, da = 0 \tag{1.11.3}$$

We now postulate that *all balance laws are valid for every part of the body and the discontinuity surface.* Applied to (1.11.3) this implies that integrands of each of the integral must vanish independently.[1] Thus

$$(\partial \rho / \partial t) + (\rho v_k)_{,k} = 0 \quad \text{in} \quad \mathscr{V} - \sigma \tag{1.11.4}$$

$$[\rho(\mathbf{v} - \boldsymbol{\nu})] \cdot \mathbf{n} = 0 \quad \text{on} \quad \sigma \tag{1.11.5}$$

[1] For nonlocal continuum theories this postulate is revoked, and only the global balance laws (valid for the entire body) are considered to be valid (cf. Eringen [1972a,b], and Eringen and Edelen [1972]).

These are the equations of local conservation of mass and the jump condition. Equation (1.11.4) is sometimes called the *equation of continuity*. It is none other than the material derivative of

$$\rho_0 = \rho j \qquad \text{or} \qquad \rho_0 = \rho(III_C)^{1/2} \tag{1.11.6}$$

Next we consider the equation of balance of momentum (1.10.5), first applying it to a tetrahedron adjacent to the surface $\mathscr{S}$ of the body. Consider a tetrahedron of volume Δv having three coordinate surfaces Δa_k in the body, with the fourth having the surface Δa on $\mathscr{S}$ (Fig. 1.11.2). By the mean value theorem, for this tetrahedron, (1.10.5) reads

$$\frac{d}{dt}(\rho^* \mathbf{v}^* \Delta v) = \mathbf{t}^*_{(\mathbf{n})} \Delta a - \mathbf{t}_k{}^* \Delta a_k + \rho^* \mathbf{f}^* \Delta v$$

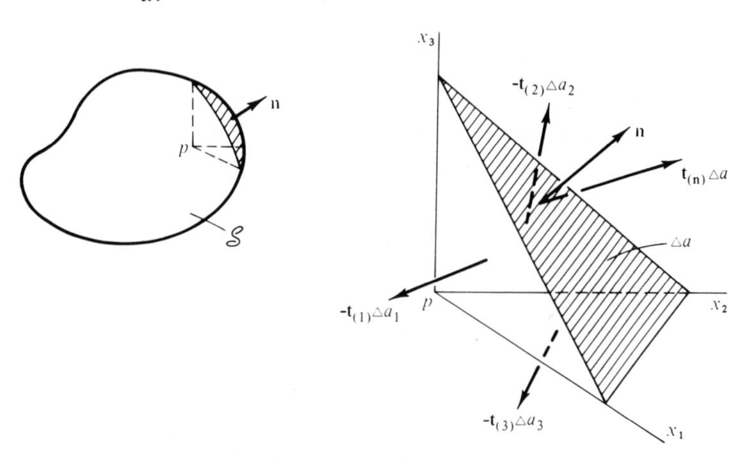

Fig. 1.11.2 Equilibrium of an infinitesimal tetrahedron near the surface $\mathscr{S}$.

where ρ^*, $\mathbf{v}^*$, and $\mathbf{f}^*$ are, respectively, the values of ρ, $\mathbf{v}$, and $\mathbf{f}$ at some interior points of the tetrahedron and $\mathbf{t}^*_{(\mathbf{n})}$ and $\mathbf{t}_k{}^*$ are the values of $\mathbf{t}_{(\mathbf{n})}$ on the surface Δa and on coordinate surfaces Δa_k (Fig. 1.11.2). In the limit as $\Delta v \to 0$ we have

$$\lim_{\Delta v \to 0} \frac{d}{dt}(\rho^* \mathbf{v}^* \Delta v) = \rho \dot{\mathbf{v}}\, dv + \mathbf{v}\, \dot{\overline{\rho\, dv}} = \rho \dot{\mathbf{v}}\, dv \tag{1.11.7}$$

since the conservation of local mass gives $\dot{\overline{\rho\, dv}} = 0$. In the limit $dv/da \to 0$, we obtain

$$\mathbf{t}_{(\mathbf{n})}\, da = \mathbf{t}_k\, da_k \tag{1.11.8}$$

Here $\mathbf{t}_k$ are called *stress vectors*.

For the tetrahedron, the area vector $d\mathbf{a}$ is equal to the sum of coordinate area vectors, i.e.,

(1.11.9)
$$d\mathbf{a} = \mathbf{n}\, da = da_k\, \mathbf{i}_k$$

Thus

(1.11.10)
$$da_k = n_k\, da$$

Substituting this into (1.11.8), we get

(1.11.11)
$$\mathbf{t}_{(\mathbf{n})} = \mathbf{t}_k n_k$$

From this, it follows that

(1.11.12)
$$\mathbf{t}_{(-\mathbf{n})} = -\mathbf{t}_{(\mathbf{n})}$$

Thus we have proved the following theorem.

Theorem. *The traction is a linear function of the normal and the tractions acting at the opposite sides of a surface are equal in magnitude and opposite in sign.*

The second part of this theorem is the counterpart of Newton's third law of motion, sometimes stated: *action is equal to reaction.*

Definition (*Stress Tensor*). *The stress tensor* t_{kl} *is the* lth *component of the stress vector* $\mathbf{t}_k$ *acting on the positive side of the* kth *coordinate surface:*

(1.11.13)
$$\mathbf{t}_k = t_{kl}\mathbf{i}_l$$

The positive components of t_{kl} on the faces of a parallelepiped built on coordinate surfaces are shown on Fig. 1.11.3. In order to avoid confusion, we have only shown the stress components on two pairs of parallel coordinate surfaces. Note that when the exterior normal of a surface is in the same direction with the coordinate axis perpendicular to the surface, the positive stress components on that surface are in the positive directions of the coordinates. When the exterior normal is opposite to the coordinate axis, then the positive stress components are in the opposite directions of the coordinates. The *components* t_{11}, t_{22}, and t_{33} *are called normal stresses*, and t_{12}, t_{23}, etc. *are called shear stresses*. The stress tensor may be arranged in a matrix form

(1.11.14)
$$\|t_{kl}\| = \begin{bmatrix} t_{11} & t_{12} & t_{13} \\ t_{21} & t_{22} & t_{23} \\ t_{31} & t_{32} & t_{33} \end{bmatrix}$$

The traction $\mathbf{t}_{(\mathbf{n})}$ is given by

(1.11.15)
$$\mathbf{t}_{(\mathbf{n})} = t_{kl} n_k \mathbf{i}_l$$

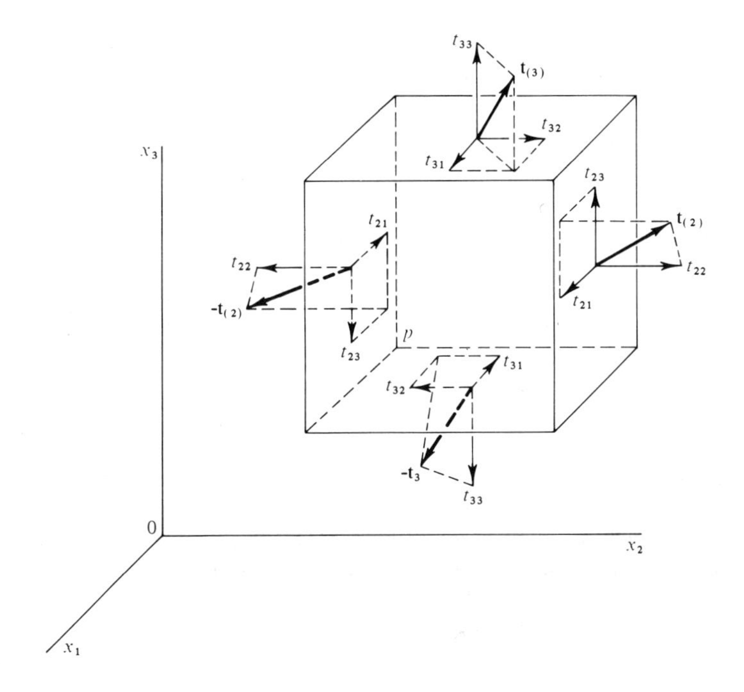

Fig. 1.11.3 Stress tensor.

The equation of global balance of momentum (1.10.5) now reads

$$\frac{d}{dt}\int_{\mathscr{V}-\sigma} \rho\mathbf{v}\, dv = \oint_{\mathscr{S}-\sigma} \mathbf{t}_k\, da_k + \int_{\mathscr{V}-\sigma} \rho\mathbf{f}\, dv$$

Using (1.11.1) and (1.11.2) with $\phi = \rho\mathbf{v}$ and $\tau_k = \mathbf{t}_k$, we obtain

(1.11.16)

$$\int_{\mathscr{V}-\sigma}\left[\frac{\partial(\rho\mathbf{v})}{\partial t} + (\rho\mathbf{v}v_k)_{,k} - \mathbf{t}_{k,k} - \rho\mathbf{f}\right] dv + \int_{\sigma}[\rho\mathbf{v}(v_k - \nu_k) - \mathbf{t}_k]n_k\, da = 0$$

This is postulated to be valid for all parts of the body. Thus the integrands vanish separately. Upon using (1.11.4), this is simplified to

$$\mathbf{t}_{k,k} + \rho(\mathbf{f} - \dot{\mathbf{v}}) = \mathbf{0} \quad \text{in} \quad \mathscr{V} - \sigma \tag{1.11.17}$$

$$[\rho\mathbf{v}(v_k - \nu_k) - \mathbf{t}_k]n_k = \mathbf{0} \quad \text{on} \quad \sigma \tag{1.11.18}$$

where

$$\dot{\mathbf{v}} \equiv (\partial\mathbf{v}/\partial t) + \mathbf{v}_{,k}v_k$$

Equation (1.11.17) is the *first law of Cauchy* expressing the *local balance of momentum*, and (1.11.18) is the associated *jump condition* on the *singular surface* σ.

Upon carrying (1.11.11) into the equation of balance of the moment of momentum (1.10.6) and using (1.11.1) and (1.11.2), for the local laws we obtain

$$\mathbf{i}_k \times \mathbf{t}_k = \mathbf{0} \quad \text{in} \quad \mathscr{V} - \sigma \tag{1.11.19}$$

where the local laws of conservation of mass and balance of momentum (1.11.4), (1.11.5), (1.11.17), and (1.11.18) are used. The associated jump condition for the moment of momentum is satisfied identically. When (1.11.13) is used, (1.11.19) gives

$$t_{kl} = t_{lk} \tag{1.11.20}$$

Thus the necessary and sufficient condition for the satisfaction of the local balance of moment of momentum is the symmetry of the stress tensor.[1] The same program can be carried out for the equation of energy balance (1.10.12) and entropy (1.10.14). Again using (1.11.11), the integral theorems (1.11.1), (1.11.2), and the local balance equations of mass and momenta, we obtain

$$\rho\dot{\varepsilon} = \mathbf{t}_k \cdot \mathbf{v}_{,k} + q_{k,k} + \rho h \qquad \text{in} \quad \mathscr{V} - \sigma \tag{1.11.21}$$

$$[(\rho\varepsilon + \tfrac{1}{2}\mathbf{v} \cdot \mathbf{v})(v_k - \nu_k) - t_{kl} v_l - q_k] n_k = 0 \qquad \text{on} \quad \sigma \tag{1.11.22}$$

for the local energy balance, and

$$\rho\gamma \equiv \rho\dot{\eta} - s_{k,k} - \rho b \geq 0 \quad \text{in} \quad \mathscr{V} - \sigma \tag{1.11.23}$$

$$[\rho\eta(\mathbf{v} - \boldsymbol{\nu}) - \mathbf{s}] \cdot \mathbf{n} \geq 0 \quad \text{on} \quad \sigma \tag{1.11.24}$$

for the local entropy inequality. Further one may write $\mathbf{s} = \mathbf{q}/\theta$ and $b \equiv h/\theta$ for a simple thermomechanical continuum (cf. Section 1.13). For details of derivations see Eringen [1967, Sections 4.2, 4.4, and 4.6].

Equations (1.11.4), (1.11.17), (1.11.20), (1.11.21), (1.11.23), and the associated jump conditions (1.11.5), (1.11.18), (1.11.22), and (1.11.24) constitute the basic local laws of continuum mechanics. Usually the energy equation (1.11.21) and the entropy inequality (1.11.23) are considered the subject matter for the thermodynamics of continua.

[1] For polar media and continua with local structure, Eq. (1.11.20) is much more complicated and is replaced by a system of partial differential equations (cf. Eringen [1962, Section 32]; see also Eringen and Şuhubi [1964]).

Equations (1.11.17), (1.11.18), and (1.11.21) by use of (1.11.13) may also be written in component forms. Collecting all balance laws, we have

(1.11.25)
$$(\partial\rho/\partial t) + (\rho v_k)_{,k} = 0 \quad \text{in} \quad \mathscr{V} - \sigma$$
$$[\rho(v_k - \nu_k]n_k = 0 \quad \text{on} \quad \sigma$$

(1.11.26)
$$t_{kl,k} + \rho(f_l - \dot{v}_l) = 0 \quad \text{in} \quad \mathscr{V} - \sigma$$
$$[\rho v_l(v_k - \nu_k) - t_{kl}]n_k = 0 \quad \text{on} \quad \sigma$$

(1.11.27)
$$t_{kl} = t_{lk} \quad \text{in} \quad \mathscr{V} - \sigma$$

(1.11.28)
$$\rho\dot{\varepsilon} = t_{kl}v_{l,k} + q_{k,k} + \rho h \quad \text{in} \quad \mathscr{V} - \sigma$$
$$[(\rho\varepsilon + \tfrac{1}{2}v^2)(v_k - \nu_k) - t_{kl}v_l - q_k]n_k = 0 \quad \text{on} \quad \sigma$$

(1.11.29)
$$\rho\dot{\eta} - s_{k,k} - \rho b \geq 0 \quad \text{in} \quad \mathscr{V} - \sigma$$
$$[\rho\eta(v_k - \nu_k) - s_k]n_k \geq 0 \quad \text{on} \quad \sigma$$

1.12 LOCAL BALANCE LAWS IN THE REFERENCE FRAME

For a large class of problems, it is convenient to employ a formulation based on the reference configuration. For example, in finite deformation theory, the shape and location of the boundary is unknown at the outset and must be determined from the solution of the differential field equations. Thus if the undeformed shape and location of the boundary of the body are known (as is usually the case) then the boundary conditions, referred to reference frame, can be written at once. This then provides a reasonable amount of simplicity in dealing with a certain class of problems even though the field equations may be somewhat more complicated.

Here we express the local balance laws in the reference frame X_K. For the equation of local conservation of mass, from (1.11.6) we have

(1.12.1)
$$\rho_0 = \rho j, \qquad j \equiv \det x_{k,K}$$

To obtain the appropriate form of the balance of momenta we introduce the stress vector $\mathbf{T}_K$ at a spatial point $\mathbf{x}$ at time t occupied by a material point $\mathbf{X}$ in the undeformed area dA_K:

(1.12.2)
$$\mathbf{t}_{(\mathbf{n})}\,da = \mathbf{t}_k\,da_k = \mathbf{T}_K\,dA_K$$

Using (1.7.2), we obtain from this

(1.12.3)
$$\mathbf{t}_k = j^{-1}x_{k,K}\mathbf{T}_K, \qquad \mathbf{T}_K = jX_{K,k}\mathbf{t}_k$$

If we now substitute the first of these into (1.11.17) and use $(1.4.6)_1$ and (1.12.1), we obtain

(1.12.4) $$\mathbf{T}_{K,K} + \rho_0(\mathbf{f} - \dot{\mathbf{v}}) = \mathbf{0}$$

This is Cauchy's equation of motion in the reference state. For component representation we introduce the Piola–Kirchhoff pseudostresses T_{Kl} and T_{KL} by

(1.12.5) $$\mathbf{T}_K = T_{Kl}\mathbf{i}_l = T_{KL}x_{l,L}\mathbf{i}_l$$

so that by (1.12.3) we have

(1.12.6) $$T_{Kl} = jX_{K,k}t_{kl}, \qquad t_{kl} = j^{-1}x_{k,K}T_{Kl} = j^{-1}x_{k,K}x_{l,L}T_{KL}$$

(1.12.7) $$T_{KL} = T_{Kl}X_{L,l} = jX_{K,k}X_{L,l}t_{kl}$$

From (1.12.5) it is clear that T_{Kl} is the stress at $\mathbf{x}$ measured per unit undeformed area at $\mathbf{X} = \mathbf{X}(\mathbf{x}, t)$.

Using (1.12.5) in (1.12.4) we obtain two different forms of the equations of motion:

(1.12.8) $$T_{Kk,K} + \rho_0(f_k - \dot{v}_k) = 0$$

(1.12.9) $$(T_{KL}x_{k,L})_{,K} + \rho_0(f_k - \dot{v}_k) = 0$$

Cauchy's second law of motion follows from $t_{kl} = t_{lk}$ upon using $(1.12.6)_2$. Hence in two different forms we have

(1.12.10) $$T_{Kk}x_{l,K} = T_{Kl}x_{k,K}$$

(1.12.11) $$T_{KL} = T_{LK}$$

By using (1.12.1), (1.12.3), and $(1.4.6)_1$ in (1.11.21) for the local balance of energy, we obtain

(1.12.12)

$$\rho_0\dot{\varepsilon} = \mathbf{T}_K \cdot \mathbf{v}_{,K} + Q_{K,K} + \rho_0 h \qquad \text{or} \qquad \rho_0\dot{\varepsilon} = T_{KL}\dot{E}_{KL} + Q_{K,K} + \rho_0 h$$

where we also introduced

(1.12.13) $$Q_K \equiv jq_k X_{K,k}$$

Similarly for the entropy inequality (1.11.23) one finds

(1.12.14) $$\rho_0\gamma \equiv \rho_0\dot{\eta} - S_{K,K} - \rho_0 b \geq 0$$

where

(1.12.15) $$S_K \equiv js_k X_{K,k}$$

To write the jump conditions in the reference frame, we first note the relations between unit exterior normals $\mathbf{n}$ and $\mathbf{N}$ of the deformed surface $\mathscr{S}$ and the undeformed surface S, respectively. From (1.7.2) we have

$$da_k = jX_{K,k}\, dA_K \tag{1.12.16}$$

but

$$n_k = da_k/(da_l\, da_l)^{1/2} = da_k/da, \qquad N_K = dA_K/(dA_L\, dA_L)^{1/2} = dA_K/dA \tag{1.12.17}$$

Hence

$$n_k = jX_{K,k}\, N_K\, dA/da \tag{1.12.18}$$

Using (1.12.16) we have

$$dA/da = j^{-1}\left(\overset{-1}{C}_{KL} N_K N_L\right)^{-1/2} \tag{1.12.19}$$

Thus

$$n_k = \left(\overset{-1}{C}_{KL} N_K N_L\right)^{-1/2} X_{M,k} N_M \tag{1.12.20}$$

Using this, (1.12.6), (1.12.13), and (1.12.15) in Eqs. (1.11.5), (1.11.18), (1.11.22), and (1.11.24), we obtain the material form of the jump conditions[1]:

$$[\rho_0(v_k - \nu_k)X_{K,k}]N_K = 0 \qquad \text{on} \quad \Sigma \tag{1.12.21}$$

$$[\rho_0\, \mathbf{v}(v_k - \nu_k)X_{K,k} - \mathbf{T}_K]N_K = 0 \qquad \text{on} \quad \Sigma \tag{1.12.22}$$

$$[\rho_0(\varepsilon + \tfrac{1}{2}v^2)(v_k - \nu_k)X_{K,k} - T_{Kl}v_l - Q_K]N_K = 0 \qquad \text{on} \quad \Sigma \tag{1.12.23}$$

$$[\rho_0(v_k - \nu_k)\eta X_{K,k} - S_K]N_K \geq 0 \qquad \text{on} \quad \Sigma \tag{1.12.24}$$

Note that $j > 0$ and $X_{K,k}$ are considered to be continuous across the discontinuity surface Σ (the image of σ in the material description). Moreover, in (1.12.24) a nonnegative factor $\left(\overset{-1}{C}_{KL} N_K N_L\right)^{-1/2}$ has been dropped.

Equations (1.12.21)–(1.12.24) can be put into other forms which will be useful in later applications. If we define

$$U \equiv (\nu_k - v_k)n_k, \qquad U_N \equiv U\left(\overset{-1}{C}_{KL} N_K N_L\right)^{1/2} \tag{1.12.25}$$

[1] Balance laws and jump conditions in the reference frame can also be derived by writing the global equations in the reference frame and using the method of Section 1.11.

Then it follows that

$$[\rho_0 U_N] = 0 \quad \text{on} \quad \Sigma \tag{1.12.26}$$

$$[\mathbf{T}_K]N_K + \rho_0 U_N[\mathbf{v}] = \mathbf{0} \quad \text{on} \quad \Sigma \tag{1.12.27}$$

$$[T_{Kk} v_k + Q_K]N_K + \rho_0 U_N[\varepsilon + \tfrac{1}{2}v^2] = 0 \quad \text{on} \quad \Sigma \tag{1.12.28}$$

$$\rho_0 U_N[\eta] + [S_K]N_K \leq 0 \quad \text{on} \quad \Sigma \tag{1.12.29}$$

The foregoing expressions may be used as a source for approximate theories in which deformation gradients, strains, or rotations are small as compared to unity. To this end we recall Eq. $(1.4.16)_2$, i.e.,

$$X_{K,k} = (\delta_{lk} - \tilde{e}_{lk} - \tilde{r}_{lk})\, \delta_{lK} \tag{1.12.30}$$

Thus *if* $\tilde{e}_{lk}$ *are small as compared to unity*, using this and $j = 1 + \tilde{e}_{kk}$ in (1.12.6) and (1.12.7) we shall have

$$T_{Kl} \simeq (t_{ml} - \tilde{r}_{mk}\, t_{kl})\, \delta_{mK} \tag{1.12.31}$$

$$T_{KL} \simeq (t_{mk} - \tilde{r}_{ml}\, t_{lk} - \tilde{r}_{kl}\, t_{lm} + \tilde{r}_{mn}\, \tilde{r}_{kl}\, t_{nl})\, \delta_{mK}\, \delta_{kL} \tag{1.12.32}$$

If *both strains* $\tilde{e}_{lk}$ and *rotations* $\tilde{r}_{lk}$ are small as compared to unity, then

$$T_{Kl} \simeq t_{ml}\, \delta_{mK}, \qquad T_{KL} \simeq t_{mk}\, \delta_{mK}\, \delta_{kL} \tag{1.12.33}$$

Therefore, we see that *when both* $\tilde{e}_{kl}$ and $\tilde{r}_{kl}$ *are small as compared to unity (the infinitesimal deformation theory), there will be no difference between Eulerian and Piola stresses.* In this case there will be no distinctions among the Lagrangian and Eulerian forms of the balance laws and jump conditions.

We further remark that in some cases some components of rotations may not be negligible as compared to unity (e.g., the Kármán–Timoshenko theory of plates). All such cases are derivable from the basic equations obtained above.

1.13 CONSTITUTIVE EQUATIONS OF THERMOELASTIC SOLIDS

The local balance laws obtained in Sections 1.11 and 1.12 are valid for all bodies irrespective of their shapes and constitutions. They constitute five differential equations [one for mass (1.11.4), three for momentum (1.11.17), and one for energy (1.11.21)] among 19 unknowns ρ, v_k, t_{kl}, q_k, ε, η, s_k, and b, since the body force $\mathbf{f}$ and heat source h are supposed to be given. Fourteen additional equations must be given in order for the system to be determinate. Such a set must reflect the nature of the material and the constitution of the body. A general constitutive theory is available for all thermomechanical materials (cf. Eringen [1965; 1967, Chapter V)]. Here we are only interested in thermoelastic solids.

A thermoelastic solid is assumed to possess a natural state in which the body is undeformed and has uniform temperature. Upon loading and heat input, the body deforms and its temperature changes. When the loads and heat are relieved, the body assumes its natural state again. Permanent deformations and flow are not allowed. This rough physical idea of the thermoelastic solid may be made precise mathematically by posing a set of general constitutive equations of the form

$$
\begin{aligned}
t_{kl} &= f_{kl}(\mathbf{x}_{,K}, \theta, \theta_{,K}, \mathbf{X}, \mathbf{x}, t), \\
q_k &= g_k(\mathbf{x}_{,K}, \theta, \theta_{,K}, \mathbf{X}, \mathbf{x}, t), \\
s_k &= h_k(\mathbf{x}_{,K}, \theta, \theta_{,K}, \mathbf{X}, \mathbf{x}, t), \\
\psi \equiv \varepsilon - \theta\eta &= \psi(\mathbf{x}_{,K}, \theta, \theta_{,K}, \mathbf{X}, \mathbf{x}, t) \\
\eta &= n(\mathbf{x}_{,K}, \theta, \theta_{,K}, \mathbf{X}, \mathbf{x}, t)
\end{aligned}
\tag{1.13.1}
$$

where introduction of the *Helmholtz free energy* ψ in place of the internal energy density proves to be convenient. Note that at the outset all *response functions* f_{kl}, g_k, h_k, ψ, and n are assumed to depend on the same set of dependent variables $\mathbf{x}_{,K}$, θ, $\theta_{,K}$, $\mathbf{X}$, $\mathbf{x}$, and t. This is known as the *rule of equipresence*. Introduction of θ, in place of b by $b = h/\theta$, represents are placement of one function with another. However, θ, called the *absolute temperature, is posited to be nonnegative*, i.e.,

$$\theta > 0, \qquad \inf \theta = 0$$

and considered to be an *absolute scalar*.

The constitutive equations are subject to several basic axioms. Essential for our purpose are the axioms of: *objectivity*, *thermomechanical admissibility*, and *material symmetry*. We discuss these axioms briefly and apply them to the set (1.13.1).

I Axiom of Objectivity. *The response functions must be form invariant under the time-dependent proper group of orthogonal transformations of the spatial frame of reference and shift of the origin of time.*

Let $Q_{kl}(t)$ be a member of the proper group of orthogonal transformations; then the general transformation of the spatial coordinates and the shift of time are expressed by

$$\bar{x}_k = Q_{kl}(t)x_l + b_k(t), \qquad \bar{t} = t - a \tag{1.13.2}$$

subject to

$$Q_{kl}\,Q_{ml} = Q_{lk}\,Q_{lm} = \delta_{km}, \qquad \det Q_{kl} = 1 \tag{1.13.3}$$

Equations (1.13.2) express general rigid motion of the spatial frame of reference and shift of the origin of time. The axiom of objectivity applied to ψ, for example, states that

$$\psi(\bar{\mathbf{x}}_{,K}, \theta, \theta_{,K}, \mathbf{X}, \bar{\mathbf{x}}, \bar{t}) = \psi(\mathbf{x}_{,K}, \theta, \theta_{,K}, \mathbf{X}, \mathbf{x}, t)$$

or, using (1.13.2),

$$\psi(\mathbf{Q}\mathbf{x}_{,K}, \theta, \theta_{,K}, \mathbf{X}, \mathbf{Q}\mathbf{x} + \mathbf{b}, t - a) = \psi(\mathbf{x}_{,K}, \theta, \theta_{,K}, \mathbf{X}, \mathbf{x}, t) \tag{1.13.4}$$

for all members of the proper orthogonal group $\{\mathbf{Q}\}$, for arbitrary $\mathbf{b}$, and arbitrary a. Consider the following three special cases:

(a) $\mathbf{Q} = \mathbf{I}$, $\mathbf{b} = \mathbf{0}$, $t = a$. This gives

$$\psi(\mathbf{x}_{,K}, \theta, \theta_{,K}, \mathbf{X}, \mathbf{x}, 0) = \psi(\mathbf{x}_{,K}, \theta, \theta_{,K}, \mathbf{X}, \mathbf{x}, a)$$

Thus ψ cannot depend *explicitly* on time

(b) $\mathbf{Q} = \mathbf{I}$, $\mathbf{x} = -\mathbf{b}$, $a = 0$. This gives

$$\psi(\mathbf{x}_{,K}, \theta, \theta_{,K}, \mathbf{X}, \mathbf{0}) = \psi(\mathbf{x}_{,K}, \theta, \theta_{,K}, \mathbf{X}, -\mathbf{b})$$

This shows that ψ cannot depend explicitly on $\mathbf{x}$.

(c) $\mathbf{Q} =$ arbitrary, $\mathbf{b} = \mathbf{0}$, $a = 0$. This gives

$$\psi(\mathbf{Q}\mathbf{x}_{,K}, \theta, \theta_{,K}, \mathbf{X}) = \psi(\mathbf{x}_{,K}, \theta, \theta_{,K}, \mathbf{X}) \tag{1.13.5}$$

for all members of the proper orthogonal group of transformations $\{\mathbf{Q}\}$. This restriction stated in another way is: ψ is an isotropic function of the three vectors $\mathbf{x}_{,K}$. According to a theorem of Cauchy (cf. Eringen [1967], App. B.6), this implies that ψ is a function of the scalars $C_{KL} \equiv \mathbf{x}_{,K} \cdot \mathbf{x}_{,L}$:

$$\psi = \psi(C_{KL}, \theta, \theta_{,K}, \mathbf{X}) \tag{1.13.6}$$

Since the general rigid motion of the spatial frame of reference is a result of application of (a)–(c) in any order, we see that (1.13.6) is necessary and sufficient for the satisfaction of the axiom of objectivity for ψ.

The application of the axiom of objectivity to tensor-valued functions (such as f_{kl} and g_k) is made as follows: First, we introduce

$$F_{KL}(\mathbf{x}_{,K}, \theta, \theta_{,K}, \mathbf{X}, \mathbf{x}, t) = f_{kl}\, x_{k,K}\, x_{l,L}$$
$$G_K(\mathbf{x}_{,K}, \theta, \theta_{,K}, \mathbf{X}, \mathbf{x}, t) = g_k\, x_{k,K}$$

Now F_{KL} and G_K are scalar functions of the spatial coordinates under transformation (1.13.2). Thus, as in the case ψ, the objectivity implies

$$F_{KL} = F_{KL}(\mathbf{C}, \theta, \theta_{,K}, \mathbf{X}), \qquad G_K = G_K(\mathbf{C}, \theta, \theta_{,K}, \mathbf{X})$$

Upon inverting, we obtain f_{kl} and g_k. Collecting these results, we have

$$(1.13.7)\quad \begin{aligned} t_{kl} &= F_{KL}(\mathbf{C}, \theta, \theta_{,K}, \mathbf{X}) X_{K,k} X_{L,l}, & \psi &= \psi(\mathbf{C}, \theta, \theta_{,K}, \mathbf{X}) \\ q_k &= G_K(\mathbf{C}, \theta, \theta_{,K}, \mathbf{X}) X_{K,k}, & \eta &= N(\mathbf{C}, \theta, \theta_{,K}, \mathbf{X}) \\ s_k &= H_K(\mathbf{C}, \theta, \theta_{,K}, \mathbf{X}) X_{K,k}, & b &= h/\theta \end{aligned}$$

More convenient forms of these are obtained if we write

$$(1.13.8)\quad \begin{aligned} F_{KL} &= j^{-1} T_{MN} C_{MK} C_{NL}, & j^{-1} &\equiv \rho/\rho_0 = (III_C)^{1/2} \\ G_L &\equiv j^{-1} Q_K C_{LK}, & H_L &\equiv j^{-1} S_K C_{LK} \end{aligned}$$

Upon using this in (1.13.7) we get

$$(1.13.9)\quad \begin{aligned} t_{kl} &= \frac{\rho}{\rho_0} T_{KL}(\mathbf{C}, \theta, \theta_{,K}, \mathbf{X}) x_{k,K} x_{l,L}, & \psi &= \psi(\mathbf{C}, \theta, \theta_{,K}, \mathbf{X}) \\ q_k &= \frac{\rho}{\rho_0} Q_K(\mathbf{C}, \theta, \theta_{,K}, \mathbf{X}) x_{k,K}, & \eta &= N(\mathbf{C}, \theta, \theta_{,K}, \mathbf{X}) \\ s_k &= \frac{\rho}{\rho_0} S_K(\mathbf{C}, \theta, \theta_{,K}, \mathbf{X}) x_{k,K}, & b &= h/\theta \end{aligned}$$

II Axiom of Thermomechanical Admissibility. *Constitutive equations must obey the balance laws and entropy inequality. The entropy inequality must be satisfied for all independent processes.*

Eliminating h between (1.11.21) and (1.11.23) after writing $\varepsilon = \psi + \theta\eta$ and $b = h/\theta$, the entropy inequality takes the form

(1.13.10)

$$\rho\gamma \equiv -\frac{\rho}{\theta}(\dot{\psi} + \dot{\theta}\eta) + \frac{1}{\theta} t_{kl} v_{l,k} + \frac{1}{\theta^2} q_k \theta_{,k} + \left(\frac{1}{\theta} q_k - s_k\right)_{,k} \geq 0 \qquad \text{in} \quad \mathscr{V} - \sigma$$

This is known as the *Clausius–Duhem inequality*. Substituting (1.13.9) into (1.13.10), we have

$$\begin{aligned} &-\frac{\rho_0}{\theta}\left(\frac{\partial \psi}{\partial \theta} + \eta\right)\dot{\theta} - \frac{\rho_0}{\theta}\frac{\partial \psi}{\partial \theta_{,K}}\dot{\theta}_{,K} + \frac{1}{2\theta}\left(T_{KL} - 2\rho_0 \frac{\partial \psi}{\partial C_{KL}}\right)\dot{C}_{KL} \\ &+ \left(\frac{1}{\theta}\frac{\partial Q_K}{\partial \theta} - \frac{\partial S_K}{\partial \theta}\right)\theta_{,K} + \frac{\partial}{\partial \theta_{,L}}\left(\frac{1}{\theta} Q_K - S_K\right)\theta_{,KL} \\ &+ \frac{\partial}{\partial C_{MN}}\left(\frac{1}{\theta} Q_K - S_K\right) C_{MN,K} + \frac{1}{\theta}\frac{\partial Q_K}{\partial X_K} - \frac{\partial S_K}{\partial X_K} \geq 0 \qquad \text{in} \quad \mathscr{V} - \sigma \end{aligned}$$

where we used (1.9.15) and $(j^{-1}x_{k,K})_{,k} = 0$, which is derivable from (1.4.6). This inequality is linear in $\dot{\theta}$, $\dot{\theta}_{,K}$, $\dot{C}_{KL}$, $C_{MN,K}$, and $\theta_{,LK}$. For arbitrary variations of these quantitites the inequality cannot hold unless the coefficients of these quantities vanish:

$$T_{KL} = 2\rho_0 \frac{\partial \psi}{\partial C_{KL}}, \qquad \frac{\partial \psi}{\partial C_{KL}} - \frac{\partial \psi}{\partial C_{LK}} = 0, \qquad \frac{\partial \psi}{\partial \theta_{,K}} = 0, \qquad \eta = -\frac{\partial \psi}{\partial \theta}$$

$$\text{(1.13.11)} \qquad \frac{\partial}{\partial C_{MN}}\left(\frac{1}{\theta}Q_K - S_K\right) + \frac{\partial}{\partial C_{NK}}\left(\frac{1}{\theta}Q_M - S_M\right) = 0$$

$$\frac{\partial}{\partial \theta_{,L}}\left(\frac{1}{\theta}Q_K - S_K\right) + \frac{\partial}{\partial \theta_{,K}}\left(\frac{1}{\theta}Q_L - S_L\right) = 0$$

$$\left(\frac{1}{\theta}\frac{\partial Q_K}{\partial \theta} - \frac{\partial S_K}{\partial \theta}\right)\theta_{,K} + \frac{1}{\theta}\frac{\partial Q_K}{\partial X_K} - \frac{\partial S_K}{\partial X_K} \geq 0$$

Of these, the fifth equation follows from

$$\frac{\partial}{\partial C_{MN}}\left(\frac{1}{\theta}Q_K - S_K\right)C_{MN,K} = 2\frac{\partial}{\partial C_{MN}}\left(\frac{1}{\theta}Q_K - S_K\right)x_{k,N}\,x_{k,MK}$$

In (1.13.11) the second equation is satisfied since $C_{KL} = C_{LK}$, and the third indicates that ψ is independent of $\theta_{,K}$. The general solutions of the fifth and sixth equations are

$$\text{(1.13.12)} \qquad \frac{1}{\theta}Q_K - S_K = \Omega_{KL}(\theta, \mathbf{X})\theta_{,L} + F_K(\theta, \mathbf{X}),$$

where $\Omega_{KL} = -\Omega_{LK}$ and F_K are functions of θ and $\mathbf{X}$ only. Substituting this into the last relation in (1.13.11) we get[1]

$$\text{(1.13.13)} \qquad \frac{1}{\theta^2}Q_K\theta_{,K} + \left(\frac{\partial \Omega_{LK}}{\partial X_L} + \frac{\partial F_K}{\partial \theta}\right)\theta_{,K} + \frac{\partial F_K}{\partial X_K} \geq 0$$

If this inequality is to be satisfied for all values of $\theta_{,K}$, we must have

$$\text{(1.13.14)} \qquad \frac{\partial F_K}{\partial X_K} \geq 0, \qquad \left(\frac{1}{\theta^2}Q_K + \frac{\partial \Omega_{LK}}{\partial X_L} + \frac{\partial F_K}{\partial \theta}\right)\theta_{,K} \geq 0$$

[1] These results were obtained in somewhat different way by Müller [1967], who also investigated possible material symmetries for which Ω_{KL} may exist. In a later paper [1971] Müller also investigated the case where $\dot{\theta}$ is included as an additional variable in the constitutive equations. In this case the energy equation can contain $\ddot{\theta}$ thus giving the interesting possibility of heat conduction with finite speeds.

If Q_K is a continuous function of $\theta_{,K}$ for the values of $\theta_{,K}$ near zero but slightly greater than zero, the factor of $\theta_{,K}$ must be nonnegative, while for the values of $\theta_{,K}$ slightly less than zero the factor of $\theta_{,K}$ must be nonpositive. Thus at $\theta_{,K} = 0$ we must have

$$\frac{1}{\theta^2} Q_K(\mathbf{C}, \mathbf{0}, \mathbf{X}, \theta) + \frac{\partial \Omega_{LK}}{\partial X_L} + \frac{\partial F_K}{\partial \theta} = 0 \tag{1.13.15}$$

For $\mathbf{C} = \mathbf{I}$ this equation becomes

$$\frac{1}{\theta^2} Q_K(\mathbf{I}, \mathbf{0}, \mathbf{X}, \theta) + \frac{\partial \Omega_{LK}}{\partial X_L} + \frac{\partial F_K}{\partial \theta} = 0 \tag{1.13.16}$$

Subtracting this from (1.13.15) we obtain

$$\frac{1}{\theta^2} [Q_K(\mathbf{C}, \mathbf{0}, \mathbf{X}, \theta) - Q_K(\mathbf{I}, \mathbf{0}, \mathbf{X}, \theta)] = 0 \tag{1.13.17}$$

It is reasonable to assume that for rigid deformation and vanishing thermal gradients there can be no heat conduction, i.e., $Q_K(\mathbf{I}, \mathbf{0}, \mathbf{X}, \theta) = 0$. Thus

$$Q_K(\mathbf{C}, \mathbf{0}, \mathbf{X}, \theta) = 0 \tag{1.13.18}$$

This is the statement and the proof of the fact that the *piezocaloric effect does not exist*. Clearly this result depends on the assumption $\mathbf{Q} = \mathbf{0}$ when $\mathbf{C} = \mathbf{I}$ and $\theta_{,K} = 0$ (the vanishing of heat with both deformation and thermal gradient).

Through (1.13.16) we now have

$$F_K = -\int^{\theta} \frac{\partial \Omega_{LK}}{\partial X_L} \, d\theta + F_K{}^0(\mathbf{X}) \tag{1.13.19}$$

Using this in (1.13.12) and (1.13.14) we obtain

$$\frac{1}{\theta} Q_K - S_K = \Omega_{KL}(\theta, \mathbf{X})\theta_{,L} + \int^{\theta} \frac{\partial \Omega_{KL}}{\partial X_L} \, d\theta + F_K{}^0(\mathbf{X}) \tag{1.13.20}$$

$$\frac{1}{\theta^2} Q_K \theta_{,K} \geq 0, \qquad \partial F_K{}^0 / \partial X_K \geq 0 \tag{1.13.21}$$

Collecting these results we have the following theorem.

Theorem 1. (i) *The thermodynamically admissible form of the entropy flux vector* $\mathbf{S}$ *is given by* (1.13.20);

(ii) *the piezocaloric effect does not exist*;

(iii) *the heat vector* Q_K *and* $F_K{}^0$ *is subject to the inequalities* (1.13.21).

Carrying the expressions of T_{KL} and η given by $(1.13.11)_{1,4}$ and using S_K obtained above [Eq. (1.13.20)] in (1.13.9), we have the next result.

***Theorem* 2.** *A thermoelastic solid is thermodynamically admissible if the constitutive equations for* $\mathbf{t}$, $\mathbf{q}$, ε, *and* η *are of the form*

$$t_{kl} = 2\rho \frac{\partial \psi}{\partial C_{KL}} x_{(k,K} x_{l),L} \qquad \varepsilon = \psi - \theta \frac{\partial \psi}{\partial \theta}$$

(1.13.22)

$$q_k = \frac{\rho}{\rho_0} Q_K(\mathbf{C}, \theta_{,K}, \theta, \mathbf{X}) x_{k,K} \qquad \eta = -\frac{\partial \psi}{\partial \theta}$$

where $\psi = \psi(\mathbf{C}, \theta, \mathbf{X})$ *is the free energy and* Q_K *is subject to conditions* (1.13.18) *and* (1.13.21).

For nonheat-conducting solids, it proves sometimes convenient to employ ε in place of ψ and η in place of θ, i.e.,

$$\varepsilon = \varepsilon(\mathbf{C}, \eta, \mathbf{X}), \qquad \theta = \theta(\mathbf{C}, \eta, \mathbf{X})$$

Through the use of

$$\psi(\mathbf{C}, \theta) = \varepsilon[\mathbf{C}, \eta(\mathbf{C}, \theta)] - \theta\eta(\mathbf{C}, \theta)$$

we calculate

$$\left.\frac{\partial \psi}{\partial \mathbf{C}}\right|_\theta = \left.\frac{\partial \varepsilon}{\partial \mathbf{C}}\right|_\eta + \left.\frac{\partial \varepsilon}{\partial \eta}\right|_\mathbf{C} \left.\frac{\partial \eta}{\partial \mathbf{C}}\right|_\theta - \theta \left.\frac{\partial \eta}{\partial \mathbf{C}}\right|_\theta$$

$$\left.\frac{\partial \psi}{\partial \theta}\right|_\mathbf{C} = \left.\frac{\partial \varepsilon}{\partial \eta}\right|_\mathbf{C} \left.\frac{\partial \eta}{\partial \theta}\right|_\mathbf{C} - \eta - \theta \left.\frac{\partial \eta}{\partial \theta}\right|_\mathbf{C}$$

Substituting from $(1.13.22)_4$ this gives

$$\left.\frac{\partial \psi}{\partial \mathbf{C}}\right|_\theta = \left.\frac{\partial \varepsilon}{\partial \mathbf{C}}\right|_\eta, \qquad \theta = \left.\frac{\partial \varepsilon}{\partial \eta}\right|_\mathbf{C}$$

Hence the constitutive equations (1.13.22) may also be expressed as

$$t_{kl} = 2\rho \frac{\partial \varepsilon}{\partial C_{KL}} x_{(k,K} x_{l),L} \qquad \varepsilon = \varepsilon(\mathbf{C}, \eta, \mathbf{X})$$

(1.13.23)

$$q_k = \frac{\rho}{\rho_0} Q_K(\mathbf{C}, \theta_{,K}, \eta, \mathbf{X}) x_{k,K}, \qquad \theta = \frac{\partial \varepsilon}{\partial \eta}$$

Finally we write the expression of the Piola–Kirchhoff stress tensor T_{Kl} which will be used in the sequel. Substituting (1.13.22) into $(1.12.6)_1$, we get

$$T_{Kl} = 2\rho_0 \frac{\partial \psi}{\partial C_{KM}} x_{l,M} \tag{1.13.24}$$

which may also be written in alternative form

$$T_{Kl} = \rho_0 \frac{\partial \psi}{\partial x_{l,K}} \tag{1.13.25}$$

Using this and $(1.12.6)_2$ we also have

$$t_{kl} = \rho \frac{\partial \psi}{\partial x_{(l,K}} x_{k),K} \tag{1.13.26}$$

To obtain (1.13.25) we consider ψ as a function of $x_{k,K}$ in place of C_{KL}. Then by the chain rule

$$\frac{\partial \psi}{\partial C_{KM}} = (\partial \psi / \partial x_{k,R})(\partial x_{k,R} / \partial C_{KM})$$

But from $(1.4.8)_2$ by differentiation

$$\delta_{KM}\, \delta_{LN} = 2(\partial x_{k,K} / \partial C_{MN}) x_{k,L}$$

or

$$\partial x_{k,K} / \partial C_{MN} = \tfrac{1}{2}\, \delta_{KM}\, X_{N,k} \tag{1.13.27}$$

Using these in (1.13.24) we obtain (1.13.25).

It is important to note that in forms (1.13.25) the axiom of objectivity is destroyed, since now $\psi = \psi(x_{k,K}, \theta_{,K}, \eta, \mathbf{X})$ which is not restricted by the objectivity. However, forms (1.13.24) and (1.13.26) obey this axiom.

The equation of heat conduction follows from the energy equation (1.11.21):

$$\rho\theta \left(\frac{\partial^2 \psi}{\partial \theta^2} \dot{\theta} + \frac{\partial^2 \psi}{\partial \theta\, \partial C_{KL}} \dot{C}_{KL} \right) + q_{k,k} + \rho h = 0 \tag{1.13.28}$$

For the *incompressible solids*, the stress constitutive equation $(1.13.22)_1$ must be further restricted, since in this case

$$\rho_0 / \rho = (III_C)^{1/2} = 1 \tag{1.13.29}$$

This implies that one of the six components of C_{KL} is determined through (1.13.29). Thus $\partial \psi / \partial C_{KL}$ are not all independent. To take care of this situation, we introduce the Lagrange multiplier p by writing

$$\psi - (p/2\rho_0)(III_C - 1)$$

in place of ψ. Substituting this and using

$$(\partial III_C/\partial C_{MN})C_{MK} = III_C\, \delta_{NK}$$

$(1.13.22)_1$ takes the form

$$t_{kl} = -p\, \delta_{kl} + 2\rho_0 \frac{\partial \psi}{\partial C_{KL}} x_{(k,K} x_{l),L} \tag{1.13.30}$$

The function $p(\mathbf{x}, t)$ introduced here is called the *pressure*. It is to be determined upon integration of differential equations and using the boundary conditions. Note that this is *not* the same as the thermodynamic pressure.

III Axiom of Material Symmetry. *Constitutive equations must be form invariant with respect to a group of orthogonal transformations* $\{\mathbf{S}\}$ *and translations* $\{\mathbf{B}\}$ *of the material coordinates. These restrictions are the result of the material symmetry conditions characterized by* $\{\mathbf{S}\}$ *and* $\{\mathbf{B}\}$ *in the material frame of reference.*

Mathematically, the response functions will be form invariant under all transformations of the form

$$\bar{X}_K = S_{KL} X_L + B_K \tag{1.13.31}$$

subject to

$$S_{KM} S_{LM} = S_{MK} S_{ML} = \delta_{KL}, \qquad \det S_{KL} = \pm 1 \tag{1.13.32}$$

for all members of $\{S_{KL}\}$ and all translations $\{B_K\}$. These conditions express geometrical symmetries represented by $\{\mathbf{S}\}$ and inhomogeneities represented by $\{\mathbf{B}\}$ at $\mathbf{X}$, in the thermomechanical properties of the body. The group $\{\mathbf{S}\}$ is a subgroup of the full orthogonal transformations. When $\{\mathbf{S}\}$ is the *full group*, the solid is called *isotropic*; when it is the *proper orthogonal group*, it is called *hemitropic*. Otherwise, the material is known as *anisotropic*. When the response functions do not depend on $\{\mathbf{B}\}$, it is called *homogeneous*; otherwise, the solid is inhomogeneous.

For the free energy function and the heat vector the material symmetry restrictions read

$$\begin{aligned} \psi(\mathbf{C}, \theta, \mathbf{X}) &= \psi(\mathbf{S}\mathbf{C}\mathbf{S}^{\mathrm{T}}, \theta, \mathbf{S}\mathbf{X} + \mathbf{B}) \\ \mathbf{S}\mathbf{Q}(\mathbf{C}, \theta, \nabla\theta, \mathbf{X}) &= \mathbf{Q}(\mathbf{S}\mathbf{C}\mathbf{S}^{\mathrm{T}}, \theta, \mathbf{S}\nabla\theta, \mathbf{S}\mathbf{X} + \mathbf{B}) \end{aligned} \tag{1.13.33}$$

We note here that out of infinitely many members of the group $\{\mathbf{S}\}$, twelve distinct ones are adequate for the description of 32 classes of crystalline materials. In the next section we obtain the consequence of these restrictions for the isotropic thermoelastic solids.

1.14 ISOTROPIC THERMOELASTIC SOLIDS

For isotropic materials, {S} is the full group of orthogonal transformations. We consider also only the homogeneous solids. Thus ψ and Q_K do not depend on $\mathbf{X}$ and are isotropic functions. This means that ψ is a function of invariants of $\mathbf{C}$, i.e.,

$$\rho_0 \psi = \Sigma(I_1, I_2, I_3, \theta) \tag{1.14.1}$$

It is well known that [cf. Spencer (1971)] a hemitropic matrix polynomial $\mathbf{P}(\mathbf{a}, \mathbf{b})$ of two matrices $\mathbf{a}$ and $\mathbf{b}$ has the following specific form when $\mathbf{P}$ and $\mathbf{a}$ are skew symmetric and $\mathbf{b}$ is symmetric:

$$\begin{aligned} \mathbf{P} = {} & K_1\mathbf{a} + K_2(\mathbf{ab} + \mathbf{ba}) + K_3(\mathbf{ab}^2 + \mathbf{b}^2\mathbf{a}) + K_4(\mathbf{a}^2\mathbf{b} - \mathbf{ba}^2) \\ & + K_5(\mathbf{a}^2\mathbf{b}^2 - \mathbf{b}^2\mathbf{a}^2) + K_6(\mathbf{ba}^2\mathbf{b}^2 - \mathbf{b}^2\mathbf{a}^2\mathbf{b}) \end{aligned} \tag{1.14.2}$$

where K_1–K_6 are polynomials in the invariants

$$\operatorname{tr} \mathbf{a}^2, \quad \operatorname{tr} \mathbf{b}, \quad \operatorname{tr} \mathbf{b}^2, \quad \operatorname{tr} \mathbf{b}^3, \quad \operatorname{tr} \mathbf{a}^2\mathbf{b}, \quad \operatorname{tr} \mathbf{a}^2\mathbf{b}^2, \quad \operatorname{tr} \mathbf{aba}^2\mathbf{b}^2 \tag{1.14.3}$$

We assume that $\mathbf{Q}$ can be expressed by a polynomial in $\mathbf{C}$ and $\nabla\theta$. Thus, if we set

$$P_{KL} \equiv e_{KLM} Q_M, \qquad a_{KL} \equiv e_{KLM} \theta_{,M}, \qquad b_{KL} \equiv C_{KL} \tag{1.14.4}$$

then (1.14.2) gives a bivector representation of the heat constitutive equation. Upon multiplying P_{KL} by e_{KLR} and using the identity

$$e_{KLM} e_{KRS} = \delta_{LR} \delta_{MS} - \delta_{LS} \delta_{MR}$$

after some manipulations (1.14.2) becomes

$$\begin{aligned} Q_K = {} & [(K_1 + K_2 C_{MM} + K_3 C_{NM} C_{MN}) \delta_{KL} - K_2 C_{KL} - K_3 C_{KM} C_{ML}]\theta_{,L} \\ & + e_{KLM}(K_4 C_{MR} + K_5 C_{RS} C_{SM})\theta_{,L} \theta_{,R} + K_6 e_{KLM} C_{MN} C_{NP} C_{RL} \theta_{,R} \theta_{,P} \end{aligned} \tag{1.14.5}$$

This is valid for hemitropic materials. For isotropic materials (1.14.5) must be form invariant under reflection of any one of the material axes. This imposes the conditions $K_4 = K_5 = K_6 = 0$. Thus for isotropic solids (1.14.5) reduces to

$$Q_K = (K_0 \delta_{KL} - K_2 C_{KL} - K_3 C_{KM} C_{ML})\theta_{,L} \tag{1.14.6}$$

where K_0, K_2, and K_3 are polynomials in invariants (1.14.3) or equivalently in the joint invariants of C_{KL} and $\theta_{,K}$:

$$I_1 \equiv C_{KK}, \qquad I_2 \equiv C_{KL}C_{LK}, \qquad I_3 \equiv C_{KL}C_{LM}C_{MK}$$
$$I_4 \equiv \theta_{,K}\theta_{,K}, \qquad I_5 \equiv C_{KL}\theta_{,K}\theta_{,L}, \qquad I_6 \equiv C_{KM}C_{ML}\theta_{,K}\theta_{,L} \tag{1.14.7}$$

Upon writing

$$C_{KL} \equiv x_{k,K}x_{k,L}, \qquad \overset{-1}{c}_{kl} \equiv X_{k,K}X_{l,K}, \qquad \theta_{,K} = \theta_{,k}x_{k,K} \tag{1.14.8}$$

the invariants (1.14.7) take the form

$$I_1 = \overset{-1}{c}_{kk}, \qquad I_2 = \overset{-1}{c}_{kl}\overset{-1}{c}_{lk}, \qquad I_3 = \overset{-1}{c}_{kl}\overset{-1}{c}_{lm}\overset{-1}{c}_{mk}$$
$$I_4 = \overset{-1}{c}_{kl}\theta_{,k}\theta_{,l}, \qquad I_5 = \overset{-1}{c}_{kl}\overset{-1}{c}_{lm}\theta_{,k}\theta_{,m}, \qquad I_0 \equiv \theta_{,k}\theta_{k,} \tag{1.14.9}$$

Here I_1 to I_5 are expressions of those listed in (1.14.7) in terms of $\overset{-1}{\mathbf{c}}$ and $\theta_{,k}$, and I_0 results from expressing I_6 in terms of I_1 to I_5 and I_0 through the use of Cayley–Hamilton theorem. Carrying (1.14.6) into $(1.13.22)_3$ we obtain the constitutive equations for heat

$$q_k = (\kappa_0\,\delta_{kl} + \kappa_1\,\overset{-1}{c}_{kl} + \kappa_2\,\overset{-1}{c}_{km}\overset{-1}{c}_{ml})\theta_{,l} \tag{1.14.10}$$

where

$$\kappa_0 = -(\rho/3\rho_0)K_3(I_3 - \tfrac{3}{2}I_1I_2 + \tfrac{1}{2}I_1{}^3)$$
$$\kappa_1 = (\rho/2\rho_0)[2K_0 + K_3(I_1{}^2 - I_2)] \tag{1.14.11}$$
$$\kappa_2 = -(\rho/\rho_0)(K_2 + K_3I_1)$$

The stress constitutive equations follow by substituting (1.14.1) into $(1.13.22)_1$:

$$t_{kl} = \frac{\rho}{\rho_0}\left\{(I_1{}^3 - 3I_1I_2 + 2I_3)\frac{\partial\Sigma}{\partial I_3}\delta_{kl} + \left[2\frac{\partial\Sigma}{\partial I_1} - 3(I_1{}^2 - I_2)\frac{\partial\Sigma}{\partial I_3}\right]\overset{-1}{c}_{kl} + \left(4\frac{\partial\Sigma}{\partial I_2} + 6I_1\frac{\partial\Sigma}{\partial I_3}\right)\overset{-1}{c}_{km}\overset{-1}{c}_{ml}\right\} \tag{1.14.12}$$

This may be expressed entirely in terms of $\overset{-1}{\mathbf{c}}$ and its principal invariants. To this end we employ the relations (1.6.10) which are valid for the tensor $\overset{-1}{\mathbf{c}}$. This gives

$$t_{kl} = h_0\,\delta_{kl} + h_1\,\overset{-1}{c}_{kl} + h_2\,\overset{-1}{c}_{km}\overset{-1}{c}_{ml} \tag{1.14.13}$$

where

(1.14.14)

$$h_0 \equiv 2(III)^{1/2}\frac{\partial\Sigma}{\partial III},\qquad h_1 \equiv \frac{2}{(III)^{1/2}}\left(\frac{\partial\Sigma}{\partial I} + I\,\frac{\partial\Sigma}{\partial II}\right),\qquad h_2 \equiv -\frac{2}{(III)^{1/2}}\frac{\partial\Sigma}{\partial II}$$

$$I \equiv I_{\overset{-1}{c}},\qquad II \equiv II_{\overset{-1}{c}},\qquad III \equiv III_{\overset{-1}{c}}$$

Another form of (1.14.13) often used is

$$t_{kl} = (h_1 + h_2 I)\,\overset{-1}{c}_{kl} + (h_0 - h_2 II)\,\delta_{kl} + h_2 IIIc_{kl} \tag{1.14.15}$$

which is obtained by replacing the factor of h_2 in (1.14.13) by use of the Cayley–Hamilton theorem. Other forms are possible (cf. Eringen [1962, Section 47]).

Similarly a second form of (1.14.10) for heat is

$$q_k = [(\kappa_1 + \kappa_2 I)\,\overset{-1}{c}_{kl} + (\kappa_0 - \kappa_2 II)\,\delta_{kl} + \kappa_2 IIIc_{kl}]\theta_{,l} \tag{1.14.16}$$

The entropy inequality (1.13.21) places restrictions on the coefficient κ_α:

$$\kappa_0 I_0 + \kappa_1 I_4 + \kappa_2 I_5 \geq 0 \tag{1.14.17}$$

The equation of heat conduction (1.13.28) takes the form

$$\frac{\rho\theta}{\rho_0}\left(\frac{\partial^2\Sigma}{\partial\theta^2}\,\dot{\theta} + \frac{\partial^2\Sigma}{\partial\theta\,\partial I_\alpha}\,\dot{I}_\alpha\right) + q_{k,k} + \rho h = 0 \tag{1.14.18}$$

Another useful form of t_{kl} and q_k for isotropic solids results from considering Σ as a function of the invariants of $\mathbf{c}$. In this case we have

$$t_{kl} = 2\,\frac{\rho}{\rho_0}\,\frac{\partial\Sigma}{\partial c_{mn}}\,\frac{\partial c_{mn}}{\partial C_{KL}}\,x_{(k,K}x_{l),L}$$

To calculate $\partial c_{mn}/\partial C_{KL}$ we note the relation

$$\partial X_{M,m}/\partial C_{KL} = -\tfrac{1}{2}\overset{-1}{C}_{ML}X_{K,m} \tag{1.14.19}$$

which is obtained by differentiating $(1.4.3)_2$ and using $(1.4.9)_2$ and (1.13.27). Using this and $(1.4.3)_2$ we obtain

$$t_{kl} = -2\,\frac{\rho}{\rho_0}\,\frac{\partial\Sigma}{\partial c_{m(k}}\,c_{l)m} = \frac{\rho}{\rho_0}\,(\delta_{km} - 2e_{km})\,\frac{\partial\Sigma}{\partial e_{lm}} \tag{1.14.20}$$

If we now substitute Σ into (1.14.20) we obtain

$$t_{kl} = a_0\,\delta_{kl} + a_1 c_{kl} + a_2\,c_{km}\,c_{ml} \tag{1.14.21}$$

$$t_{kl} = b_0\,\delta_{kl} + b_1 e_{kl} + b_2\,e_{km}\,e_{ml} \tag{1.14.22}$$

where

$$a_0(I_c, II_c, III_c, \theta, \mathbf{X}) \equiv -2III_c^{3/2}\, \partial\Sigma/\partial III_c$$

(1.14.23) $$a_1(I_c, II_c, III_c, \theta, \mathbf{X}) \equiv -2III_c^{1/2}\left(\frac{\partial\Sigma}{\partial I_c} + I_c \frac{\partial\Sigma}{\partial II_c}\right)$$

$$a_2(I_c, II_c, III_c, \theta, \mathbf{X}) = 2III_c^{1/2}\frac{\partial\Sigma}{\partial II_c}$$

$$b_0(I_e, II_e, III_e, \theta, \mathbf{X}) = \frac{\rho}{\rho_0}\left[\frac{\partial\Sigma}{\partial I_e} + I_e\frac{\partial\Sigma}{\partial II_e} + (II_e - 2III_e)\frac{\partial\Sigma}{\partial III_e}\right]$$

(1.14.24) $$b_1(I_e, II_e, III_e, \theta, \mathbf{X}) = -\frac{\rho}{\rho_0}\left[2\frac{\partial\Sigma}{\partial I_e} + (1 + 2I_e)\frac{\partial\Sigma}{\partial II_e} + I_e\frac{\partial\Sigma}{\partial III_e}\right]$$

$$b_2(I_e, II_e, III_e, \theta, \mathbf{X}) = \frac{\rho}{\rho_0}\left(2\frac{\partial\Sigma}{\partial II_e} + \frac{\partial\Sigma}{\partial III_e}\right)$$

$$\rho/\rho_0 = (1 - 2I_e + 4II_e - 8III_e)^{1/2}$$

The expression of q_k involving $\mathbf{c}$ (or $\mathbf{e}$) has the same form as (1.4.10) in which $\overset{-1}{\mathbf{c}}$ is replaced by $\mathbf{c}$ (or $\mathbf{e}$) and κ_α are now considered to be functions of the joint invariants of $\mathbf{c}$ (or $\mathbf{e}$) and $\theta_{,k}$.

In the *undistorted* state $\mathbf{c} = \mathbf{I}$, $\mathbf{e} = \mathbf{0}$. From any one of (1.14.13), (1.14.15), (1.14.21), and (1.14.22), we arrive at the following theorem.

Theorem. *In an isotropic elastic solid the state of stress in the reference configuration (undistorted) is always hydrostatic.*

Incompressible solids. For the incompressible solids $\rho_0 = \rho$, $III = 1$, and therefore Σ is independent of III, i.e., $\Sigma = \Sigma(I, II, \theta, \mathbf{X})$. In this case, we add an arbitrary pressure term $-p(\mathbf{x}, t)$ on the right-hand side of (1.14.13) and (1.14.15). Thus

(1.14.25)

$$t_{kl} = -p\,\delta_{kl} + 2\left(\frac{\partial\Sigma}{\partial I} + I\frac{\partial\Sigma}{\partial II}\right)\overset{-1}{c}_{kl} - 2\frac{\partial\Sigma}{\partial II}\overset{-1}{c}_{km}\overset{-1}{c}_{ml} \qquad \text{(incompressible)}$$

(1.14.26) $$t_{kl} = -p\,\delta_{kl} + 2\frac{\partial\Sigma}{\partial I}\overset{-1}{c}_{kl} - 2\frac{\partial\Sigma}{\partial II}c_{kl} \qquad \text{(incompressible)}$$

The constitutive equations (1.14.10) and (1.14.16) do not change forms except that κ_α do not depend on III (=1). The equation of heat conduction (1.14.18) now reads

(1.14.27)

$$-\theta\left(\frac{\partial^2\Sigma}{\partial\theta^2}\dot{\theta}+\frac{\partial^2\Sigma}{\partial\theta\,\partial I}\dot{I}+\frac{\partial^2\Sigma}{\partial\theta\,\partial II}\dot{II}\right)+q_{k,k}+\rho h=0 \qquad \text{(incompressible)}$$

The foregoing equations are all exact. Various order approximate theories can be obtained from these by assuming that strains and temperature gradients are small. In the following two sections we present two such approximations, namely, the linear and quadratic theories.

Finally, we examine the expression of entropy **S** given by (1.13.20). Both Q_K and S_K are invariant under the full group of orthogonal transformations $\{S_{KL}\}$ of the material frame of reference as expressed by (1.13.31) and (1.13.32). Since Ω_{KL} is a skew-symmetric tensor, it follows that $\Omega_{KL}=0$, $F_K{}^0=0$. Thus we have the following theorem.

Theorem. *For isotropic thermoelastic solids the entropy flux* **S** *is given by*

(1.14.28)
$$\mathbf{S}=\mathbf{Q}/\theta, \qquad \mathbf{s}=\mathbf{q}/\theta$$

This is the classical expression of the entropy flux.

1.15 LINEAR THEORY

For approximate theories employing the order of magnitude of the strain tensors, it is convenient to write the constitutive equations in terms of the Lagrangian strain measure

$$E_{KL}\equiv\tfrac{1}{2}(C_{KL}-\delta_{KL})$$

Equation (1.13.22) in terms of **E** reads

(1.15.1)

$$t_{kl}=\frac{\rho}{\rho_0}\frac{\partial\Sigma}{\partial E_{KL}}x_{(k,K}x_{l),L}, \qquad \varepsilon=\frac{1}{\rho_0}\left(\Sigma-\theta\frac{\partial\Sigma}{\partial\theta}\right)$$

$$q_k=\frac{\rho}{\rho_0}Q_K(\mathbf{E},\theta_{,K}\theta,\mathbf{X})x_{k,K}, \qquad \eta=-\frac{1}{\rho_0}\frac{\partial\Sigma}{\partial\theta}$$

where

(1.15.2)
$$\rho_0^{-1}\Sigma(\mathbf{E},\theta,\mathbf{X})=\psi$$

is the free energy. The function Q_K is subject to

(1.15.3)
$$Q_K\theta_{,K}\geq 0, \qquad Q_K(\mathbf{E},\mathbf{0},\theta,\mathbf{X})=0$$

of which the inequality expresses the fact that *heat flows from hot to cold*, and $(1.15.3)_2$ means *that the heat vanishes with the temperature gradient*. Equations (1.15.1) are valid for anisotropic inhomogeneous solids. We now proceed to linearize these equations. Compatible with this, the strain measure $\mathbf{E}$ is to be replaced by the infinitesimal strain tensor $\tilde{\mathbf{E}}$, and we employ the infinitesimal rotation tensor $\tilde{\mathbf{R}}$:

$$\tilde{E}_{KL} \equiv \tfrac{1}{2}(U_{K,L} + U_{L,K}), \qquad \tilde{R}_{KL} \equiv \tfrac{1}{2}(U_{K,L} - U_{L,K}) \tag{1.15.4}$$

Linear theory requires that we write Σ as a quadratic polynomial in $\tilde{\mathbf{E}}$ and Q_K as a linear function in $\tilde{\mathbf{E}}$ and $\theta_{,K}$, i.e.,

(1.15.5)

$$\Sigma = \Sigma_0 + \Sigma_{KL}\tilde{E}_{KL} + \tfrac{1}{2}\Sigma_{KLMN}\tilde{E}_{KL}\tilde{E}_{MN}, \qquad Q_K = B_K + B_{KM}\theta_{,M} + B_{KLM}\tilde{E}_{LM}$$

where Σ_0, Σ_{KL}, Σ_{KLMN}, B_K, B_{KM}, and B_{KLM} are functions of θ and $\mathbf{X}$ and satisfy the symmetry conditions

(1.15.6)

$$\Sigma_{KL} = \Sigma_{LK}, \qquad \Sigma_{KLMN} = \Sigma_{MNKL} = \Sigma_{LKMN} = \Sigma_{KLNM}, \qquad B_{KLM} = B_{KML}$$

which result from the symmetry of the strain tensor. Since, according to $(1.15.3)_2$ Q_K must vanish with $\theta_{,M}$, we have $B_K = 0$, $B_{KLM} = 0$. Substituting (1.15.5) into $(1.15.1)_{1,2}$ and using expression (1.4.16), i.e.,

$$x_{k,K} = (\delta_{MK} + \tilde{E}_{MK} + \tilde{R}_{MK})\,\delta_{Mk} \tag{1.15.7}$$

we obtain

(1.15.8)

$$t_{kl} = (\rho/\rho_0)[\Sigma_{KL} + \Sigma_{LM}(\tilde{E}_{KM} + \tilde{R}_{KM}) + \Sigma_{KM}(\tilde{E}_{LM} + \tilde{R}_{LM}) + \Sigma_{KLMN}\tilde{E}_{MN}]\,\delta_{Kk}\,\delta_{Ll}$$

$$q_k = (\rho/\rho_0)B_{KM}\,\theta_{,M}\,\delta_{Kk} \tag{1.15.9}$$

In linear theory we have

(1.15.10)

$$\tilde{E}_{KL} \simeq \tilde{e}_{kl}\,\delta_{Kk}\,\delta_{Ll}, \qquad \tilde{R}_{KL} \simeq \tilde{r}_{kl}\,\delta_{Kk}\,\delta_{Ll}, \qquad (\rho/\rho_0) \simeq 1 - \tilde{E}_{KK} \simeq 1 - \tilde{e}_{kk}$$

Using these in (1.15.7) and (1.15.8) we have the spatial forms

(1.15.11)

$$t_{kl} = \sigma_{kl}(1 - \tilde{e}_{mm}) + \sigma_{lm}(\tilde{e}_{km} + \tilde{r}_{km}) + \sigma_{km}(\tilde{e}_{lm} + \tilde{r}_{lm}) + \sigma_{klmn}\,\tilde{e}_{mn}$$

$$q_k = b_{km}\,\theta_{,m} \tag{1.15.12}$$

where

(1.15.13)

$$\sigma_{kl} \equiv \Sigma_{KL}\,\delta_{kK}\,\delta_{lL}, \qquad \sigma_{klmn} \equiv \Sigma_{KLMN}\,\delta_{kK}\,\delta_{lL}\,\delta_{mM}\,\delta_{nN}, \qquad b_{kl} \equiv B_{KL}\,\delta_{kK}\,\delta_{lL}$$

are functions of θ and $\mathbf{X}$ and subject to symmetry conditions

$$\sigma_{kl} = \sigma_{lk}, \qquad \sigma_{klmn} = \sigma_{mnkl} = \sigma_{lkmn} = \sigma_{klnm}, \tag{1.15.14}$$

Inequality (1.15.3) implies that

$$b_{km}\theta_{,k}\theta_{,m} \geq 0 \tag{1.15.15}$$

which states that the *heat conduction matrix* b_{km} *is a nonnegative* form.

The stress potential Σ takes the form

$$\Sigma = \Sigma_0 + \sigma_{kl}\tilde{e}_{kl} + \tfrac{1}{2}\sigma_{klmn}\tilde{e}_{kl}\tilde{e}_{mn} \tag{1.15.16}$$

The entropy given by $(1.15.1)_4$ takes the form

$$\eta = -\frac{1}{\rho_0}\left(\frac{\partial \Sigma_0}{\partial \theta} + \frac{\partial \sigma_{kl}}{\partial \theta}\tilde{e}_{kl} + \frac{1}{2}\frac{\partial \sigma_{klmn}}{\partial \theta}\tilde{e}_{kl}\tilde{e}_{mn}\right) \tag{1.15.17}$$

The heat conduction equation (1.13.28) becomes

$$\frac{\rho\theta}{\rho_0}\left(\frac{\partial^2 \Sigma_0}{\partial \theta^2}\dot{\theta} + \frac{\partial^2 \sigma_{kl}}{\partial \theta^2}\dot{\theta}\tilde{e}_{kl} + \frac{\partial \sigma_{kl}}{\partial \theta}\dot{\tilde{e}}_{kl}\right) + (b_{kl}\theta_{,l})_{,k} + \rho h = 0 \tag{1.15.18}$$

Note that all these equations are still nonlinear with respect to temperature. Further linearization may be made by taking

$$\theta = T_0 + T, \qquad |T| \ll T_0, \qquad 0 < T_0 \tag{1.15.19}$$

where T is the temperature change from the temperature T_0 of the natural state. In this case, we may write

$$\Sigma_0 = S_0 - \rho_0\eta_0 T - (\rho_0\gamma/2T_0)T^2, \qquad \sigma_{kl} = \alpha_{kl} - \beta_{kl}T \tag{1.15.20}$$

so that

$$\Sigma = S_0 - \rho_0\eta_0 T - (\rho_0\gamma/2T_0)T^2 + \alpha_{kl}\tilde{e}_{kl} - \beta_{kl}\tilde{e}_{kl}T + \tfrac{1}{2}\sigma_{klmn}\tilde{e}_{kl}\tilde{e}_{mn} \tag{1.15.21}$$

where S_0, η_0, γ, α_{kl}, β_{kl}, and σ_{klmn} are functions of $\mathbf{X}$ only. The constitutive equations now become

$$t_{kl} = \alpha_{kl}(1 - \tilde{e}_{mm}) - \beta_{kl}T + \alpha_{lm}(\tilde{e}_{km} + \tilde{r}_{km}) + \alpha_{km}(\tilde{e}_{lm} + \tilde{r}_{lm}) + \sigma_{klmn}\tilde{e}_{mn} \tag{1.15.22}$$

$$q_k = b_{kl}\theta_{,l} \tag{1.15.23}$$

$$\rho_0\varepsilon = S_0 + \rho_0\eta_0 T_0 + \rho_0\gamma[T + (T^2/2T_0)] + (\alpha_{kl} + T_0\beta_{kl})\tilde{e}_{kl} + \tfrac{1}{2}\sigma_{klmn}\tilde{e}_{kl}\tilde{e}_{mn} \tag{1.15.24}$$

$$\eta = \eta_0 + (\gamma/T_0)T + (1/\rho_0)\beta_{kl}\tilde{e}_{kl} \tag{1.15.25}$$

The linear equation of heat conduction follows from (1.15.18)

(1.15.26) $$-\rho_0 \gamma \dot{T} - T_0 \beta_{kl} \dot{\tilde{e}}_{kl} + (b_{kl} T_{,l})_{,k} + \rho h = 0$$

When the natural state is stress free, then $\alpha_{kl} = 0$, and (1.15.22) in this case reads

(1.15.27) $$t_{kl} = -\beta_{kl} T + \sigma_{klmn} \tilde{e}_{mn}$$

with corresponding reduction in ε by setting $\alpha_{kl} = 0$.

Equations (1.15.22)–(1.15.25) *are the constitutive equations of the linear theory of anisotropic, inhomogeneous thermoelasticity, and* (1.15.26) *is the equation of heat conduction.* Of these, (1.15.22) is known as the generalized *Hooke's law* and (1.15.23) is *Fourier's law* of heat conduction. These equations are subject to material symmetry conditions. For anisotropic materials possessing various material symmetries, the number of constitutive coefficients α_{kl}, β_{kl}, σ_{klmn}, and b_{kl} are reduced. For various types of anisotropic elastic solids, the reader is referred to books in classical elasticity (e.g., Love [1944], Sokolnikoff [1956], and Eringen [1967, Section 6]). Here we give the equations for the isotropic solids.

Isotropic solids. In this case, α_{kl}, β_{kl}, b_{kl}, and σ_{klmn} are isotropic tensors. Thus they have the general forms:

(1.15.28) $$\alpha_{kl} = \alpha\, \delta_{kl}, \qquad \beta_{kl} = \beta\, \delta_{kl}, \qquad b_{kl} = \kappa\, \delta_{kl}$$
$$\sigma_{klmn} = \lambda\, \delta_{kl}\, \delta_{mn} + \mu(\delta_{km}\, \delta_{ln} + \delta_{kn}\, \delta_{lm})$$

With these, (1.15.22)–(1.15.25) become

(1.15.29)
$$t_{kl} = \alpha(1 - \tilde{e}_{mm})\, \delta_{kl} - \beta T\, \delta_{kl} + \lambda \tilde{e}_{mm}\, \delta_{kl} + 2(\mu + \alpha)\tilde{e}_{kl}, \qquad q_k = \kappa T_{,k}$$
$$\rho_0 \varepsilon = S_0 + \rho_0 \eta_0 T_0 + \rho_0 \gamma[T + (T^2/2T_0)] + (\alpha + \beta T_0) I_{\tilde{e}} + \tfrac{1}{2}(\lambda + 2\mu) I_{\tilde{e}}^{\,2} - 2\mu II_{\tilde{e}}$$
$$\eta = \eta_0 + (\gamma/T_0)T + (\beta/\rho_0) I_{\tilde{e}}$$

The equation of heat conduction (1.15.26) reduces to

(1.15.30) $$-\rho_0 \gamma \dot{T} - \beta T_0 \dot{u}_{k,k} + (\kappa T_{,k})_{,k} + \rho h = 0$$

When the natural state is stress free then $\alpha = 0$.

For *incompressible* solids we have $\tilde{e}_{rr} = 0$, and the above equations are modified to

(1.15.31) $$t_{kl} = (-p + \alpha - \beta T)\, \delta_{kl} + 2(\mu + \alpha)\tilde{e}_{kl}$$

(1.15.32) $$q_k = \kappa T_{,k}$$

(1.15.33) $$\rho_0 \varepsilon = S_0 + \rho_0 \eta_0 T_0 + \rho_0 \gamma[T + (T^2/2T_0)] - 2\mu II_{\tilde{e}}$$

(1.15.34) $$\eta = \eta_0 + (\gamma/T_0)T$$

The equation of heat conduction is

$$-\rho_0 \gamma \dot{T} + (\kappa T_{,k})_{,k} + \rho_0 h = 0 \tag{1.15.35}$$

In this case, we have the incompressibility condition

$$\dot{e}_{kk} = \dot{u}_{k,k} = 0 \tag{1.15.36}$$

1.16 QUADRATIC THEORY

To obtain stress and heat constitutive equations which are polynomials of second degree in the deformation and temperature gradients, $U_{K,L}$ and $\theta_{,K}$, we include the next order terms in (1.15.5), i.e.,

(1.16.1)
$$\begin{aligned}\Sigma &= \Sigma_0 + \Sigma_{KL} E_{KL} + \tfrac{1}{2}\Sigma_{KLMN} E_{KL} E_{MN} + \tfrac{1}{3}\Sigma_{KLMNPQ} E_{KL} E_{MN} E_{PQ} \\ Q_K &= B_K + B_{KM}\theta_{,M} + \bar{B}_{KLM} E_{LM} + B_{KLM}\theta_{,L}\theta_{,M} + B_{KLMN} E_{LM}\theta_{,N} \\ &\quad + B_{KMNPQ} E_{MN} E_{PQ}\end{aligned}$$

Here the constitutive coefficients Σ_0, Σ_{KL}, Σ_{KLMN}..., and B_{KMNPQ} are functions of θ and $\mathbf{X}$ only. For homogeneous materials they are independent of $\mathbf{X}$. If the stress vanishes in the natural state then $\Sigma_{KL} = 0$. Condition (1.13.18) implies that $B_K = \bar{B}_{KLM} = B_{KMNPQ} = 0$. Condition $(1.13.21)_1$ now gives

$$B_{KM}\theta_{,K}\theta_{,M} + B_{KLM}\theta_{,K}\theta_{,L}\theta_{,M} + B_{KLMN}\theta_{,K} E_{LM}\theta_{,N} \geq 0$$

which must be satisfied for all independent variations of $\theta_{,K}$ and E_{LM}. Since this inequality is linear in E_{LM}, clearly it is not possible to maintain it unless $B_{KLMN} = 0$. Further, it is not difficult to see that the remaining third-degree polynomial cannot be made nonnegative for all values of $\theta_{,K}$ unless $B_{KLM} = 0$. Hence we have

$$Q_K = B_{KM}\theta_{,M} \tag{1.16.2}$$

as in the linear theory.

From the composition $(1.16.1)_1$, it is clear that the following symmetry conditions hold for the material moduli:

(1.16.3)
$$\begin{aligned}&\Sigma_{KL} = \Sigma_{LK}, \qquad \Sigma_{KLMN} = \Sigma_{MNKL} = \Sigma_{LKMN} = \Sigma_{KLNM} \\ &\Sigma_{KLMNPQ} = \Sigma_{MNKLPQ} = \Sigma_{KLPQMN} = \Sigma_{LKMNPQ} = \Sigma_{KLNMPQ} = \Sigma_{KLMNQP}\end{aligned}$$

Substituting $(1.16.1)_1$ and (1.16.2) into $(1.15.1)_1$ we get

(1.16.4)
$$t_{kl} = (\rho/\rho_0)(\Sigma_{KL} + \Sigma_{KLMN} E_{MN} + \Sigma_{KLMNPQ} E_{MN} E_{PQ}) x_{(k,K} x_{l),L}$$

which contains terms of higher than second order in $U_{K,L}$. To obtain the final form of (1.16.4) involving terms not higher than the second degree in $U_{K,L}$ we further substitute

$$
\begin{aligned}
& x_{k,K} = \delta_{Kk} + U_{M,K}\,\delta_{Mk} \\
(1.16.5)\qquad & \rho/\rho_0 = j^{-1} = (1 + 2I_E + 4II_E + 8III_E)^{-1/2} \\
& = 1 - I_{\tilde{E}} - 2II_{\tilde{E}} + \tfrac{3}{2}I_{\tilde{E}}^{\,2} - \tfrac{1}{2}U_{P,M}\,U_{P,M} + 0(\tilde{\mathbf{E}}^3)
\end{aligned}
$$

and drop all products of $U_{K,L}$ whose degrees are greater than two. Hence

$$
\begin{aligned}
(1.16.6)\qquad t_{kl} = {} & \{\Sigma_{KL}(1 - I_{\tilde{E}} - 2II_{\tilde{E}} + \tfrac{3}{2}I_{\tilde{E}}^{\,2} - \tfrac{1}{2}U_{P,M}\,U_{P,M}) \\
& + (1 - I_{\tilde{E}})(\Sigma_{KM}\,U_{L,M} + \Sigma_{ML}\,U_{K,M}) + \Sigma_{MN}\,U_{K,M}\,U_{L,N} \\
& + [\Sigma_{KLMN}(1 - I_{\tilde{E}}) + \Sigma_{PLMN}\,U_{K,P} + \Sigma_{KPMN}\,U_{L,P}]\tilde{E}_{MN} \\
& + \tfrac{1}{2}\Sigma_{KLMN}\,U_{P,M}\,U_{P,N} + \Sigma_{KLMNPQ}\,\tilde{E}_{MN}\,\tilde{E}_{PQ}\}\delta_{Kk}\,\delta_{Ll}
\end{aligned}
$$

Similarly, the expression of q_k is obtained as

$$
(1.16.7)\qquad q_k = B_{KM}[(1 - I_{\tilde{E}})\,\delta_{Kk} + U_{N,K}\,\delta_{Nk}]\theta_{,M}
$$

The constitutive equations for ε and η follow from $(1.15.1)_{2,4}$ by substituting $(1.16.1)_1$:

(1.16.8)

$$
\begin{aligned}
\varepsilon = {} & \frac{1}{\rho_0}\left[\Sigma_0 - \theta\,\frac{\partial\Sigma_0}{\partial\theta} + \left(\Sigma_{KL} - \theta\,\frac{\partial\Sigma_{KL}}{\partial\theta}\right)E_{KL} + \left(\Sigma_{KLMN} - \theta\,\frac{\partial\Sigma_{KLMN}}{\partial\theta}\right)E_{KL}\,E_{MN}\right. \\
& \left. + \frac{1}{2}\left(\Sigma_{KLMNPQ} - \theta\,\frac{\partial\Sigma_{KLMNPQ}}{\partial\theta}\right)E_{KL}\,E_{MN}\,E_{PQ}\right]
\end{aligned}
$$

(1.16.9)

$$
\eta = -\frac{1}{\rho_0}\left(\frac{\partial\Sigma_0}{\partial\theta} + \frac{\partial\Sigma_{KL}}{\partial\theta}\,E_{KL} + \frac{1}{2}\,\frac{\partial\Sigma_{KLMN}}{\partial\theta}\,E_{KL}\,E_{MN} + \frac{1}{3}\,\frac{\partial\Sigma_{KLMNPQ}}{\partial\theta}\,E_{KL}\,E_{MN}\,E_{PQ}\right)
$$

Constitutive equations (1.16.6)–(1.16.9) are valid for arbitrary values of θ. If a quadratic approximation is desired in the temperature change T from the temperature of the natural state T_0, then we must further substitute

(1.16.10)

$$
\begin{aligned}
& \theta = T_0 + T, \qquad |T| \ll T_0, \qquad 0 < T_0 \\
& \Sigma_0 = \Sigma_0{}^0 + \Sigma_0{}^1 T + \Sigma_0{}^2 T^2 + \Sigma_0{}^3 T^3, \quad \Sigma_{KL} = \Sigma_{KL}^0 + \Sigma_{KL}^1\,T + \Sigma_{KL}^2\,T^2 \\
& \Sigma_{KLMN} = \Sigma_{KLMN}^0 + \Sigma_{KLMN}^1\,T, \qquad\qquad B_{KL} = B_{KL}^0 + B_{KL}^1\,T
\end{aligned}
$$

and drop terms that are higher than degree two in the products of $U_{K,L}$, $T_{,K}$, and T. The new material moduli $\Sigma_0{}^\alpha$, Σ_{KL}^α, ..., B_{KL}^α are functions of $\mathbf{X}$ only and, of course, they possess the same symmetry conditions as (1.16.3).

(i) *Isotropic solids.* For isotropic solids the material moduli are isotropic tensors. Under the symmetry conditions (1.16.3) the forms of various tensors are given by

(1.16.11)

$$\Sigma_{KL} = \alpha_E \, \delta_{KL}, \qquad \Sigma_{KLMN} = \lambda_E \, \delta_{KL} \, \delta_{MN} + \mu_E(\delta_{KM} \, \delta_{LN} + \delta_{KN} \, \delta_{LM})$$

$$\begin{aligned} \Sigma_{KLMNPQ} = {} & \tfrac{1}{2}(6l_E + 3m_E + n_E) \, \delta_{KL} \, \delta_{MN} \, \delta_{PQ} \\ & - \tfrac{1}{4}(m_E + n_E)[\delta_{KL}(\delta_{MP} \, \delta_{NQ} + \delta_{MQ} \, \delta_{NP}) \\ & + \delta_{PQ}(\delta_{KM} \, \delta_{LN} + \delta_{LM} \, \delta_{KN}) + \delta_{MN}(\delta_{KP} \, \delta_{LQ} + \delta_{KQ} \, \delta_{LP})] \\ & + (n_E/8)[\delta_{LM}(\delta_{KP} \, \delta_{NQ} + \delta_{KQ} \, \delta_{NP}) + \delta_{KM}(\delta_{LP} \, \delta_{NQ} + \delta_{LQ} \, \delta_{NP}) \\ & + \delta_{KN}(\delta_{LP} \, \delta_{MQ} + \delta_{LQ} \, \delta_{MP}) + \delta_{LN}(\delta_{KP} \, \delta_{MQ} + \delta_{KQ} \, \delta_{MP})] \end{aligned}$$

where α_E, λ_E, μ_E, l_E, m_E, and n_E are the six material elastic moduli of the second-degree theory.

Substituting (1.16.11) into (1.16.6) and (1.16.7) we obtain the constitutive equations for stress and heat for isotropic second-degree thermoelastic solids:

(1.16.12)

$$\begin{aligned} t_{kl} = {} & \{[\alpha_E + (\lambda_E - \alpha_E)I_{\tilde{E}} + (m_E + n_E - 2\alpha_E)II_{\tilde{E}} + \tfrac{1}{2}(3\alpha_E - 2\lambda_E + 6l_E + 2m_E)I_{\tilde{E}}^2 \\ & + \tfrac{1}{2}(\lambda_E - \alpha_E)U_{P,M} \, U_{P,M}] \, \delta_{KL} + [2(\mu_E + \alpha_E) \\ & + (2\lambda_E - 2\mu_E - 2\alpha_E - m_E - n_E)I_{\tilde{E}}]\tilde{E}_{KL} + \mu_E \, U_{M,K} \, U_{M,L} \\ & + \alpha_E \, U_{K,M} \, U_{L,M} + 2\mu_E(\tilde{E}_{ML} \, U_{K,M} + \tilde{E}_{KM} \, U_{L,M}) + n_E \, \tilde{E}_{KM} \, \tilde{E}_{ML}\} \, \delta_{Kk} \, \delta_{Ll} \end{aligned}$$

(1.16.13)

$$q_k = \kappa[(1 - I_{\tilde{E}}) \, \delta_{KM} + U_{M,K}] \, \delta_{Mk} \, \theta_{,K}$$

The stress potential, for this approximation follows from $(1.16.1)_1$ by use of $(1.16.11)_1$. Thus[1]

(1.16.14)

$$\Sigma = \Sigma_0 + \alpha_E \, I_E + \tfrac{1}{2}(\lambda_E + 2\mu_E)I_E^2 - 2\mu_E \, II_E + l_E \, I_E^3 + m_E \, I_E \, II_E + n_E \, III_E$$

The expression of entropy follows from substituting (1.16.14) into $(1.13.22)_4$.

Again if the quadratic approximation in the temperature change is desired, we further expand Σ_0, α_E, λ_E, μ_E, l_E, m_E, and n_E into power series of T about $\theta = T_0$ and retain only the second-degree products of T, $U_{K,L}$, and $T_{,L}$ in (1.16.12) and (1.16.14).

Of course, results (1.16.12)–(1.16.14) may also be derived directly from (1.14.10) and (1.14.12) in which one employs (1.16.14) for Σ and writes κ_α as

[1] Note that (1.16.14) contains higher-order terms than the third in $U_{K,L}$, since the invariants II_E and III_E do. There is no particular difficulty in writing the exact third-order approximation by replacing these quantities by the invariants of $\tilde{E}_{KL}$ and products of $U_{K,L}$.

polynomials in the invariants of **E**. Below we derive the expressions of t_{kl} and q_k in terms of the Eulerian strain measures in this fashion. For this purpose, we write Σ in the form

(1.16.15)
$$\Sigma = \Sigma_0 + \alpha_e I_e + \tfrac{1}{2}(\lambda_e + 2\mu_e)I_e^2 - 2\mu_e II_e + l_e I_e^3 + m_e I_e II_e + n_e III_e$$

Substituting this into (1.14.22) and retaining the terms up to and including second degree in $u_{k,l}$ we obtain[1]

(1.16.16)
$$\begin{aligned} t_{kl} = {} & [\alpha_e + (\lambda_e - \alpha_e)I_{\tilde{e}} - \tfrac{1}{2}(\lambda_e - \alpha_e)u_{p,q}u_{p,q} + (3l_e + m_e - \lambda_e - \tfrac{1}{2}\alpha_e)I_{\tilde{e}}^2 \\ & + (m_e + n_e + 2\alpha_e)II_{\tilde{e}}]\,\delta_{kl} + [2(\mu_e - \alpha_e) - (m_e + n_e + 2\lambda_e \\ & + 2\mu_e - 2\alpha_e)I_{\tilde{e}}]\tilde{e}_{kl} + (\alpha_e - \mu_e)u_{p,k}u_{p,l} + (n_e - 4\mu_e)\tilde{e}_{km}\,\tilde{e}_{ml} \end{aligned}$$

To obtain the heat conduction law in Eulerian coordinates we first write

(1.16.17)
$$q_k = \frac{\rho}{\rho_0}\kappa\theta_{,K}x_{k,K} = \frac{\rho}{\rho_0}\kappa\theta_{,l}x_{l,K}x_{k,K} = \frac{\rho}{\rho_0}\kappa\,\overset{-1}{c}_{kl}\,\theta_{,l}$$

The entropy inequality $q_k\theta_{,k} \geq 0$ is satisfied if $\kappa \geq 0$. Using (1.16.5) in this we obtain the second-degree approximation:

(1.16.18)
$$q_k = \kappa[(1 - I_{\tilde{e}})\,\delta_{kl} + 2\tilde{e}_{kl}]\theta_{,l}$$

To find the relations between the Eulerian and Lagrangian thermoelastic moduli by (1.6.20), (1.6.21), and (1.6.24), we express invariants of **E** in terms of those of **e**. When these are used in (1.16.15) and the result is compared with (1.16.14) we obtain

(1.16.19)
$$\begin{aligned} &\alpha_e = \alpha_E, \quad \lambda_e = \lambda_E, \quad \mu_e = \mu_E + 2\alpha_E \\ &l_e = l_E + 2\lambda_E + 4\mu_E + 4\alpha_E, \quad m_e = m_E - 4\lambda_E - 12\mu_E - 12\alpha_E \\ &n_e = n_E + 12\mu_E + 12\alpha_E \end{aligned}$$

The constitutive equations ε and η follow from substituting (1.16.15) into $(1.15.1)_{3,4}$. If a quadratic approximation in T is desired these coefficients must be expanded into a power series in T, and the products of T, e_{kl}, and $T_{,k}$ exceeding second degree must be dropped from the constitutive equations (1.16.16) and (1.16.17).

(ii) *Incompressible solids.* The incompressibility condition $\rho = \rho_0$ in quadratic approximation is given by

(1.16.20)
$$-I_{\tilde{e}} + 2II_{\tilde{e}} - \tfrac{1}{2}I_{\tilde{e}}^2 + \tfrac{1}{2}u_{p,k}u_{p,k} = 0$$

[1] Equations (60.4) and (60.2) of Eringen [1962] corresponding to (1.16.16) and (1.16.19) contain some numerical errors.

Using this we eliminate $II_{\tilde{e}}$ from (1.16.16) and add the pressure term on the right-hand side, thus obtaining

(1.16.21)

$$
\begin{aligned}
t_{kl} = {} & -p\,\delta_{kl} + [\alpha_e + \tfrac{1}{2}(2\lambda_e + m_e + n_e)I_{\tilde{e}} - \tfrac{1}{4}(2\lambda_e + m_e + n_e)u_{p,q}u_{p,q} \\
& + \tfrac{1}{4}(12l_e + 5m_e + n_e - 4\lambda_e)I_{\tilde{e}}^{\,2}]\,\delta_{kl} \\
& - (m_e + n_e + 2\lambda_e + 2\mu_e - 2\alpha_e)I_{\tilde{e}}\,\tilde{e}_{kl} \\
& + 2(\mu_e - \alpha_e)(\tilde{e}_{kl} - \tfrac{1}{2}u_{p,k}u_{p,l}) + (n_e - 4\mu_e)\tilde{e}_{km}\,\tilde{e}_{ml} \qquad \text{(incompressible)}
\end{aligned}
$$

(1.16.22)
$$q_k = \kappa(\delta_{kl} + 2\tilde{e}_{kl})\theta_{,l} \qquad \text{(incompressible)}$$

valid to a second-degree approximation.

1.17 FIELD EQUATIONS

The field equations of the nonlinear theory of thermoelasticity are the union of the balance laws and the constitutive equations. These may be expressed in Lagrangian or Eulerian forms. Here we give the Lagrangian form for nonlinear theory. For linear theory the two forms are identical.

Cauchy's equations of motion and the equation of energy balance are, respectively, given by (1.12.9) and (1.12.12):

(1.17.1)
$$(T_{KL}x_{k,L})_{,K} + \rho_0(f_k - \ddot{x}_k) = 0$$

(1.17.2)
$$\rho_0\,\dot{\varepsilon} - T_{KL}\dot{x}_{l,L}x_{l,K} - Q_{K,K} - \rho_0\,h = 0$$

The constitutive equations are given by

(1.17.3)
$$
\begin{gathered}
T_{KL} = 2\,\frac{\partial\Sigma}{\partial C_{KL}}, \qquad Q_K = Q_K(\mathbf{C}, \theta_{,K}, \theta, \mathbf{X}) \\
\Sigma = \Sigma(\mathbf{C}, \theta, \mathbf{X}), \qquad \eta = -\frac{1}{\rho_0}\frac{\partial\Sigma}{\partial\theta}, \qquad \varepsilon = \frac{1}{\rho_0}\left(\Sigma - \theta\,\frac{\partial\Sigma}{\partial\theta}\right)
\end{gathered}
$$

where for the deformation tensor $\mathbf{C}$ we have

(1.17.4)
$$C_{KL} = x_{k,K}x_{k,L}$$

Substituting these into (1.17.1) and (1.17.2) we obtain

(1.17.5)
$$A_{KkLl}x_{l,KL} + A_{Kk}\theta_{,K} + A_k + \rho_0(f_k - \ddot{x}_k) = 0$$

(1.17.6)
$$\rho_0 c\dot{\theta} - \theta A_{Kk}\dot{x}_{k,K} - H_{KkL}x_{k,KL} - H_{KL}\theta_{,LK} - H_K\theta_{,K} - H - \rho_0 h = 0$$

where A_{KkLl}, A_{Kk}, and A_k, called *elasticities*, H_{KkL}, H_{KL}, and H_K, called *conductivities*, and c, called *specific heat*, are given by

$$
\begin{aligned}
A_{KkLl}(\mathbf{C}, \theta, \mathbf{X}) &\equiv 4 \frac{\partial^2 \Sigma}{\partial C_{KM}\, \partial C_{LN}} x_{k,M}\, x_{l,N} + 2 \frac{\partial \Sigma}{\partial C_{KL}} \delta_{kl} \\
A_{Kk}(\mathbf{C}, \theta, \mathbf{X}) &\equiv 2 \frac{\partial^2 \Sigma}{\partial C_{KL}\, \partial \theta} x_{k,L} \\
A_k(\mathbf{C}, \theta, \mathbf{X}) &\equiv 2 \frac{\partial^2 \Sigma}{\partial C_{KL}\, \partial X_K} x_{k,L} \\
c(\mathbf{C}, \theta, \mathbf{X}) &\equiv -\frac{\theta}{\rho_0} \frac{\partial^2 \Sigma}{\partial \theta^2} \\
H_{KkL}(\mathbf{C}, \theta_{,K}, \theta, \mathbf{X}) &\equiv 2 \frac{\partial Q_K}{\partial C_{LN}} x_{k,N} \\
H_{KL}(\mathbf{C}, \theta_{,K}, \theta, \mathbf{X}) &\equiv \frac{\partial Q_K}{\partial \theta_{,L}} \\
H_K(\mathbf{C}, \theta_{,K}, \theta, \mathbf{X}) &\equiv \frac{\partial Q_K}{\partial \theta} \\
H(\mathbf{C}, \theta_{,K}, \theta, \mathbf{X}) &\equiv \frac{\partial Q_K}{\partial X_K}
\end{aligned}
\tag{1.17.7}
$$

Given the body force f_k and the heat source h, field equations (1.17.5) and (1.17.6) constitute a system of four nonlinear partial differential equations for the determination of four unknown functions $x_k(\mathbf{X}, t)$ and $\theta(\mathbf{X}, t)$. When the reference configuration is *homogeneous* then

$$
A_k = 0, \qquad H = 0 \tag{1.17.8}
$$

When the state of body is *isothermal*, $\theta = \text{const}$. Since $\mathbf{Q}$ vanishes with $\theta_{,K} = 0$, in this case (1.17.5) and (1.17.6) reduce to

$$
A_{KkLl}\, x_{l,KL} + A_k + \rho_0(f_k - \ddot{x}_k) = 0, \qquad \theta A_{Kk}\, \dot{x}_{k,K} - \rho_0 h = 0 \tag{1.17.9}
$$

If the body is not heat conducting (*adiabatic*), $\mathbf{Q} = \mathbf{0}$, the heat source vanishes, $h = 0$, and (1.17.6) gives $\eta = \text{const}$. In this case simplicity is achieved if the constitutive equations (1.13.23) employing η (in place of θ) are used. Then (1.17.5) is replaced by

$$
A_{KkLl}\, x_{l,KL} + A_k + \rho_0(f_k - \ddot{x}_k) = 0 \tag{1.17.10}
$$

where A_{KkLl} and A_k are now given by

(1.17.11)
$$A_{KkLl}(\mathbf{C}, \eta, \mathbf{X}) = 4\rho_0 \frac{\partial^2 \varepsilon}{\partial C_{KM}\, \partial C_{LN}} x_{k,M} x_{l,N} + 2\rho_0 \frac{\partial \varepsilon}{\partial C_{KL}} \delta_{kl}$$
$$A_k(\mathbf{C}, \eta, \mathbf{X}) = 2\rho_0 \frac{\partial^2 \varepsilon}{\partial C_{KL}\, \partial X_K} x_{k,L} \qquad \text{(adiabatic)}$$

For isothermal and adiabatic states the deformation is uncoupled from the temperature field, and we need to solve three equations $(1.17.9)_1$ or (1.17.10) to determine the motion $x_k(\mathbf{X}, t)$.

If Σ is considered as a function of $x_{k,K}$ in place of C_{KL}, expression $(1.17.7)_1$ takes a particularly simple form:

(1.17.12)
$$A_{KkLl} = \partial^2 \Sigma / \partial x_{k,K}\, \partial x_{l,L} = \partial T_{Kk} / \partial x_{l,L}$$

To obtain this result we employ the chain rule and use (1.13.27), i.e.,

$$\frac{\partial \Sigma}{\partial C_{KM}} = \frac{\partial \Sigma}{\partial x_{r,R}} \frac{\partial x_{r,R}}{\partial C_{KM}} = \frac{1}{2} \frac{\partial \Sigma}{\partial x_{r,K}} X_{M,r}$$

Differentiating this once more with respect to C_{LN} and using (1.14.19) and (1.4.8) we obtain $(1.17.12)_1$. Noting (1.13.25) we get $(1.17.12)_2$.

For isotropic solids Σ may be considered as a function of $\overset{-1}{\mathbf{c}}$. For future purposes we give the explicit form of (1.17.12) for isotropic solids. In this case, from (1.17.12), (1.13.25), and $(1.12.6)_1$ we have

$$A_{KkLl} = \partial (j X_{K,r} t_{kr}) / \partial x_{l,L}$$

Carrying out the differentiation and using $(1.4.6)_2$ and

(1.17.13)
$$\partial X_{K,k} / \partial x_{l,L} = -X_{K,l} X_{L,k}$$

obtained by differentiating $(1.4.3)_2$, we obtain

(1.17.14)
$$A_{KkLl} = 2j(t_{km} X_{[K,|m|} X_{L],l} + \tfrac{1}{2} X_{K,m}\, \partial t_{km} / \partial x_{l,L})$$

which is still valid for anisotropic solids. For isotropic solids, Σ and hence t_{km} may be considered functions of $\overset{-1}{\mathbf{c}}$. Thus calculating

$$\frac{\partial t_{km}}{\partial x_{l,L}} = \frac{\partial t_{km}}{\partial \overset{-1}{c}_{rs}} \frac{\partial \overset{-1}{c}_{rs}}{\partial x_{l,L}} = \frac{\partial t_{km}}{\partial \overset{-1}{c}_{rs}} (\delta_{rl} x_{s,L} + \delta_{sl} x_{r,L})$$

(1.17.14) becomes

(1.17.15)
$$A_{KkLl} = 2j(t_{km} X_{[K,|m|} X_{L],l} + F_{kmln} x_{n,L} X_{K,m})$$

where by use of (1.14.13) and (1.14.14), F_{klmn} is found to be

$$
\begin{aligned}
F_{klmn} = \frac{\partial t_{kl}}{\partial \overset{-1}{c}_{mn}} &= \frac{h_1}{2}(\delta_{km}\delta_{ln} + \delta_{lm}\delta_{kn}) \\
&+ \frac{h_2}{2}\left(\delta_{km}\overset{-1}{c}_{ln} + \delta_{kn}\overset{-1}{c}_{lm} + \delta_{lm}\overset{-1}{c}_{kn} + \delta_{ln}\overset{-1}{c}_{km}\right) \\
&+ \delta_{kl}\left[\frac{\partial h_0}{\partial I}\delta_{mn} + \frac{\partial h_0}{\partial II}\left(I\delta_{mn} - \overset{-1}{c}_{mn}\right) + III\frac{\partial h_0}{\partial III}c_{mn}\right] \\
&+ \overset{-1}{c}_{kl}\left[\frac{\partial h_1}{\partial I}\delta_{mn} + \frac{\partial h_1}{\partial II}\left(I\delta_{mn} - \overset{-1}{c}_{mn}\right) + III\frac{\partial h_1}{\partial III}c_{mn}\right] \\
&+ \overset{-1}{c}_{kr}\overset{-1}{c}_{rl}\left[\frac{\partial h_2}{\partial I}\delta_{mn} + \frac{\partial h_2}{\partial II}\left(I\delta_{mn} - \overset{-1}{c}_{mn}\right) + III\frac{\partial h_2}{\partial III}c_{mn}\right]
\end{aligned}
\tag{1.17.16}
$$

We note the following symmetry relations

$$
A_{KkLl} = A_{KlLk} = A_{LkKl} = A_{LlKk}, \qquad F_{klmn} = F_{lkmn} = F_{klnm} \tag{1.17.17}
$$

of which the first set follows (1.17.5) and $(1.17.6)_1$, which is valid for anisotropic solids, and the second set is clear from (1.17.16).

For nonlinear theory the nature of *reasonable* boundary conditions is not easy to decide. Two physically meaningful boundary conditions are the prescription of the displacement field or the traction field. For the thermal condition, temperature or heat may be prescribed. Expressed as a *mixed boundary* value problem we may be given

$$
\begin{aligned}
T_{KL}N_K &= \bar{T}_K, && \mathbf{X} \in S_T \\
\mathbf{x}(\mathbf{X}, t) &= \bar{\mathbf{x}}, && \mathbf{X} \in S_x \equiv S - S_T
\end{aligned}
\tag{1.17.18}
$$

$$
\begin{aligned}
Q_K N_K &= \bar{Q}, && \mathbf{X} \in S_Q \\
\theta &= \bar{\theta}, && \mathbf{X} \in S_\theta \equiv S - S_Q
\end{aligned}
\tag{1.17.19}
$$

Of course, other possibilities exist, but in all cases ultimately it is necessary to know if the *uniqueness* theorem is satisfied. Unfortunately, the uniqueness theorem for nonlinear theory has not been proven.

For the initial conditions, usually employed conditions are

$$
\begin{aligned}
\mathbf{x}(\mathbf{X}, 0) &= \mathbf{x}_0, && \mathbf{X} \in B \\
\dot{\mathbf{x}}(\mathbf{X}, 0) &= \mathbf{v}_0, && \mathbf{X} \in B
\end{aligned}
\tag{1.17.20}
$$

$$
\theta(\mathbf{X}, 0) = \theta_0, \qquad \mathbf{X} \in B \tag{1.17.21}
$$

In finite deformation theory the uniqueness of the solution may fail. Especially for the problems in which surface tractions are prescribed, the uniqueness may neither be expected nor desired. For example, a circular cylindrical tube or a spherical shell may be turned inside out corresponding to vanishing forces on its inside and outside surfaces (cf. Eringen [1962, Section 54]).

For static problems in finite elasticity a condition often used is the *strong ellipticity* of differential equations (1.17.10) with $\ddot{\mathbf{x}} = \mathbf{0}$. This is expressed by

$$A_{KkLl}\, a_k\, a_l\, B_K\, B_L > 0 \tag{1.17.22}$$

for nonvanishing vectors a_k and B_K. This condition is often also retained for the dynamical problems.

It is of course possible to formulate the problems in Eulerian forms. In this case additional problems arise regarding the boundary conditions. If the place of the deformed boundary is not known in advance (as is usually the case) then the boundary conditions depend on the solution, since the boundary displacements are required to fix the points on the deformed surface of the body.

For static problems, in addition, the global equilibrium conditions must be satisfied. These are

$$\oint_{\mathscr{S}} \mathbf{t}_k\, da_k + \int_{\mathscr{V}} \rho \mathbf{f}\, dv = \mathbf{0}, \qquad \oint_{\mathscr{S}} \mathbf{p} \times \mathbf{t}_k\, da_k + \int_{\mathscr{V}} \rho \mathbf{p} \times \mathbf{f}\, dv = \mathbf{0} \tag{1.17.23}$$

When discontinuity surfaces (e.g., shock waves) sweep the body, in addition to the solution of the boundary–initial value problem on two sides of the surface of discontinuity, the jump conditions of the form (1.12.21)–(1.12.24) must be satisfied. In Chapter II we give an extensive discussion on the propagation of singular surfaces.

Linear Theory

In linear theory there is no difference between Eulerian and Lagrangian descriptions. Substituting the stress constitutive equations (1.15.27) into Cauchy's law of motion $(1.11.26)_1$ we obtain

$$-(\beta_{kl} T)_{,k} + (\sigma_{klmn} u_{m,n})_{,k} + \rho(f_l - \ddot{u}_l) = 0 \tag{1.17.24}$$

The equation of heat conduction is already given by (1.15.26):

$$-\rho_0 \gamma \dot{T} - T_0 \beta_{kl} \dot{u}_{k,l} + (b_{kl} T_{,l})_{,k} + \rho h = 0 \tag{1.17.25}$$

Equations (1.17.24) and (1.17.25) are known as *Navier–Duhamel equations* of linear anisotropic thermoelasticity.

For the *isotropic solids* these equations take the form:

$$-(\beta T)_{,l} + (\lambda u_{k,k})_{,l} + (\mu u_{k,l})_{,k} + (\mu u_{l,k})_{,k} + \rho(f_l - \ddot{u}_l) = 0 \tag{1.17.26}$$

$$-\rho\gamma \dot{T} - \beta T_0\, \dot{u}_{k,k} + (\kappa T_{,k})_{,k} + \rho h = 0 \tag{1.17.27}$$

For *homogeneous and isotropic solids* β, λ, μ, and κ are independent of $\mathbf{x}$, and we have

$$-\beta T_{,l} + (\lambda + \mu)u_{k,kl} + \mu u_{l,kk} + \rho(f_l - \ddot{u}_l) = 0 \tag{1.17.28}$$

$$-\rho\gamma\dot{T} - \beta T_0 \dot{u}_{k,k} + \kappa T_{,kk} + \rho h = 0 \tag{1.17.29}$$

For *homogeneous, incompressible, isotropic* solids we have

$$-\beta T_{,l} - p_{,l} + \mu u_{l,kk} + \rho(f_l - \ddot{u}_l) = 0 \tag{1.17.30}$$

$$-\rho\gamma\dot{T} + \kappa T_{,kk} + \rho h = 0 \tag{1.17.31}$$

$$\dot{u}_{k,k} = 0 \tag{1.17.32}$$

The boundary and initial conditions to be considered in linear theory are much more clear. We possess uniqueness theorems (cf. Vol. II, Section 5.7). A *mixed* boundary–initial value problem requires that

$$\begin{aligned} t_{kl} n_k &= \bar{t}_k, && \mathbf{x} \in \mathscr{S}_t \\ \mathbf{u}(\mathbf{x}, t) &= \bar{\mathbf{u}}, && \mathbf{x} \in \mathscr{S}_u \equiv \mathscr{S} - \mathscr{S}_t \end{aligned} \tag{1.17.33}$$

$$\begin{aligned} q_k n_k &= \bar{q}, && \mathbf{x} \in \mathscr{S}_q \\ \theta &= \bar{\theta}, && \mathbf{x} \in \mathscr{S}_\theta \equiv \mathscr{S} - \mathscr{S}_q \end{aligned} \tag{1.17.34}$$

$$\begin{aligned} \mathbf{u}(\mathbf{x}, 0) &= \mathbf{u}_0, && \mathbf{x} \in \mathscr{B} \\ \dot{\mathbf{u}}(\mathbf{x}, 0) &= \mathbf{v}_0, && \mathbf{x} \in \mathscr{B} \end{aligned} \tag{1.17.35}$$

$$\theta(\mathbf{x}, 0) = \theta_0, \qquad \mathbf{x} \in \mathscr{B} \tag{1.17.36}$$

Of course other possibilities exist. For example, certain components of the traction and remaining component of displacement (e.g., t_1, t_2, u_3) may be specified on a part of the surface. Similarly, instead of $\bar{q}$ a combination of heat and temperature may be specified (the radiation boundary condition). The *mixed-mixed* type boundary value problems (cf. Eringen [1954]) often make the solution tractable. Nevertheless, all such cases must obey the uniqueness theorem.

1.18 RESTRICTIONS ON MATERIAL MODULI

In the linear theory of thermoelasticity, the restrictions arising from Clausius–Duhem inequality (1.15.3) are the nonnegative definiteness of the heat conduction coefficients (1.15.15), i.e.,

$$b_{km}\theta_{,k}\theta_{,m} \geq 0, \qquad \kappa \geq 0 \tag{1.18.1}$$

for any nonvanishing vector $\theta_{,k}$. The thermoelastic moduli σ_{klmn}, β_{kl}, and γ for anisotropic solids (and λ, μ for the isotropic solids) are left with no other restriction beyond their symmetry regulation (1.15.14), i.e.,

$$\sigma_{klmn} = \sigma_{mnkl} = \sigma_{lkmn} = \sigma_{klnm}, \qquad \beta_{kl} = \beta_{lk} \tag{1.18.2}$$

However, physical observations and certain mathematical considerations may be used to restrict these coefficients further. Here we give a discussion of these conditions for the linear isotropic solids. Later we present briefly some of the conditions, occasionally used, in nonlinear theory. The physical considerations are based on certain special types of static loads which cause certain special states of strain. Mathematical considerations include work theorems, stability criteria, the uniqueness theorem, the existence of real wave speeds, etc.

The constitutive equations of linear, isotropic thermoelastic solids were obtained in Section 1.15 (1.15.29):

(1.18.3)

$$t_{kl} = -\beta T\,\delta_{kl} + \lambda\,\tilde{e}_{mm}\,\delta_{kl} + 2\mu\,\tilde{e}_{kl}, \qquad q_k = \kappa T_{,k}$$

$$\rho_0\,\varepsilon = S_0 + \rho_0\,\eta_0\,T_0 + \rho_0\,\gamma[T + (T^2/2T_0)] + \beta T_0\,I_{\tilde{e}} + \tfrac{1}{2}(\lambda + 2\mu)I_{\tilde{e}}^{\,2} - 2\mu II_{\tilde{e}}$$

$$\eta = \eta_0 + (\gamma/T_0)T + (\beta/\rho_0)I_{\tilde{e}}$$

If we assume that $(1.18.3)_1$ is soluble for $\tilde{e}_{kl}$, then by setting $k = l$ and then using this result we obtain

$$t_{kk} \equiv I_t = -3\beta T + (3\lambda + 2\mu)I_{\tilde{e}}$$

$$\tilde{e}_{kl} = \frac{1}{2\mu(3\lambda + 2\mu)}(2\mu\,\beta T - \lambda t_{mm})\,\delta_{kl} + \frac{1}{2\mu}\,t_{kl} \tag{1.18.4}$$

provided that

$$\mu \neq 0, \qquad 3\lambda + 2\mu \neq 0 \tag{1.18.5}$$

and in order not to have zero strain for finite stress or temperature we must have

$$|\mu| < \infty, \qquad |3\lambda + 2\mu| < \infty \tag{1.18.6}$$

(i) In any state of strain, the shearing strain on any plane has the same direction as the shear if and only if

$$\mu > 0 \tag{1.18.7}$$

(ii) From (1.18.4), it follows that for vanishing temperature, a pressure $\bar{p} = -t_{kk}/3$ applied to a specimen causes volume decrease if and only if

$$3\lambda + 2\mu > 0 \tag{1.18.8}$$

(iii) For vanishing stress, a temperature rise ($T > 0$) causes a volume increase if and only if

$$\beta > 0 \tag{1.18.9}$$

(iv) For an arbitrary temperature change and state of strain the internal energy change is nonnegative if and only if

$$\gamma \geq 0, \qquad 3\lambda + 2\mu \geq 0, \qquad \mu \geq 0 \tag{1.18.10}$$

To prove this theorem we write $(1.18.3)_3$ in the form

(1.18.11)
$$\rho_0 \varepsilon - S_0 - \rho_0 \eta_0 T_0 = \rho_0 \gamma([T + (T^2/2T_0)] + \tfrac{1}{6}(3\lambda + 2\mu)I_{\tilde{e}}^{\,2} + (\mu/3)J_2$$

where

(1.18.12)
$$J_2 \equiv (\tilde{e}_{11} - \tilde{e}_{22})^2 + (\tilde{e}_{22} - \tilde{e}_{33})^2 + (\tilde{e}_{33} - \tilde{e}_{11})^2 + 6\tilde{e}_{12}^2 + 6\tilde{e}_{23}^2 + 6\tilde{e}_{31}^2$$

Now since $\theta = T_0 + T \geq 0$, we have $-T_0 \leq T$. Each of the terms on the right-hand side of (1.18.11) is independent. Thus for arbitrary variations of $T \geq -T_0$ and $\tilde{e}_{kl}$ the energy increase is nonnegative if and only if (1.18.10) is satisfied.

Conditions (1.18.7) and (1.18.8) can be shown to be sufficient but not necessary for the global stability of nonheat-conducting elastic solids, i.e., for a displacement field satisfying the conditions of equilibrium corresponding to a minimal stored energy. The necessary and sufficient conditions are given by

$$\lambda + 2\mu > 0, \qquad \mu > 0 \tag{1.18.13}$$

We shall also see that (cf. Vol. II, Section 5.3) these conditions give positive wave speeds in elastic solids.

We shall also find that (1.18.7) and (1.18.8) are sufficient for a unique solution of mixed problems in elastostatics; they are also sufficient but not necessary for the uniqueness of solution of the initial value problems of elastodynamics. (For details, see Vol. II, Section 5.7.)

In the theory of elasticity other material constants are often used in place of the Lamé constants λ and μ. Following are relations among some of these constants:

(1.18.14)
$$\begin{aligned} &\lambda \equiv E\nu/(1 + \nu)(1 - 2\nu) = 2G\nu/(1 - 2\nu), && \mu \equiv G = E/2(1 + \nu) \\ &E = \mu(3\lambda + 2\mu)/(\lambda + \mu), && \nu = \lambda/2(\lambda + \mu) \\ &k \equiv \lambda + \tfrac{2}{3}\mu = E/3(1 - 2\nu) && \end{aligned}$$

where E is called the *Young's modulus*, G the *shear modulus*, ν *Poisson's ratio*, and k the *bulk modulus*. The inequalities

$$0 < G, \qquad -1 < \nu < \tfrac{1}{2} \tag{1.18.15}$$

are equivalent to (1.18.7) and (1.18.8). The solution of the static traction boundary value problem is unique in arbitrary smooth regions *if and only if*

$$-1 < \nu < 1 \tag{1.18.16}$$

and the displacement boundary value problem is unique *if and only if*

$$\nu < \tfrac{1}{2} \quad \text{or} \quad 1 < \nu \tag{1.18.17}$$

Nonlinear Theory

A number of special inequalities have been proposed based on certain "reasonable" physical expectations when an isotropic (nonheat-conducting) elastic solid is subjected to pressure, tensions, and shears. For a discussion of these we refer the reader to Truesdell and Noll [1965]. We now discuss briefly three of these inequalities.

(i) *Baker–Ericksen* [1954] *inequalities*. If it is assumed that in an isotropic body greatest (least) tension occurs in the direction corresponding to the greatest (least) stretch, then

$$t_\alpha > t_\beta \qquad \text{whenever} \quad \lambda_\alpha > \lambda_\beta \tag{1.18.18}$$

From (1.14.15) we have

$$t_\alpha = (h_1 + h_2 I)\lambda_\alpha^{\,2} + (h_0 - h_2 I) + h_2 III\lambda_\alpha^{-2} \tag{1.18.19}$$

Using (1.18.18) and (1.14.14) we derive

$$\begin{aligned} &\frac{\partial \Sigma}{\partial I} + \lambda_\alpha^{\,2}\frac{\partial \Sigma}{\partial II} > 0, \qquad \text{if} \quad \lambda_\beta \neq \lambda_\gamma \\ &\frac{\partial \Sigma}{\partial I} + \lambda_\alpha^{\,2}\frac{\partial \Sigma}{\partial II} \geq 0, \qquad \text{if} \quad \lambda_\beta = \lambda_\gamma \end{aligned} \qquad (\alpha, \beta, \gamma \text{ unequal}) \tag{1.18.20}$$

In linear theory these reduce to $\mu > 0$

(ii) *Compressibility condition.* If the volume of an isotropic elastic solid is to be increased (decreased) when the body is subjected to a uniform tension (compression) then the tension t (pressure p) must be a strictly increasing (decreasing) function of the stretch $\lambda_1 = \lambda_2 = \lambda_3 \equiv \lambda$. Mathematically,

$$(\bar{t} - t)(\bar{\lambda} - \lambda) > 0, \qquad \bar{t} \neq t$$

where $t(\lambda) \equiv t_{kk}/3$. If the body is subjected to different uniform tensions on its principal planes the volume is again expected to increase. If this is accepted to be the case for the principal forces t_α/λ_α, then

$$\sum_{\alpha=1}^{3} t_\alpha(\lambda_\alpha - 1)/\lambda_\alpha > 0, \qquad \text{when not all} \quad \lambda_\alpha = 1 \tag{1.18.21}$$

where the natural state ($\lambda_\alpha = 1$) is assumed to be force free, i.e., $t_\alpha = 0$. This condition, in the linear limit, is equivalent to $\mu > 0$, $3\lambda + 2\mu > 0$.

(iii) *Thermostatic inequality.* For vanishing thermal gradients the equation of local entropy production (1.12.14) may be written as

$$-\rho_0 \theta \gamma \equiv \dot{\Sigma} + \rho_0 \dot{\theta} \eta - \tfrac{1}{2} T_{KL} \dot{C}_{KL} \tag{1.18.22}$$

where we used (1.12.1), (1.12.6), and $(1.13.1)_4$. If T_{KL} and η are fixed, this integrates into

(1.18.23)

$$-\int_t^{\bar{t}} \rho_0 \theta \gamma \, dt = \Sigma(\bar{\mathbf{C}}, \bar{\theta}) - \Sigma(\mathbf{C}, \theta) - \tfrac{1}{2} T_{KL}(\bar{C}_{KL} - C_{KL}) + \rho_0 \eta(\bar{\theta} - \theta)$$

where $\bar{\mathbf{C}}$ and $\bar{\theta}$ are the values of $\mathbf{C}$ and θ at time $\bar{t}$. Physically, a material will not deform or acquire temperature changes unless the force (T_{KL}) and/or entropy (η) change. Thus, in this hypothetical process the entropy production must be negative, i.e.,

$$\Sigma(\bar{\mathbf{C}}, \theta) - \Sigma(\mathbf{C}, \theta) - \frac{\partial \Sigma}{\partial C_{KL}} (\bar{C}_{KL} - C_{KL}) - \frac{\partial \Sigma}{\partial \theta} (\bar{\theta} - \theta) > 0 \tag{1.18.24}$$

where we also used the constitutive equations $(1.13.11)_1$ and $(1.13.11)_4$. Thus *under the conditions* $\mathbf{T}$ *and* η *fixed, the stress potential of nonheat-conducting elastic solids must satisfy the inequality* (1.18.24).[1]

Since $\mathbf{C}$ and $\bar{\mathbf{C}}$ are positive-definite symmetric tensors, there exists a positive-definite symmetric tensor Q_{KL} such that

$$\bar{\mathbf{C}} = \mathbf{Q}\mathbf{C}\mathbf{Q}^{\mathrm{T}} \tag{1.18.25}$$

Therefore, we can state that (1.18.24) is valid for all positive-definite symmetric tensors $\mathbf{Q} \neq \mathbf{1}$. In particular, for $\bar{\theta} = \theta$ we obtain

$$\Sigma(\bar{\mathbf{C}}, \theta) - \Sigma(\mathbf{C}, \theta) - T_{KL}(\bar{C}_{KL} - C_{KL}) > 0 \tag{1.18.26}$$

[1] An inequality similar to this was given by Coleman and Noll [1959]. However, they used the internal energy function ϵ, the Piola–Kirchhoff stress T_{Kk}, and considered ϵ as a function of the deformation gradient $\mathbf{x}_{,K}$ and η. The present inequality eliminates some of the extra conditions since the rotation and translation dependence are naturally excluded from Σ by the axiom of objectivity.

Various other inequalities can be derived from this. For example, interchanging $\mathbf{C}$ and $\bar{\mathbf{C}}$ and adding the results we obtain

$$(\bar{T}_{KL} - T_{KL})(\bar{C}_{KL} - C_{KL}) > 0 \tag{1.18.27}$$

a condition equivalent to the condition proposed by Coleman and Noll [1964]. In linear theory this condition gives the classical results (1.18.7) and (1.18.8). Other special cases are discussed by Truesdell and Noll [1965, Sections 52 and 87].

1.19 RESUME OF BASIC EQUATIONS OF THERMOELASTICITY

The theory of thermoelasticity is based upon the following equations:

A. Equations of Balance

$$\rho_0/\rho = (III)^{1/2} \tag{1.19.1}$$

$$t_{kl,k} + \rho(f_l - \dot{v}_l) = 0 \tag{1.19.2}$$

$$t_{kl} = t_{lk} \tag{1.19.3}$$

$$\rho\theta\dot{\eta} = q_{k,k} + \rho h \qquad \text{in} \quad \mathscr{V} - \sigma \tag{1.19.4}$$

B. Equations of Jump Conditions

If there is a discontinuity surface $\sigma(t)$ sweeping the body with velocity $\mathbf{v}$ in the direction of its positive normal, then the following jump conditions must be satisfied on σ:

$$[\rho(\mathbf{v} - \mathbf{v})] \cdot \mathbf{n} = 0 \tag{1.19.5}$$

$$-[t_{kl}]n_k + [\rho v_l(\mathbf{v} - \mathbf{v})] \cdot \mathbf{n} = 0 \tag{1.19.6}$$

$$[\rho(\varepsilon + \tfrac{1}{2}\mathbf{v} \cdot \mathbf{v})(\mathbf{v} - \mathbf{v})] \cdot \mathbf{n} - [t_{kl} v_l + q_k]n_k = 0 \tag{1.19.7}$$

$$[\rho\eta(\mathbf{v} - \mathbf{v}) - \theta^{-1}q] \cdot \mathbf{n} \geq 0 \qquad \text{on} \quad \sigma(t) \tag{1.19.8}$$

C. Constitutive Equations (Compressible Solids)

(i) *Nonlinear isotropic solids*

$$t_{kl} = \frac{2\rho}{\rho_0}\left[III\frac{\partial\Sigma}{\partial III}\delta_{kl} + \left(\frac{\partial\Sigma}{\partial I} + I\frac{\partial\Sigma}{\partial II}\right)\overset{-1}{c}_{kl} - \frac{\partial\Sigma}{\partial II}\overset{-1}{c}_{kr}\overset{-1}{c}_{rl}\right] \tag{1.19.9}$$

$$q_k = \left(\kappa_0\,\delta_{kl} + \kappa_1\overset{-1}{c}_{kl} + \kappa_2\overset{-1}{c}_{kr}\overset{-1}{c}_{rl}\right)\theta_{,l} \tag{1.19.10}$$

$$\psi \equiv \varepsilon - \theta\eta = \rho_0^{-1}\Sigma(I, II, III, \theta, \mathbf{X}) \tag{1.19.11}$$

$$\eta = -\rho_0^{-1}\,\partial\Sigma/\partial\theta \tag{1.19.12}$$

where κ_0, κ_1, κ_2 are functions of **X** and the invariants (1.14.9) of which I_1, I_2, and I_3 may be replaced by *I*, *II*, and *III*. Also these functions are subject to inequality (1.13.17), i.e.,

$$\kappa_0 I_0 + \kappa_1 I_4 + \kappa_2 I_5 \geq 0 \tag{1.19.13}$$

(For other forms of constitutive equations see Section 1.14; for anisotropic solids see Section 1.13.)

(ii) *Linear isotropic solids*

$$t_{kl} = (-\beta T + \lambda I_{\tilde{e}})\, \delta_{kl} + 2\mu \tilde{e}_{kl} \tag{1.19.14}$$

$$q_k = \kappa T_{,k}, \qquad \kappa \geq 0 \tag{1.19.15}$$

$$\begin{aligned} \rho_0 \varepsilon = S_0 + \rho_0 \eta_0 T_0 + \rho_0 \gamma[T + (T^2/2T_0)] + \beta T_0 I_{\tilde{e}} \\ + \tfrac{1}{2}(\lambda + 2\mu) I_{\tilde{e}}^2 - 2\mu II_{\tilde{e}} \end{aligned} \tag{1.19.16}$$

$$\eta = \eta_0 + (\gamma/T_0)T + (\beta/\rho_0) I_{\tilde{e}} \tag{1.19.17}$$

(iii) *Quadratic theory*. (See Section 1.16.)

D. Constitutive Equations (Incompressible Solids)

(i) *Nonlinear isotropic solids*

$$t_{kl} = -p\, \delta_{kl} + 2\left(\frac{\partial \Sigma}{\partial I} + I \frac{\partial \Sigma}{\partial II}\right) \overset{-1}{c}_{kl} - \frac{\partial \Sigma}{\partial II} \overset{-1}{c}_{kr} \overset{-1}{c}_{rl} \tag{1.19.18}$$

$$q_k = \left(\kappa_0\, \delta_{kl} + \kappa_1 \overset{-1}{c}_{kl} + \kappa_2 \overset{-1}{c}_{kr} \overset{-1}{c}_{rl}\right) \theta_{,l} \tag{1.19.19}$$

$$\psi \equiv \varepsilon - \theta\eta = \rho_0^{-1} \Sigma(I, II, \theta, \mathbf{X}) \tag{1.19.20}$$

$$\eta = -\rho_0^{-1}\, \partial \Sigma / \partial \theta \tag{1.19.21}$$

where κ_α is a function of θ, **X**, and the invariants *I*, *II* and I_4, I_5, I_6, and I_0 are given by (1.14.9) and are subject to inequality (1.19.13).

(ii) *Linear isotropic solids*

$$t_{kl} = -(\beta T + p)\, \delta_{kl} + 2\mu \tilde{e}_{kl} \tag{1.19.22}$$

$$q_k = \kappa T_{,k}, \qquad 0 \leq \kappa \tag{1.19.23}$$

$$\rho_0 \varepsilon = S_0 + \rho_0 \eta_0 T_0 + \rho_0 \gamma[T + (T^2/2T_0)] - 2\mu II_{\tilde{e}} \tag{1.19.24}$$

$$\eta = \eta_0 + (\gamma/T_0)T \tag{1.19.25}$$

E. Kinematical Relations

(1.19.26) $\dot{v}_k = (\partial v_k/\partial t) + v_{k,l}v_l$

(1.19.27) $\overset{-1}{c}_{kl} \equiv X_{K,k}X_{K,l}$

(1.19.28) $2e_{kl} \equiv u_{k,l} + u_{l,k} - u_{m,k}u_{m,l}$

(1.19.29) $C_{KL} \equiv x_{k,K}x_{k,L}, \qquad 2E_{KL} = U_{K,L} + U_{L,K} + U_{M,K}U_{M,L}$

(1.19.30)
$$I \equiv I_{\overset{-1}{c}} = \overset{-1}{c}_{kk}, \qquad II = II_{\overset{-1}{c}} = \tfrac{1}{2}\left(\overset{-1}{c}_{kk}\overset{-1}{c}_{ll} - \overset{-1}{c}_{kl}\overset{-1}{c}_{lk}\right)$$
$$III \equiv III_{\overset{-1}{c}} = \det \overset{-1}{c}_{kl}$$

where u_k and U_K are the Eulerian and Lagrangian displacement fields, respectively, and v_k is the velocity vector. For linear theory in (1.19.26) and (1.19.29), the nonlinear terms drop out and

(1.19.31) $v_k \simeq \partial u_k/\partial t, \qquad \dot{v}_k \simeq \partial^2 u_k/\partial t^2 \qquad$ (linear)

F. Field Equations (Linear Theory)

For linear theory, by substituting $(1.19.31)_2$ and the expressions of linear strain tensors into constitutive equations (1.19.14)–(1.19.17) for compressible solids, or (1.19.22)–(1.19.25) for the incompressible solids, and the result into the balance laws (1.19.2), and (1.19.4), upon linearization, we obtain the field equations [cf. Eqs. (1.17.28) and (1.17.29)]

(1.19.32) $(\lambda + \mu)u_{l,lk} + \mu u_{k,ll} - (\beta T)_{,k} + \rho(f_k - \ddot{u}_k) = 0$

(1.19.33) $-\rho_0 \gamma \dot{T} - \beta T_0 \dot{u}_{k,k} + (\kappa T_{,k})_{,k} + \rho h = 0$

(compressible)

(1.19.34)
$$-p_{,k} + \mu u_{k,ll} - (\beta T)_{,k} + \rho(f_k - \ddot{u}_k) = 0, \qquad u_{k,k} = 0$$

(1.19.35) $-\rho_0 \gamma \dot{T} + (\kappa T_{,k})_{,k} + \rho_0 h = 0$

(incompressible)

These equations are known as the *Navier–Duhamel* equations of thermoelasticity.

G. Boundary Conditions

(1.19.36) $t_{lk}n_l = \bar{t}_k \qquad$ on $\mathscr{S}_t - \sigma$

(1.19.37) $u_k = \bar{u}_k \qquad$ on $\mathscr{S}_u - \sigma \equiv \mathscr{S} - \mathscr{S}_t - \sigma$

(1.19.38) $\mathbf{q} \cdot \mathbf{n} = \bar{q}_{(\mathbf{n})} \qquad$ on $\mathscr{S}_q - \sigma$

(1.19.39) $T = \bar{T} \qquad$ on $\mathscr{S}_T - \sigma \equiv \mathscr{S} - \mathscr{S}_q - \sigma$

where $\bar{t}_k$ is the surface traction prescribed on $\mathscr{S}_t$ of the surface of body $\mathscr{S} = \mathscr{S}_t + \mathscr{S}_u$, and $\bar{u}_k$ is the displacement field prescribed on the remaining part $\mathscr{S}_u$ of the surface not overlapping $\mathscr{S}_t$. Similarly, $\bar{q}_{(\mathbf{n})}$ is prescribed on a part $\mathscr{S}_q$ of the surface, while the temperature is prescribed on the remaining part $\mathscr{S}_T$. Other mixed combinations of the boundary conditions are possible. For any set of boundary conditions to be admissible together with the initial conditions, it must not violate the uniqueness theorem of the theory.

For the heat radiation, another important boundary condition replacing one of (1.19.38) or (1.19.39) is

(1.19.40) $$\mathbf{q} \cdot \mathbf{n} + a(T - T_1) = 0$$

where $a \geq 0$ is an appropriate function of the surface coordinates and T_1 is the temperature outside of the body near its surface $\mathscr{S}$.

H. Initial Conditions

(1.19.41) $$u_k(\mathbf{x}, 0) = u_k{}^0(\mathbf{x}) \quad \text{in} \quad \mathscr{V} - \sigma$$

(1.19.42) $$\dot{u}_k(\mathbf{x}, 0) = v_k{}^0(\mathbf{x}) \quad \text{in} \quad \mathscr{V} - \sigma$$

(1.19.43) $$T(\mathbf{x}, 0) = T^0(\mathbf{x}) \quad \text{in} \quad \mathscr{V} - \sigma$$

where $u_k{}^0$, $v_k{}^0$, and T^0 are prescribed in the body.

The linear theory of thermoelasticity is based on the Navier–Duhamel equations (1.19.32) and (1.19.33) for compressible solids or (1.19.34) and (1.19.35) for incompressible solids, subject to boundary conditions of the type given above.

For anisotropic solids, the field equations are (see Section 1.17):

(1.19.44) $$(\sigma_{klmn} u_{m,n})_{,k} - (\beta_{kl} T)_{,k} + \rho(f_l - \ddot{u}_l) = 0 \quad \text{in} \quad \mathscr{V} - \sigma$$

(1.19.45) $$-\rho_0 \gamma \dot{T} - T_0 \beta_{kl} \dot{u}_{k,l} + (b_{kl} T_{,l})_{,k} + \rho h = 0 \quad \text{in} \quad \mathscr{V} - \sigma$$

which follow from (1.19.2) and (1.19.4) by use of the linear constitutive equations (1.15.22) and (1.15.23) with the natural state being assumed stress free ($\alpha_{kl} = 0$) [cf. (1.17.24) and (1.17.25)].

For nonlinear theory one must employ the balance laws (1.19.1)–(1.19.4) and the constitutive equations (1.19.9)–(1.19.12) for compressible solids and (1.19.18)–(1.19.21) for the incompressible solids.

A more specific form of the equation of heat conduction replacing (1.19.4) is given by (1.19.45) for compressible solids. For the incompressible solids, Eq. (1.19.45) remains valid with $\dot{I}_1 = 0$.

For general nonlinear theory, the field equations are given by (1.17.5) and (1.17.6). For the restrictions placed on the material moduli, see Section 1.18.

1.20 CURVILINEAR COORDINATES

For a large class of boundary value problems the use of an appropriate curvilinear coordinate system simplifies the boundary conditions, thus allowing a tractable solution. Here we make the passage to curvilinear coordinates by some simple observations: Let x^k be a set of curvilinear coordinates, and $g_{kl}(\mathbf{x})$ be the metric tensor with reciprocal $g^{kl}(\mathbf{x})$. Thus, the square of arc length is given by (cf. Appendix A)

$$ds^2 = g_{kl}\, dx^k\, dx^l \tag{1.20.1}$$

The passage from rectangular coordinates to curvilinear coordinates may be made by observing the following two simple rules:

(a) The partial derivative symbol (,) must be replaced by the covariant partial differentiation sign (;); and

(b) The repeated indices must be on the diagonal position.

Applying these rules to Cauchy's equations of motion (1.19.2) and the energy equation (1.19.4), we obtain its curvilinear form

$$t^k{}_{l;k} + \rho(f_l - \dot{v}_l) = 0 \tag{1.20.2}$$

$$\rho\theta\dot{\eta} - q^k{}_{;k} - \rho h = 0 \tag{1.20.3}$$

where $t^k{}_l$ and q^k are, respectively, the mixed components of the stress tensor and the contravariant component of the heat vector, and an index following a semicolon indicates covariant partial differentiation, e.g.,

$$t^k{}_{l;r} \equiv t^k{}_{l,r} + \begin{Bmatrix} k \\ mr \end{Bmatrix} t^m{}_l - \begin{Bmatrix} m \\ rl \end{Bmatrix} t^k{}_m \tag{1.20.4}$$

$$q^k{}_{;l} = q^k{}_{,l} + \begin{Bmatrix} k \\ mr \end{Bmatrix} q^r \tag{1.20.5}$$

where the *Christoffel symbol of the second kind* is given by

$$\begin{Bmatrix} k \\ ml \end{Bmatrix} \equiv g^{kr}[ml, r], \qquad [ml, r] \equiv \tfrac{1}{2}(g_{mr,l} + g_{lr,m} - g_{ml,r}) \tag{1.20.6}$$

the last of which is the Christoffel symbol of the first kind.

Deformation tensors are given by

$$\begin{aligned} C_{KL} &\equiv g_{kl}\, x^k{}_{,K}\, x^l{}_{,L}, & c_{kl} &\equiv G_{KL} X^K{}_{,k}\, X^L{}_{,l} \\ \overset{-1}{C}{}^{KL} &= g^{kl} X^K{}_{,k}\, X^L{}_{,l}, & \overset{-1}{c}{}^{kl} &= G^{KL} x^k{}_{,K}\, x^l{}_{,L} \end{aligned} \tag{1.20.7}$$

where G_{KL} is the metric tensor in the curvilinear material coordinates X^K, and G^{KL} is its reciprocal.

The strain tensors **E** and **e** are given by

$$2E_{KL} = C_{KL} - \delta_{KL} = U_{K;L} + U_{L;K} + G^{MN} U_{M;K} U_{N;L}$$
$$2e_{kl} = \delta_{kl} - c_{kl} = u_{k;l} + u_{l;k} - g^{mn} u_{m;k} u_{n;l} \tag{1.20.8}$$

where

$$U_{K;L} \equiv U_{K,L} - \begin{Bmatrix} M \\ KL \end{Bmatrix} U_M$$
$$u_{k;l} \equiv u_{k,l} - \begin{Bmatrix} m \\ kl \end{Bmatrix} u_m \tag{1.20.9}$$

The Christoffel symbols $\{{}^{M}_{KL}\}$ of the material frame have the same form as (1.20.6) with the metric tensors G_{KL} and G^{KL} replacing g_{kl} and g^{kl}.

Often these equations are expressed in terms of the *physical components* of the vectors and tensors involved. The physical components $t^{(k)}{}_{(l)}$ and $u^{(k)}$ of $t^k{}_l$ and u^k are related to each other by (cf. Appendix A)

$$t^k{}_l = t^{(k)}{}_{(l)}(g_{\underline{ll}}/g_{\underline{kk}})^{1/2}, \qquad u^k = u^{(k)}/(g_{\underline{kk}})^{1/2} \tag{1.20.10}$$

where underscores are used to suspend the summation on repeated indices. We now give expressions of Cauchy's law of motion (1.20.2), the energy equation (1.20.3) and the Eulerian strain tensor **e** in *orthogonal curvilinear* coordinates. In this case $g_{kl} = 0$ when $k \neq l$, and

$$ds^2 = g_{11}(dx^1)^2 + g_{22}(dx^2)^2 + g_{33}(dx^3)^2$$
$$g^{\underline{kk}} = 1/g_{\underline{kk}}, \qquad g \equiv \det g_{kl} = g_{11} g_{22} g_{33}$$
$$\begin{Bmatrix} k \\ lm \end{Bmatrix} = \frac{1}{2g_{\underline{kk}}} \left(\frac{\partial g_{\underline{kk}}}{\partial x^m} \delta_{kl} + \frac{\partial g_{\underline{mm}}}{\partial x^l} \delta_{km} - \frac{\partial g_{\underline{ll}}}{\partial x^k} \delta_{lm} \right) \tag{1.20.11}$$

Using these and (1.20.10) in (1.20.2) and (1.20.3), we get

$$\sum_{k=1}^{3} \left\{ \frac{1}{\sqrt{g}} \frac{\partial}{\partial x^k} \left[t^{(k)}{}_{(l)} \frac{\sqrt{g}}{(g_{\underline{kk}})^{1/2}} \right] + \frac{1}{(g_{\underline{kk}} g_{\underline{ll}})^{1/2}} \frac{\partial (g_{\underline{ll}})^{1/2}}{\partial x^k} t^{(k)}{}_{(l)} - \frac{1}{(g_{\underline{kk}} g_{\underline{ll}})^{1/2}} \frac{\partial (g_{\underline{kk}})^{1/2}}{\partial x^l} t^{(k)}{}_{(k)} \right\} - \frac{1}{(g_{\underline{ll}})^{1/2}} \frac{\partial (\beta T)}{\partial x^l} + \rho(f^{(l)} - \dot{v}^{(l)}) = 0 \tag{1.20.12}$$

$$\rho\theta\dot{\eta} - \rho h - \sum_{k=1}^{3} \frac{1}{\sqrt{g}} \left(\frac{\sqrt{g}}{(g_{\underline{kk}})^{1/2}} q^{(k)} \right)_{,k} = 0 \tag{1.20.13}$$

where $f^{(l)} = f^l (g_{\underline{ll}})^{1/2}$ and $q^{(l)} = q^l (g_{\underline{ll}})^{1/2}$ are, respectively, the physical components of the body force and the heat vector. To express the Eulerian strain tensor in terms of the physical components of u^k, we write it as

$$2e^k{}_l = u^k{}_{;l} + g_{nl}\, g^{km} u^n{}_{;m} + g^{kr} g_{mn}\, u^m{}_{;r}\, u^n{}_{;l} \tag{1.20.14}$$

where

$$u^k = g^{kl} u_l, \qquad u^k{}_{;l} = u^k{}_{,l} + \begin{Bmatrix} k \\ lm \end{Bmatrix} u^m \tag{1.20.15}$$

Now we use the format (1.20.10) and (1.20.11) to obtain the components of the displacement gradient in terms of its physical components. Thus

$$u^k{}_{;l} = \frac{1}{(g_{\underline{kk}})^{1/2}}\, u^{(k)}{}_{,l} - \frac{u^{(l)}}{2g_{\underline{kk}}(g_{\underline{ll}})^{1/2}} \frac{\partial g_{\underline{ll}}}{\partial x^k} + \frac{\delta_{kl}}{2g_{\underline{kk}}} \sum_{r=1}^{3} \frac{u^{(r)}}{(g_{\underline{rr}})^{1/2}} \frac{\partial g_{\underline{kk}}}{\partial x^r} \tag{1.20.16}$$

For orthogonal curvilinear coordinates the physical components of $e^k{}_l$ have the form

$$e^{(k)}{}_{(l)} = e^k{}_l \frac{(g_{\underline{kk}})^{1/2}}{(g_{\underline{ll}})^{1/2}} = \frac{(g_{\underline{kk}})^{1/2}}{2\,(g_{\underline{ll}})^{1/2}} \left(u^k{}_{;l} + \frac{g_{\underline{ll}}}{g_{\underline{kk}}}\, u^l{}_{;k} + \frac{1}{g_{\underline{kk}}} \sum_{n=1}^{3} g_{\underline{nn}}\, u^n{}_{;k}\, u^n{}_{,l} \right) \tag{1.20.17}$$

Substitution of $u^k{}_{;l}$ from (1.20.16) into (1.20.17) gives the physical components of $\mathbf{e}$ in terms of the gradients of the physical components of the displacement vector $\mathbf{u}$.

For linear theory these read:

$$\begin{aligned} \tilde{e}^{(k)}{}_{(l)} = \tilde{e}^k{}_l \frac{(g_{\underline{kk}})^{1/2}}{(g_{\underline{ll}})^{1/2}} = \frac{1}{2} \Bigg[& \frac{(g_{\underline{kk}})^{1/2}}{(g_{\underline{ll}})^{1/2}} \frac{\partial}{\partial x^l} \left(\frac{u^{(k)}}{(g_{\underline{kk}})^{1/2}} \right) \\ & + \frac{(g_{\underline{ll}})^{1/2}}{(g_{\underline{kk}})^{1/2}} \frac{\partial}{\partial x^k} \left(\frac{u^{(l)}}{(g_{\underline{ll}})^{1/2}} \right) + \frac{\delta_{kl}}{(g_{\underline{kk}}\, g_{\underline{ll}})^{1/2}} \sum_{r=1}^{3} \frac{\partial g_{\underline{kk}}}{\partial x^r} \frac{u^{(r)}}{(g_{\underline{rr}})^{1/2}} \Bigg] \end{aligned} \tag{1.20.18}$$

Similarly, for the infinitesimal rotation tensors and vectors we get

(1.20.19)

$$\tilde{r}^{(k)}{}_{(l)} = \tilde{r}^k{}_l \frac{(g_{\underline{kk}})^{1/2}}{(g_{\underline{ll}})^{1/2}} = \frac{1}{2(g_{\underline{kk}}\, g_{\underline{ll}})^{1/2}} \left\{ \frac{\partial}{\partial x_l} [(g_{\underline{kk}})^{1/2} u^{(k)}] - \frac{\partial}{\partial x^k} [(g_{\underline{ll}})^{1/2} u^{(l)}] \right\}$$

$$\tilde{r}^k = \tfrac{1}{2} \sum_{k,l} e^{mlk} \tilde{r}^{(k)}{}_{(l)} \qquad \text{or} \qquad \tilde{\mathbf{r}} = \tfrac{1}{2} \nabla \times \mathbf{u} \tag{1.20.20}$$

The stress constitutive equations of the linear theory of isotropic solids are

$$t^{(k)}{}_{(l)} = (-\beta T + \lambda \tilde{e}^{(m)}{}_{(m)})\, \delta^k{}_l + 2\mu \tilde{e}^{(k)}{}_{(l)} \tag{1.20.21}$$

Fourier's law is

$$q^{(k)} = \frac{\kappa}{(g_{\underline{kk}})^{1/2}}\, \theta_{,k}, \qquad \kappa \geq 0 \tag{1.20.22}$$

The Navier–Duhamel equations for homogeneous solids can be written in terms of vector operators. To this end, we first write these equations in the tensorial forms:

(1.20.23) $$(\lambda + \mu)u^l{}_{;lk} + \mu u_{k;}{}^l{}_l - \beta T_{,k} + \rho(f_k - \ddot{u}_k) = 0$$

(1.20.24) $$T_{;}{}^k{}_k - \beta T_0 \dot{u}^k{}_{;k} - \rho_0 \gamma \dot{T} + \rho h = 0$$

where

(1.20.25) $$u_{k;}{}^l = g^{lr} u_{k;r}, \qquad T_{;}{}^k \equiv g^{kr} T_{,r}, \qquad T_{;}{}^k{}_k \equiv g^{kr} T_{;kr}$$

We now recall the following vector identities

(1.20.26) $$u_{k;}{}^l{}_l = -(\nabla \times \nabla \times \mathbf{u})_k + (\nabla\nabla \cdot \mathbf{u})_k, \qquad u^l{}_{;kl} = (\nabla\nabla \cdot \mathbf{u})_k$$

where ∇ is the gradient operator (cf. Appendix A). Using these in (1.20.23) and (1.20.24), we get

(1.20.27) $$(\lambda + 2\mu)\nabla\nabla \cdot \mathbf{u} - \mu\nabla \times \nabla \times \mathbf{u} - \beta\nabla T + \rho(\mathbf{f} - \ddot{\mathbf{u}}) = \mathbf{0}$$

(1.20.28) $$\kappa\nabla^2 T - \beta T_0 \nabla \cdot \dot{\mathbf{u}} - \rho_0 \gamma \dot{T} + \rho h = 0$$

By substituting the expressions of gradient, divergence, and curl, these equations can be expressed in any curvilinear coordinate system. We next give the specific forms of these equations in cylindrical and spherical coordinates.

(i) *Cylindrical coordinates.* Cylindrical coordinates (r, θ, z) are related to rectangular coordinates (x, y, z) by (Fig. 1.20.1)

(1.20.29) $$x = r\cos\theta, \qquad y = r\sin\theta, \qquad z = z$$

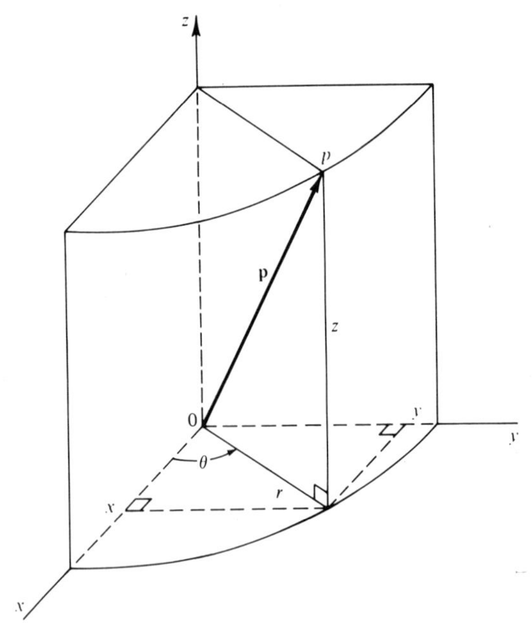

Fig. 1.20.1 Cylindrical coordinates.

The square of arc length and metric tensor g_{kl} are given by

$$ds^2 = dr^2 + r^2\, d\theta^2 + dz^2$$
$$g_{11} = 1, \qquad g_{22} = r^2, \qquad g_{33} = 1, \qquad g_{kl} = 0 \qquad (k \neq l) \tag{1.20.30}$$

Cauchy's equations of motion (1.20.12), the energy equation (1.20.13), and the infinitesimal strains and rotations (1.20.18) and (1.20.20) become

$$\frac{\partial t_{rr}}{\partial r} + \frac{1}{r}\frac{\partial t_{r\theta}}{\partial \theta} + \frac{\partial t_{rz}}{\partial z} + \frac{1}{r}(t_{rr} - t_{\theta\theta}) - \frac{\partial}{\partial r}(\beta T) + \rho(f_r - \ddot{u}_r) = 0$$

$$\frac{\partial t_{r\theta}}{\partial r} + \frac{1}{r}\frac{\partial t_{\theta\theta}}{\partial \theta} + \frac{\partial t_{\theta z}}{\partial z} + \frac{2}{r} t_{\theta r} - \frac{1}{r}\frac{\partial}{\partial \theta}(\beta T) + \rho(f_\theta - \ddot{u}_\theta) = 0 \tag{1.20.31}$$

$$\frac{\partial t_{rz}}{\partial r} + \frac{1}{r}\frac{\partial t_{\theta z}}{\partial \theta} + \frac{\partial t_{zz}}{\partial z} + \frac{1}{r} t_{rz} - \frac{\partial}{\partial z}(\beta T) + \rho(f_z - \ddot{u}_z) = 0$$

$$\rho\theta\dot{\eta} - \rho h - \frac{1}{r}\frac{\partial}{\partial r}(r q_r) - \frac{1}{r}\frac{\partial q_\theta}{\partial \theta} - \frac{\partial q_z}{\partial z} = 0 \tag{1.20.32}$$

(1.20.33)

$$\tilde{e}_{rr} = \frac{\partial u_r}{\partial r}, \qquad e_{\theta\theta} = \frac{1}{r}\frac{\partial u_\theta}{\partial \theta} + \frac{u_r}{r}, \qquad \tilde{e}_{zz} = \frac{\partial u_z}{\partial z}$$

$$2\tilde{e}_{\theta z} = \frac{\partial u_\theta}{\partial z} + \frac{1}{r}\frac{\partial u_z}{\partial \theta}, \qquad 2e_{zr} = \frac{\partial u_r}{\partial z} + \frac{\partial u_z}{\partial r}, \qquad 2\tilde{e}_{r\theta} = \frac{1}{r}\frac{\partial u_r}{\partial \theta} + \frac{\partial u_\theta}{\partial r} - \frac{u_\theta}{r}$$

$$2\tilde{r}_r = \frac{1}{r}\frac{\partial u_z}{\partial \theta} - \frac{\partial u_\theta}{\partial z}, \qquad 2\tilde{r}_\theta = \frac{\partial u_r}{\partial z} - \frac{\partial u_z}{\partial r}, \qquad 2\tilde{r}_z = \frac{1}{r}\frac{\partial}{\partial r}(r u_\theta) - \frac{1}{r}\frac{\partial u_r}{\partial \theta}$$

where $(t_{rr}, t_{r\theta}, \ldots, t_{zz})$, $(\tilde{e}_{rr}, \tilde{e}_{r\theta}, \ldots, \tilde{e}_{zz})$, (u_r, u_θ, u_z), and $(\tilde{r}_r, \tilde{r}_\theta, \tilde{r}_z)$ denote, respectively, the physical components of stress, strain, displacement, and rotation in cylindrical coordinates.

The Navier–Duhamel equations (1.20.27) and (1.20.28) read

$$(\lambda + 2\mu)\frac{\partial I_{\tilde{e}}}{\partial r} - 2\mu\left(\frac{1}{r}\frac{\partial \tilde{r}_z}{\partial \theta} - \frac{\partial \tilde{r}_\theta}{\partial z}\right) - \beta\frac{\partial T}{\partial r} + \rho(f_r - \ddot{u}_r) = 0$$

$$(\lambda + 2\mu)\frac{1}{r}\frac{\partial I_{\tilde{e}}}{\partial \theta} - 2\mu\left(\frac{\partial \tilde{r}_r}{\partial z} - \frac{\partial \tilde{r}_z}{\partial r}\right) - \frac{\beta}{r}\frac{\partial T}{\partial \theta} + \rho(f_\theta - \ddot{u}_\theta) = 0 \tag{1.20.34}$$

$$(\lambda + 2\mu)\frac{\partial I_{\tilde{e}}}{\partial z} - \frac{2\mu}{r}\left[\frac{\partial}{\partial r}(r\tilde{r}_\theta) - \frac{\partial \tilde{r}_r}{\partial \theta}\right] - \beta\frac{\partial T}{\partial z} + \rho(f_z - \ddot{u}_z) = 0$$

$$\kappa\left(\frac{\partial^2 T}{\partial r^2} + \frac{1}{r}\frac{\partial T}{\partial r} + \frac{1}{r^2}\frac{\partial^2 T}{\partial \theta^2} + \frac{\partial^2 T}{\partial z^2}\right) - \beta T_0 \dot{I}_{\tilde{e}} - \rho_0 \gamma \dot{T} + \rho h = 0 \tag{1.20.35}$$

where

$$I_{\tilde{e}} \equiv \operatorname{div} \mathbf{u} = \frac{1}{r}\frac{\partial}{\partial r}(ru_r) + \frac{1}{r}\frac{\partial u_\theta}{\partial \theta} + \frac{\partial u_z}{\partial z} \tag{1.20.36}$$

(ii) *Spherical coordinates.* Spherical coordinates (r, θ, ϕ) are related to rectangular coordinates (x, y, z) by (Fig. 1.20.2)

$$x = r \sin\theta \cos\phi, \qquad y = r \sin\theta \sin\phi, \qquad z = r\cos\theta \tag{1.20.37}$$

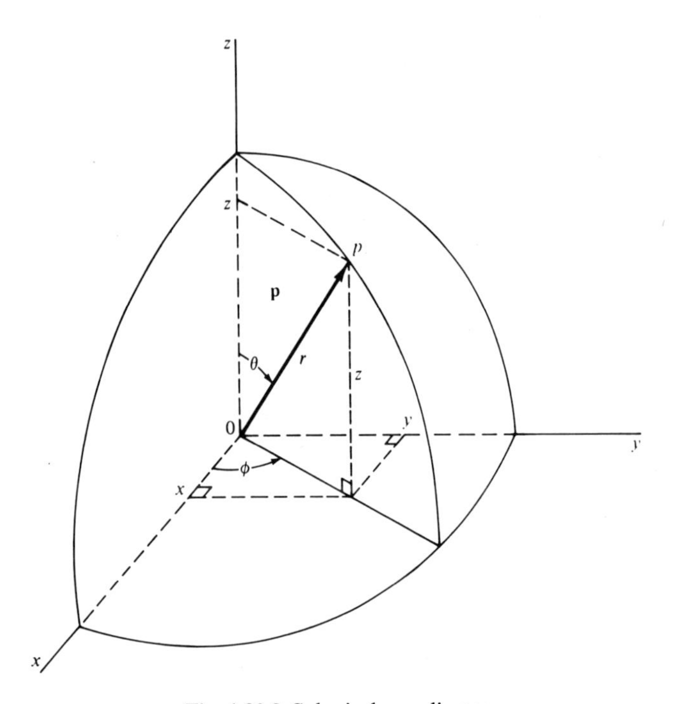

Fig. 1.20.2 Spherical coordinates.

The square of the arc length and the metric tensor g_{kl} are given by

$$\begin{gathered} ds^2 = dr^2 + r^2\, d\theta^2 + r^2 \sin^2\theta\, d\phi^2 \\ g_{11} = 1, \qquad g_{22} = r^2, \qquad g_{33} = r^2 \sin^2\theta, \qquad g_{kl} = 0 \qquad (k \neq l) \end{gathered} \tag{1.20.38}$$

Cauchy's equations of motion (1.20.12), the energy equation (1.20.13), and the infinitesimal strains and rotations are

$$\frac{\partial t_{rr}}{\partial r} + \frac{1}{r}\frac{\partial t_{r\theta}}{\partial \theta} + \frac{1}{r\sin\theta}\frac{\partial t_{r\phi}}{\partial \phi} + \frac{1}{r}(2t_{rr} - t_{\theta\theta} - t_{\phi\phi} + t_{r\theta}\cot\theta) - \beta\frac{\partial T}{\partial r} + \rho(f_r - \ddot{u}_r) = 0$$

$$\frac{\partial t_{r\theta}}{\partial r} + \frac{1}{r}\frac{\partial t_{\theta\theta}}{\partial r} + \frac{1}{r\sin\theta}\frac{\partial t_{\theta\phi}}{\partial \phi} + \frac{1}{r}[3t_{r\theta} + (t_{\theta\theta} - t_{\phi\phi})\cot\theta] - \frac{\beta}{r}\frac{\partial T}{\partial \theta} + \rho(f_\theta - \ddot{u}_\theta) = 0 \tag{1.20.39}$$

$$\frac{\partial t_{r\phi}}{\partial r} + \frac{1}{r}\frac{\partial t_{\theta\phi}}{\partial \theta} + \frac{1}{r\sin\theta}\frac{\partial t_{\phi\phi}}{\partial \phi} + \frac{1}{r}(3t_{r\phi} + 2t_{\theta\phi}\cot\theta) - \frac{\beta}{r\sin\theta}\frac{\partial T}{\partial \phi} + \rho(f_\phi - \ddot{u}_\phi) = 0$$

$$\rho\theta\dot{\eta} - \rho h - \frac{1}{r^2}\frac{\partial}{\partial r}(r^2 q_r) - \frac{1}{r\sin\theta}\frac{\partial}{\partial\theta}(q_\theta\sin\theta) - \frac{1}{r\sin\theta}\frac{\partial q_\phi}{\partial\phi} = 0 \tag{1.20.40}$$

$$\tilde{e}_{rr} = \frac{\partial u_r}{\partial r}, \qquad \tilde{e}_{\theta\theta} = \frac{1}{r}\frac{\partial u_\theta}{\partial\theta} + \frac{u_r}{r}, \qquad \tilde{e}_{\phi\phi} = \frac{1}{r\sin\theta}\frac{\partial u_\phi}{\partial\phi} + \frac{u_r}{r} + \frac{u_\theta}{r}\cot\theta$$

$$2\tilde{e}_{\theta\phi} = \frac{1}{r}\frac{\partial u_\phi}{\partial\theta} + \frac{1}{r\sin\theta}\frac{\partial u_\theta}{\partial\phi} - \frac{u_\phi}{r}\cot\theta, \qquad 2\tilde{e}_{\phi r} = \frac{1}{r\sin\theta}\frac{\partial u_r}{\partial\phi} + \frac{\partial u_\phi}{\partial r} - \frac{u_\phi}{r}$$

(1.20.41)

$$2\tilde{e}_{r\theta} = \frac{1}{r}\frac{\partial u_r}{\partial\theta} + \frac{\partial u_\theta}{\partial r} - \frac{u_\theta}{r}, \qquad 2\tilde{r}^r = \frac{1}{r\sin\theta}\left[\frac{\partial}{\partial\theta}(u_\phi\sin\theta) - \frac{\partial u_\theta}{\partial\phi}\right]$$

$$2\tilde{r}_\theta = \frac{1}{r\sin\theta}\frac{\partial u_r}{\partial\phi} - \frac{1}{r}\frac{\partial}{\partial r}(ru_\phi), \qquad 2\tilde{r}_\phi = \frac{1}{r}\frac{\partial}{\partial r}(ru_\theta) - \frac{1}{r}\frac{\partial u_r}{\partial\theta}$$

where $(t_{rr}, t_{r\theta}, \ldots, t_{\phi\phi})$, $(\tilde{e}_{rr}, \tilde{e}_{r\theta}, \ldots, \tilde{e}_{\phi\phi})$, (u_r, u_θ, u_ϕ), and $(\tilde{r}_r, \tilde{r}_\theta, \tilde{r}_\phi)$ denote, respectively, the physical components of the stress tensor, strain tensor, displacement vector, and rotation vector.

The Navier–Duhamel equations (1.20.27) and (1.20.28) take the form

(1.20.42)

$$(\lambda + 2\mu)\frac{\partial I_{\tilde{e}}}{\partial r} - \frac{2\mu}{r\sin\theta}\left[\frac{\partial}{\partial\theta}(\tilde{r}_{\phi}\sin\theta) - \frac{\partial\tilde{r}_{\theta}}{\partial\phi}\right] - \beta\frac{\partial T}{\partial r} + \rho(f_r - \ddot{u}_r) = 0$$

$$(\lambda + 2\mu)\frac{1}{r}\frac{\partial I_{\tilde{e}}}{\partial\theta} - \frac{2\mu}{r}\left[\frac{1}{\sin\theta}\frac{\partial\tilde{r}_r}{\partial\phi} - \frac{\partial}{\partial r}(r\tilde{r}_{\phi})\right] - \frac{\beta}{r}\frac{\partial T}{\partial\theta} - \rho(f_{\theta} - \ddot{u}_{\theta}) = 0$$

$$\frac{\lambda + 2\mu}{r\sin\theta}\frac{\partial I_{\tilde{e}}}{\partial\phi} - \frac{2\mu}{r}\left[\frac{\partial}{\partial r}(r\tilde{r}_{\theta}) - \frac{\partial\tilde{r}_r}{\partial\theta}\right] - \frac{\beta}{r\sin\theta}\frac{\partial T}{\partial\phi} + \rho(f_{\phi} - \ddot{u}_{\phi}) = 0$$

(1.20.43)

$$\frac{\kappa}{r^2}\frac{\partial}{\partial r}\left(r^2\frac{\partial T}{\partial r}\right) + \frac{\kappa}{r^2\sin\theta}\frac{\partial}{\partial\theta}\left(\sin\theta\frac{\partial T}{\partial\theta}\right) + \frac{\kappa}{r^2\sin\theta}\frac{\partial^2 T}{\partial\phi^2} - \beta T_0\dot{I}_{\tilde{e}} - \rho_0\gamma\dot{T} + \rho h = 0$$

where

(1.20.44)

$$I_{\tilde{e}} \equiv \operatorname{div}\mathbf{u} = \frac{1}{r^2}\frac{\partial(r^2u_r)}{\partial r} + \frac{1}{r\sin\theta}\frac{\partial}{\partial\theta}(u_{\theta}\sin\theta) + \frac{1}{r\sin\theta}\frac{\partial u_{\phi}}{\partial\phi}$$

is the dilatation (the first invariant of $\tilde{\mathbf{e}}$).

CHAPTER II

Propagation of Singular Surfaces

2.1 SCOPE OF THE CHAPTER

The study of the propagation of finite amplitude waves in elastic materials is tedious and often presents insurmountable mathematical difficulties. In order to gain an insight into such problems, one may conceive of some limiting forms of the motion from which important information may be extracted regarding the actual motion. A limiting form of a wave motion may be thought to be a moving wave front across which the motion displays certain discontinuities. This mathematical abstraction represents a good approximation for the propagation of a signal of very narrow width or a wave of very high frequency.

In Section 2.2 we present a précis of the differential geometry leading to fundamental formulas relevant to moving surfaces. Singular surfaces are introduced in Section 2.3 and compatibility conditions across a singular surface are derived in Sections 2.4 and 2.5. In Sections 2.6–2.8 singular surfaces associated with the motion of the material body, dynamical compatibility conditions, and classification of singular surfaces are discussed. Section 2.9 is devoted to the formulation of shock waves. The general theory is specialized to incompressible material in Section 2.10 and to weak shock waves in Section 2.11. The theory of acceleration waves is developed in Section 2.12, and in Sections 2.13 and 2.14 it is specialized to isotropic and incompressible solids. The growth and decay of acceleration waves (Section 2.15) concludes the chapter.

2.2 FUNDAMENTAL FORMULAS FOR MOVING SURFACES

In a three-dimensional Euclidean space E_3 a one-parameter family of surfaces $\sigma(t)$ may be represented by the Gaussian forms

(2.2.1)
$$\mathbf{x} = x_i \mathbf{i}_i = \mathbf{x}(p^\alpha, t) \quad \text{or} \quad x_i = x_i(p^\alpha, t), \quad t \in [a, b], \quad i = 1, 2, 3, \quad \alpha = 1, 2$$

where $p^\alpha(p^1, p^2)$ are the curvilinear coordinates of the surfaces, x_i the rectangular coordinates of a point on the surface, and $\mathbf{i}_i$ the Cartesian unit vectors. If we identify the parameter t as time, then (2.2.1) represents a moving surface, and the family of surfaces $\sigma(t)$ is the set of all spatial configurations of the same surface in a given closed time interval $[a, b]$, which may be of infinite extent. We assume that the functions (2.2.1) are continuously differentiable with respect to their arguments and $\sigma(t)$ is smooth in the sense that the matrix $\|\partial x_k/\partial p^\alpha\|$ always has rank two. This condition is sufficient for the existence of a tangent plane at each point of $\sigma(t)$ and permits us to eliminate p^α in (2.2.1) to obtain a representation of the moving surface in the form

$$f(\mathbf{x}, t) = 0 \tag{2.2.2}$$

An infinitesimal tangent vector on $\sigma(t)$ is given by

$$d\mathbf{x} = \mathbf{a}_\alpha \, dp^\alpha \quad \text{or} \quad dx_i = x_{i\alpha} \, dp^\alpha, \qquad i = 1, 2, 3, \quad \alpha = 1, 2 \tag{2.2.3}$$

where $\partial x_i/\partial p^\alpha \equiv x_{i,\alpha} \equiv x_{i\alpha}$ and

$$\mathbf{a}_\alpha(\mathbf{p}, t) \equiv \partial \mathbf{x}(\mathbf{p}, t)/\partial p^\alpha = (\partial x_i/\partial p^\alpha)\mathbf{i}_i = x_{i\alpha} \mathbf{i}_i, \qquad \alpha = 1, 2 \tag{2.2.4}$$

are the *base vectors* tangent to the curvilinear coordinates on the surface (Fig. 2.2.1). It is clear that the Cartesian components $x_{i\alpha}$ of $\mathbf{a}_\alpha$ constitute a *two-point tensor field*, which transforms as an absolute tensor under the transformation of both coordinates x_i and p^α (cf. Section A.7).

The element of arc length ds on the surface is given by

$$ds^2 = a_{\alpha\beta} \, dp^\alpha \, dp^\beta \tag{2.2.5}$$

where the symmetric surface tensor

$$a_{\alpha\beta}(\mathbf{p}, t) \equiv \mathbf{a}_\alpha \cdot \mathbf{a}_\beta = x_{i\alpha} x_{i\beta} \tag{2.2.6}$$

is the *metric tensor of the surface*. The elements $a^{\alpha\beta}$ of the inverse of 2×2 matrix $\|a_{\alpha\beta}\|$ are given by

$$a^{\alpha\beta} \equiv (\text{cofactor } a_{\alpha\beta})/a, \qquad a \equiv \det a_{\alpha\beta} = |\mathbf{a}_1 \times \mathbf{a}_2|^2 \tag{2.2.7}$$

Thus

$$a^{\alpha\beta} a_{\beta\gamma} = \delta^\alpha{}_\gamma \tag{2.2.8}$$

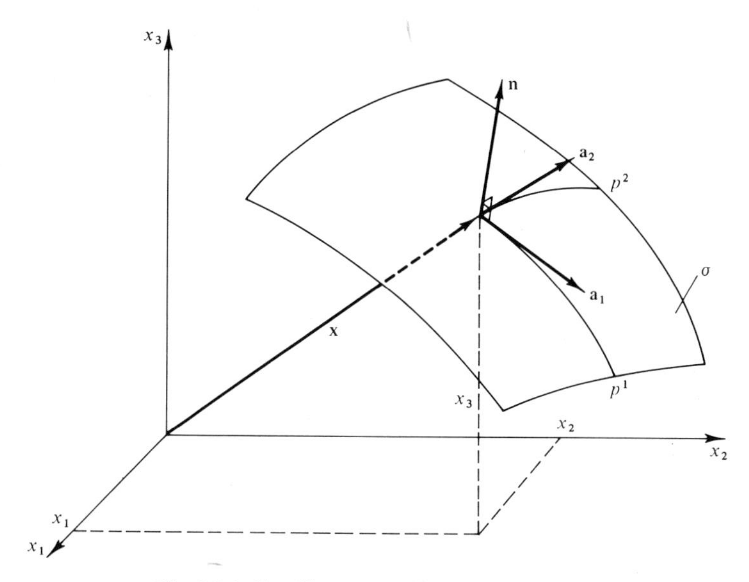

Fig. 2.2.1 Curvilinear coordinates on a surface.

If **u** is a vector field on the surface having covariant components u_α and contravariant components u^α, then

$$\mathbf{u} = u^\alpha \mathbf{a}_\alpha = u_\alpha \mathbf{a}^\alpha \tag{2.2.9}$$

where

$$\mathbf{a}^\alpha \equiv a^{\alpha\beta} \mathbf{a}_\beta \tag{2.2.10}$$

are the *reciprocal bases* to $\mathbf{a}_\beta$. We have

$$\mathbf{a}^\alpha \cdot \mathbf{a}_\beta = a^{\alpha\gamma} a_{\gamma\beta} = \delta^\alpha{}_\beta \tag{2.2.11}$$

Thus by taking the scalar product of (2.2.9) with $\mathbf{a}^\beta$ and $\mathbf{a}_\beta$, we obtain

$$u^\alpha = a^{\alpha\beta} u_\beta, \qquad u_\alpha = a_{\alpha\beta} u^\beta \tag{2.2.12}$$

This is known as the raising and lowering of indices pertinent to the surface (cf. Section A.3). This process is valid for higher-order tensors, e.g.,

$$t^{\alpha\beta} = a^{\alpha\gamma} t_\gamma{}^\beta = a^{\alpha\gamma} a^{\beta\delta} t_{\gamma\delta}$$

The unit normal **n** to the surface at a point **x** can be expressed by means of the two tangent vectors x_{i1} and x_{i2} (Fig. 2.2.1):

$$\mathbf{n} = (\mathbf{a}_\alpha \times \mathbf{a}_\beta)/|\mathbf{a}_\alpha \times \mathbf{a}_\beta| \quad \text{or} \quad n_i = \tfrac{1}{2}\varepsilon^{\alpha\beta} e_{ijk} x_{j\alpha} x_{k\beta}, \qquad \alpha \neq \beta \tag{2.2.13}$$

where e_{ijk} is the usual three-dimensional permutation symbol,

$$\varepsilon^{\alpha\beta} \equiv e^{\alpha\beta}/\sqrt{a}, \qquad \varepsilon_{\alpha\beta} \equiv e_{\alpha\beta}\sqrt{a} \tag{2.2.14}$$

with $e_{\alpha\beta}$ and $e^{\alpha\beta}$ the two-dimensional permutation symbols, i.e.,

$$e_{11} = e_{22} = e^{11} = e^{22} = 0, \qquad e_{12} = -e_{21} = e^{12} = -e^{21} = 1 \tag{2.2.15}$$

We note that

$$\varepsilon^{\alpha\beta}\varepsilon_{\gamma\beta} = \delta^{\alpha}{}_{\gamma} \tag{2.2.16}$$

From (2.2.13) it is clear that $\mathbf{n}$ is a unit vector, i.e.,

$$\mathbf{n} \cdot \mathbf{n} = 1 \tag{2.2.17}$$

and since $\mathbf{n}$ is perpendicular to $\mathbf{a}_\alpha$, we have

$$\mathbf{n} \cdot \mathbf{a}_\alpha = 0 \qquad \text{or} \qquad n_i x_{i\alpha} = 0 \tag{2.2.18}$$

The following identity will be particularly useful in subsequent developments:

$$a^{\alpha\beta} x_{i\alpha} x_{j\beta} = \delta_{ij} - n_i n_j \tag{2.2.19}$$

which follows from $(2.2.13)_2$, with the use of the identities (cf. A.5.17)

$$a^{\alpha\beta} = \varepsilon^{\alpha\gamma}\varepsilon^{\beta\delta} a_{\gamma\delta}, \qquad e_{ijk} e_{lmn} = \begin{vmatrix} \delta_{il} & \delta_{im} & \delta_{in} \\ \delta_{jl} & \delta_{jm} & \delta_{jn} \\ \delta_{kl} & \delta_{km} & \delta_{kn} \end{vmatrix} \tag{2.2.20}$$

To determine the partial derivatives of surface vectors we need the partial derivatives of $\mathbf{a}_\alpha$ and $\mathbf{a}^\alpha$. Since any vector on the surface is linearly dependent on $\mathbf{a}_\alpha$ or $\mathbf{a}^\alpha$ and (2.2.11) is valid, we can write (Section A.5)

$$\begin{aligned} \mathbf{a}_{\alpha,\beta} &\equiv \partial \mathbf{a}_\alpha / \partial p^\beta = \begin{Bmatrix} \gamma \\ \alpha\beta \end{Bmatrix} \mathbf{a}_\gamma \\ \mathbf{a}^\alpha{}_{,\beta} &\equiv \partial \mathbf{a}^\alpha / \partial p^\beta = -\begin{Bmatrix} \alpha \\ \beta\gamma \end{Bmatrix} \mathbf{a}^\gamma \end{aligned} \tag{2.2.21}$$

where $\left\{ {\gamma \atop \alpha\beta} \right\}$ is called the *Christoffel symbol* of the *second kind*, which is related to the *Christoffel symbol* of the *first kind* $[\alpha\beta, \delta]$ by

$$\begin{Bmatrix} \gamma \\ \alpha\beta \end{Bmatrix} = a^{\gamma\delta}[\alpha\beta, \delta] \tag{2.2.22}$$

To obtain the explicit expressions of Christoffel symbols from $(2.2.21)_1$, we write

$$\begin{Bmatrix} \gamma \\ \alpha\beta \end{Bmatrix} = \mathbf{a}^\gamma \cdot \mathbf{a}_{\alpha,\beta} = a^{\gamma\delta} \mathbf{a}_\delta \cdot \mathbf{a}_{\alpha,\beta}$$

Thus

$$[\alpha\beta, \gamma] = \mathbf{a}_\gamma \cdot \mathbf{a}_{\alpha,\beta} = x_{i,\gamma} x_{i,\alpha\beta} \tag{2.2.23}$$

where we used (2.2.4). By substituting (2.2.6) we can verify that

$$[\alpha\beta, \gamma] = \tfrac{1}{2}(a_{\alpha\gamma,\beta} + a_{\beta\gamma,\alpha} - a_{\alpha\beta,\gamma}) \tag{2.2.24}$$

Thus both Christoffel symbols are expressible in terms of the partial derivatives of the metric tensor of the surface.

If $\mathbf{v}$ is a vector on the surface then

$$\mathbf{v}_{,\beta} = (v^\alpha \mathbf{a}_\alpha)_{,\beta} = v^\alpha{}_{,\beta}\,\mathbf{a}_\alpha + v^\alpha \mathbf{a}_{\alpha,\beta}$$

or on using (2.2.21) this may be written as

$$\mathbf{v}_{,\beta} = v^\alpha{}_{;\beta}\,\mathbf{a}_\alpha, \qquad \mathbf{v}_{,\beta} = v_{\alpha;\beta}\,\mathbf{a}^\alpha \tag{2.2.25}$$

where here, and throughout, a semicolon is used to indicate *covariant partial differentiations* defined by

$$v^\alpha{}_{;\beta} \equiv v^\alpha{}_{,\beta} + \begin{Bmatrix} \alpha \\ \beta\gamma \end{Bmatrix} v^\gamma, \qquad v_{\alpha;\beta} \equiv v_{\alpha,\beta} - \begin{Bmatrix} \gamma \\ \alpha\beta \end{Bmatrix} v_\gamma \tag{2.2.26}$$

Two basic formulas from the theory of surfaces are

$$x_{i\alpha;\beta} = b_{\alpha\beta} n_i, \qquad n_{i;\alpha} = -a^{\beta\gamma} b_{\alpha\beta} x_{i\gamma} = -b^\gamma{}_\alpha x_{i\gamma} \tag{2.2.27}$$

where the symmetric surface tensor $b_{\alpha\beta}$ is usually introduced through the *second fundamental form* of the surface

$$B = -dn_i\, dx_i = -n_{i;\alpha} x_{i\beta}\, dp^\alpha\, dp^\beta = b_{\alpha\beta}\, dp^\alpha\, dp^\beta \tag{2.2.28}$$

It may be seen that this definition gives $(2.2.27)_2$, for if we multiply $(2.2.27)_2$ by $x_{i\beta}$ we obtain $b_{\alpha\beta}$ defined by $(2.2.28)_1$. To obtain (2.2.27), we first write

$$x_{i\alpha;\beta} = x_{i\alpha,\beta} - \begin{Bmatrix} \delta \\ \alpha\beta \end{Bmatrix} x_{i\delta}$$

Using (2.2.22) and (2.2.23) this reads

$$\begin{aligned} x_{i\alpha;\beta} &= x_{i,\alpha\beta} - a^{\gamma\delta} x_{j\gamma} x_{j,\alpha\beta} x_{i\delta} \\ &= x_{i,\alpha\beta} - (\delta_{ij} - n_i n_j) x_{j,\alpha\beta} = n_i n_j x_{j,\alpha\beta} \\ &= -\, n_i n_{j,\alpha} x_{j\beta} \end{aligned}$$

where (2.2.18) and (2.2.19) are used. Since $n_{j;\alpha} = n_{j,\alpha}$ this last result is identical to $(2.2.27)_1$. Multiplying both sides of $(2.2.27)_1$ by n_i and using (2.2.8) and (2.2.13), we obtain

$$b_{\alpha\beta} = n_i x_{i\alpha;\beta} = n_i x_{i\alpha,\beta} = -\, n_{i;\alpha} x_{i\beta} = \tfrac{1}{2}\varepsilon^{\gamma\delta} e_{ijk} x_{i,\alpha\beta} x_{j\gamma} x_{k\delta} \tag{2.2.29}$$

Two surface scalars which are the first and second invariants of the tensor $b^\alpha{}_\beta$ are defined by

$$2\Omega \equiv a^{\alpha\beta} b_{\alpha\beta} = b^\alpha{}_\alpha = \operatorname{tr} \mathbf{b}, \qquad 2K \equiv (b^\alpha{}_\alpha)^2 - b^\alpha{}_\beta b^\beta{}_\alpha = (\operatorname{tr} \mathbf{b})^2 - \operatorname{tr} \mathbf{b}^2 \tag{2.2.30}$$

the former of which is called the *mean curvature* of the surface and the latter the *total* or *Gaussian curvature* of the surface. For additional accounts on the theory of surfaces the reader may consult McConnell [1957], Thomas [1961a], and Eringen [1971].

For a moving surface all the geometrical properties considered above are functions of time. For a particular choice of the time parameter t these equations describe the geometrical properties of a particular surface at time t coinciding with the corresponding configuration of the moving surface. We now briefly discuss the kinematics of a moving surface. The surface velocity $\mathbf{u}(\mathbf{p}, t)$ is defined by

$$\mathbf{u}(\mathbf{p}, t) = \left.\frac{\partial \mathbf{x}}{\partial t}\right|_{\mathbf{p}} \tag{2.2.31}$$

It is clear that (2.2.31) gives the velocity of a point on the surface, along its trajectory

$$x_i = x_i(\mathbf{p}, t)\Big|_{\mathbf{p}} \tag{2.2.32}$$

If the representation (2.2.2) is used, we have $df = 0$, hence

$$(\partial f/\partial t) + \mathbf{u} \cdot \nabla f = 0$$

where ∇ denotes the gradient operator. The normal velocity of the surface, or *the speed of displacement*, is defined by

$$u_n \equiv \mathbf{u} \cdot \mathbf{n} = -(\partial f/\partial t)/|\nabla f| = n_i\, \partial x_i/\partial t \tag{2.2.33}$$

The speed of displacement is clearly independent of a specific choice of the surface coordinates p^α in (2.2.1), while the velocity $\mathbf{u}$, in general, depends on a particular choice of the surface coordinates. For all $\mathbf{u}$ corresponding to infinitely many sets of surface coordinates, the normal velocity of the surface points is the same. Sometimes it is advantageous to select the surface coordinates in such a way that $\mathbf{u} = u_n \mathbf{n}$. This selection of the surface coordinates is tantamount to selecting the trajectories of surface points as orthogonal trajectories of the one-parameter family of surfaces $\sigma(t)$. If u_n is independent of the surface coordinates, then the configurations of the moving surface constitute a family of parallel surfaces.

The *δ-time derivative* (Thomas [1957, 1961b]) or the *displacement derivative* (Truesdell and Toupin [1960]) of a surface tensor $\phi(\mathbf{x}, t)$ is defined by

$$\delta\phi/\delta t = \lim_{\Delta t \to 0} \Delta\phi/\Delta t \tag{2.2.34}$$

where ϕ is evaluated at a point $\mathbf{x}$ on $\sigma(t)$ and $\phi + \Delta\phi$ at the intersection point $\mathbf{x} + \Delta\mathbf{x}$, on $\sigma(t + \Delta t)$, of the normal $\mathbf{n}$ erected to $\sigma(t)$ at the point $\mathbf{x}$ (Fig. 2.2.2). This derivative is evidently the time rate of change of ϕ as apparent

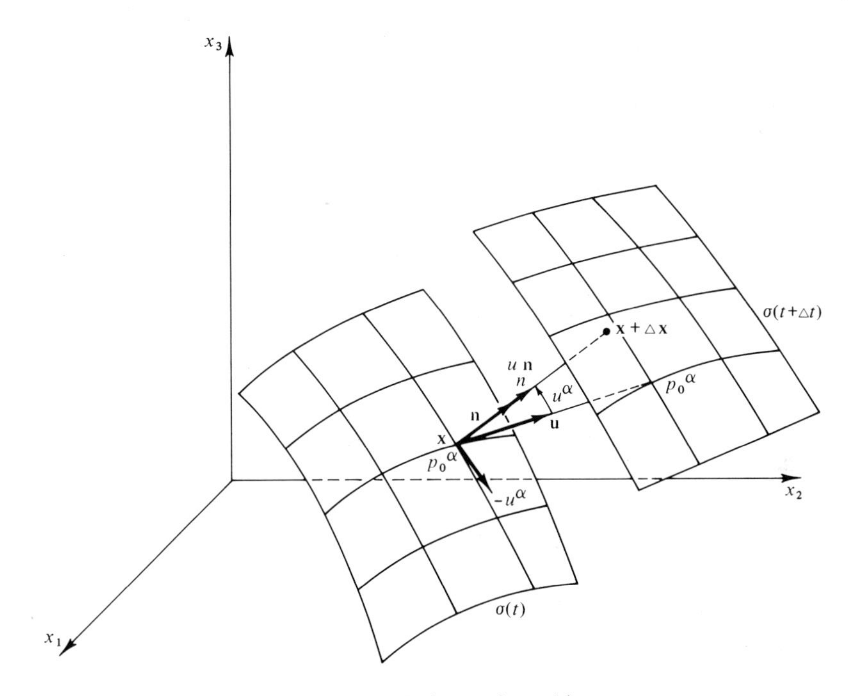

Fig. 2.2.2 Moving surface $\sigma(t)$.

to an observer moving with the normal velocity of the surface. It can be seen that the δ-derivative *obeys the rules of ordinary differentiation.* We have

(2.2.35) $$\delta\phi/\delta t = (\partial\phi/\partial t) + \phi_{,i}\,\delta x_i/\delta t$$

Now clearly

(2.2.36) $$\delta x_i/\delta t = u_n n_i$$

Therefore, (2.2.35) can be put into the form

(2.2.37) $$\delta\phi/\delta t = (\partial\phi/\partial t) + u_n \phi_{,i} n_i$$

If the parametrization of the surface is not the *normal parametrization,* then $\delta\mathbf{x}/\delta t$ is not equal to $\partial\mathbf{x}/\partial t$, but it can only differ by a tangential surface velocity vector. If $-u^\alpha$ is the component of this vector along the surface tangent vector $\mathbf{a}_\alpha$, then we can write at once

(2.2.38) $$\delta x_i/\delta t = u_n n_i = (\partial x_i/\partial t) + u^\alpha x_{i\alpha}$$

Since $x_i = x_i(p^\alpha, t)$, we also have

$$\delta x_i/\delta t = (\partial x_i/\partial t) + (\delta p^\alpha/\delta t)x_{i\alpha}$$

from these we deduce that

$$u^{\alpha} = \delta p^{\alpha}/\delta t \tag{2.2.39}$$

This may be referred to as the *tangential velocity of the parametrization.* At first glance it may appear that u^{α} should vanish since surface coordinates are assumed to be time independent. However, if $\mathbf{u} \neq u_n \mathbf{n}$, an observer moving with the normal velocity of the surface will encounter surface coordinates with varying numerical values during his motion and hence will see a continuous change in surface coordinates along his way (Fig. 2.2.2).

On multiplying (2.2.38) by $x_{i\beta}$, we obtain

$$u_{\alpha} = a_{\alpha\beta} u^{\beta} = -(\partial x_i/\partial t) x_{i\alpha} \tag{2.2.40}$$

The insertion of (2.2.38) into (2.2.37) yields a surface description of the displacement derivative:

$$\frac{\delta \phi}{\delta t} = \frac{\partial \phi}{\partial t} + \left(\frac{\partial x_i}{\partial t} + u^{\alpha} x_{i\alpha}\right) \phi_{,i} = \frac{\partial \phi}{\partial t} + \frac{\partial x_i}{\partial t} \phi_{,i} + u^{\alpha} \phi_{;\alpha}$$

The surface description of $\phi(\mathbf{x}, t)$ may be denoted by

$$\Phi(\mathbf{p}, t) = \phi[\mathbf{x}(p^{\alpha}, t), t]$$

from which we get

$$\partial \Phi/\partial t = (\partial \phi/\partial t) + \phi_{,i}\, \partial x_i/\partial t$$

If we use the same symbol to specify the same function but recall that it is now a function of the surface coordinates p^{α} and time t, we obtain the following formula for the displacement derivative:

$$\delta \phi/\delta t = (\partial \phi/\partial t) + u^{\alpha} \phi_{;\alpha} \tag{2.2.41}$$

where it should be understood that ϕ is given as a function of the surface coordinates and time instead of the space coordinates and time.

We can now derive an important formula which gives the displacement derivative of the unit normal $\mathbf{n}(p^{\alpha}, t)$ to the surface. To this end we first calculate the variation of the normal velocity $u_n(p^{\alpha}, t)$ along the surface $\sigma(t)$. From (2.2.33) it follows that

$$u_{n,\alpha} = \frac{\partial x_{i\alpha}}{\partial t} n_i + \frac{\partial x_i}{\partial t} n_{i,\alpha} = -x_{i\alpha} \frac{\partial n_i}{\partial t} + \frac{\partial x_i}{\partial t} n_{i,\alpha}$$

where in the extreme right we used (2.2.18). Multiplying both sides of this equation by $a^{\alpha\beta} x_{j\beta}$ and using (2.2.19) and $(2.2.27)_2$ we obtain

$$a^{\alpha\beta} x_{j\beta} u_{n,\alpha} = -(\delta_{ij} - n_i n_j)(\partial n_i/\partial t) - a^{\alpha\beta} b^{\gamma}{}_{\alpha} x_{j\beta} x_{i\gamma} (\partial x_i/\partial t) \tag{2.2.42}$$

But, from (2.2.17) we have

$$n_i\, \partial n_i/\partial t = 0$$

Thus on using this and the symmetry of the tensor **b**, (2.2.42) reduces to

$$-a^{\alpha\beta} x_{j\beta}\, u_{n,\alpha} = (\partial n_j/\partial t) - a^{\alpha\beta} b^{\gamma}{}_{\alpha} u_{\gamma}\, x_{j\beta} = (\partial n_j/\partial t) - b^{\beta}{}_{\gamma} u^{\gamma} x_{j\beta}$$

With the help of $(2.2.27)_2$ we finally have

$$-a^{\alpha\beta} x_{j\beta}\, u_{n,\alpha} = (\partial n_j/\partial t) + u^{\alpha} n_{j;\alpha}$$

the right-hand side of which is the displacement derivative of **n** [cf. (2.2.41)]. Thus

$$\delta n_i/\delta t = -a^{\alpha\beta} x_{i\alpha}\, u_{n,\beta} \tag{2.2.43}$$

From this equation it follows that *a necessary and sufficient condition for parallel propagation of a moving surface* $\sigma(t)$ *is*

$$u_n = u_n(t)$$

In this case the normal trajectories to the family of surfaces $\sigma(t)$ are all straight lines.

Equation (2.2.43) can be modified for a spatial description. For this we write $u_{n,\beta} = u_{n,j}\, x_{j\beta}$ and make use of (2.2.19). Thus[1]

$$\delta n_i/\delta t = (n_i n_j - \delta_{ij}) u_{n,j} \tag{2.2.44}$$

where both **n** and u_n are supposed to be functions of the surface points **x** and time t. Finally, if we multiply both sides of (2.2.43) by $x_{i\gamma}$ we get

$$u_{n,\alpha} = -x_{i\alpha}\, \delta n_i/\delta t \tag{2.2.45}$$

Equation (2.2.43) was stated by Hayes [1957] without proof. The proof was given by Thomas [1957].

2.3 SINGULAR SURFACES

A regular surface $\sigma(t)$ moving in a spatial region R, at time t divides this region into two subregions $R^+(t)$ and $R^-(t)$. We take the positive unit normal to σ directed toward the region $R^+(t)$ (Fig. 2.3.1). The side of the surface which is in touch with the region $R^+(t)$ will be labeled with a superscript $(+)$, while the other side will be labeled with $(-)$.

Consider a tensor-valued function $\phi(\mathbf{x}, t)$ (ϕ may be components of a tensorial quantity) which is continuous in $R^+(t)$ and $R^-(t)$ and possesses the limiting values ϕ^+ and ϕ^- as any point **x** on the surface $\sigma(t)$ is approached

[1] The operator $a^{\alpha\beta} x_{i\alpha}(\ )_{;\beta} = (\delta_{ij} - n_i n_j)(\ )_{,j}$ is called the surface gradient operator.

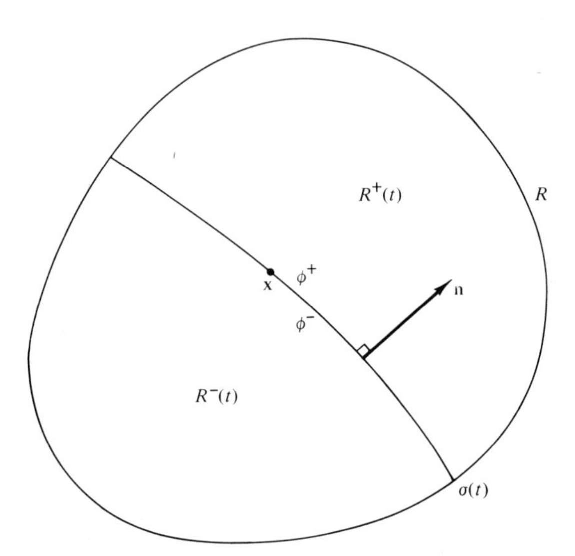

Fig. 2.3.1 Singular surface $\sigma(t)$ sweeping a region R.

from $R^+(t)$ and $R^-(t)$, respectively. When ϕ is continuous across the surface, these two values are identical. Otherwise, there will be a jump across $\sigma(t)$ at a surface point $\mathbf{x}$ given by[1]

$$[\phi] = \phi^+ - \phi^- \tag{2.3.1}$$

The quantity $[\phi]$ is obviously a function of positions on the surface $\sigma(t)$ and time t. Therefore it is expressible in surface coordinates and time only. *When* $[\phi] \neq 0$ *the surface* $\sigma(t)$ *is called a singular surface relative to the field* ϕ. This definition may be extended to include the spatial and temporal derivatives of ϕ. Thus, for example, if ϕ is continuous across σ but some of its derivatives are discontinuous, we shall still name the surface as a singular surface relative to ϕ.

The *order* of a singular surface is usually defined as the lowest order $p + q$ of the derivative $\partial^q \phi_{,i_1 i_2 \cdots i_p}/\partial t^q$ which suffers a finite jump across the surface. Therefore, the zeroth-order singular surface is such that the field quantity ϕ itself suffers a discontinuity across it.

The general and invariant theory of singular surfaces and its application to continuum mechanics may be attributed to Hadamard, although some fundamental ideas were discussed by Riemann and Hugoniot in their analysis of shock waves in perfect gases. The entire theory of singular surfaces can be constructed systematically starting from a lemma due to Hadamard [1903].[2]

[1] This notation is attributed to Christoffel.

[2] A more elaborate proof of this lemma is given in Lichtenstein [1929].

Hadamard's Lemma. *Let a tensor-valued function ϕ be defined and continuously differentiable in regions R^+ and R^- which are separated by a smooth surface σ, and let ϕ and $\phi_{,i}$ tend to finite limits $(\phi^+, (\phi_{,i})^+)$ and $(\phi^-, (\phi_{,i})^-)$ as σ is approached on paths interior to R^+ and R^-, respectively. If $\mathbf{x} = \mathbf{x}(s)$ is a smooth curve on σ, and ϕ^+ and ϕ^- are differentiable along this path, then*

$$d\phi^+/ds = (\phi_{,i})^+ \, dx_i/ds, \qquad d\phi^-/ds = (\phi_{,i})^- \, dx_i/ds \tag{2.3.2}$$

In other words, the theorem of total differentiation holds true for the limiting values as σ is approached from one side only. The proof is quite similar to the one in classical analysis. Considering that ϕ^+, ϕ^-,$(\phi_{,i})^+$, and $(\phi_{,i})^-$ are functions of surface coordinates in (2.3.2) and writing $\phi_{,i}^+$ and $\phi_{,i}^-$ for $(\phi_{,i})^+$ and $(\phi_{,i})^-$, respectively, it follows from (2.3.2) that

$$\phi_{;\alpha}^+ = \phi_{,i}^+ x_{i\alpha}, \qquad \phi_{;\alpha}^- = \phi_{,i}^- x_{i\alpha} \tag{2.3.3}$$

Hadamard's lemma imposes certain restrictions called *compatibility conditions* among various orders of discontinuities of ϕ. The next two sections are devoted to the development of these relations.

2.4 GEOMETRICAL CONDITIONS OF COMPATIBILITY

If we subtract the second of relations (2.3.3) from the first, we obtain

$$[\phi]_{;\alpha} = [\phi_{,i}]x_{i\alpha} \tag{2.4.1}$$

Multiplying both sides of (2.4.1) by $a^{\alpha\beta}x_{j\beta}$ and using (2.2.19) we find

$$a^{\alpha\beta}x_{j\beta}[\phi]_{;\alpha} = (\delta_{ij} - n_i n_j)[\phi_{,i}] \tag{2.4.2}$$

Henceforth we employ the following abbreviations:

$$A \equiv [\phi], \qquad B \equiv [\phi_{,i}]n_i \tag{2.4.3}$$

Note that A and B are not scalars if ϕ is not a scalar. With these definitions we deduce from (2.4.2)[1]

$$[\phi_{,i}] = Bn_i + a^{\alpha\beta}A_{;\alpha}x_{i\beta} \tag{2.4.4}$$

which determines the jumps in the three first-order partial derivatives of ϕ with respect to space variables on σ in terms of A and B. Therefore, we need to prescribe only the jump in the function itself and the jump in the normal derivative of the function to determine *completely* all jumps in the first-order partial derivatives. In other words, three jumps $[\phi_{,i}]$ cannot be assigned arbitrarily across the surface. They must satisfy the *first-order geometrical*

[1] The second term in (2.4.4) is the surface gradient of A (cf. footnote, p. 85).

compatibility condition (2.4.4). Equation (2.4.4) is also valid if space coordinates are curvilinear. In this case partial derivatives with respect to space variables should be replaced by covariant derivatives and covariant derivatives with respect to surface coordinates by the *total covariant derivatives.*

If ϕ is continuous across $\sigma(t)$, or more generally if $[\phi]$ is independent of surface coordinates, (2.4.4) reduces to

$$[\phi_{,i}] = Bn_i = [\phi_{,j} n_j] n_i \tag{2.4.5}$$

In the general expressions (2.4.4) we may replace ϕ by $\phi_{,i}$ and, modeled after (2.4.3), we define

$$A_i \equiv [\phi_{,i}], \qquad B_i \equiv [\phi_{,ij}] n_j \tag{2.4.6}$$

which gives

$$[\phi_{,ij}] = B_i n_j + a^{\alpha\beta} A_{i;\alpha} x_{j\beta} \tag{2.4.7}$$

where A_i are expressible in terms of A and B through (2.4.4). Since the left-hand side of (2.4.7) is symmetric in the indices i and j, the right-hand side should also be symmetric in these two indices. Hence we have

$$B_i n_j + a^{\alpha\beta} A_{i;\alpha} x_{j\beta} = B_j n_i + a^{\alpha\beta} A_{j;\alpha} x_{i\beta} \tag{2.4.8}$$

Now multiplying (2.4.7) by $n_i n_j$ and using (2.2.18) we obtain

$$B_i n_i = C \tag{2.4.9}$$

where

$$C \equiv [\phi_{,ij} n_i n_j] = [\phi_{,ij}] n_i n_j \tag{2.4.10}$$

Next we multiply (2.4.8) by n_j and use (2.4.9) to obtain

$$B_i = C n_i + a^{\alpha\beta} A_{j;\alpha} x_{i\beta} n_j \tag{2.4.11}$$

Differentiating (2.4.4), noting that $a^{\alpha\beta}{}_{;\gamma} \equiv 0$, and recalling relations (2.2.27), we find

$$A_{j;\alpha} = B_{;\alpha} n_j - B a^{\beta\gamma} b_{\alpha\beta} x_{j\gamma} + a^{\beta\gamma}(A_{;\alpha\beta} x_{j\gamma} + A_{;\beta} b_{\alpha\gamma} n_j)$$

If we introduce this into (2.4.11) and the resulting expression into (2.4.7), we arrive at the *second-order compatibility conditions*

$$\begin{aligned}[\phi_{,ij}] = {} & C n_i n_j + a^{\alpha\beta}(B_{;\alpha} + a^{\gamma\delta} b_{\alpha\gamma} A_{;\delta})(n_i x_{j\beta} + n_j x_{i\beta}) \\ & + a^{\alpha\beta} a^{\gamma\delta}(A_{;\alpha\gamma} - B b_{\alpha\gamma}) x_{i\beta} x_{j\delta}\end{aligned} \tag{2.4.12}$$

which determine the jumps in the second-order spatial partial derivatives across the singular surface in terms of the derivatives of the lesser-order discontinuities along the surface and a single jump in the second derivative along the normal direction to the surface. Since (2.4.12) is symmetric in the indices i and j, it can be put into the form

$$[\phi_{,ij}] = Cn_i n_j + 2a^{\alpha\beta} n_{(i} x_{j)\beta}(B_{;\alpha} + b^{\gamma}{}_{\alpha} A_{;\gamma}) + a^{\alpha\beta} a^{\gamma\delta} x_{(i|\beta|} x_{j)\delta}(A_{;\alpha\gamma} - b_{\alpha\gamma} B) \tag{2.4.13}$$

where parentheses enclosing indices indicate symmetrization excluding the indices separated by vertical bars (cf. Section A.3), e.g.,

$$a_{(ij)} \equiv \tfrac{1}{2}(a_{ij} + a_{ji}), \qquad x_{(i|\beta|} x_{j)\delta} \equiv \tfrac{1}{2}(x_{i\beta} x_{j\delta} + x_{j\beta} x_{i\delta})$$

Similarly, a bracket will be used to denote antisymmetrization, e.g.,

$$a_{[ij]} \equiv \tfrac{1}{2}(a_{ij} - a_{ji})$$

Note that $A_{;\alpha\gamma}$ is symmetric in indices α and γ since A is, by definition, a spatial quantity, and its second covariant derivative in surface coordinates given by

$$A_{;\alpha\gamma} = A_{,\alpha\gamma} - \begin{Bmatrix} \beta \\ \alpha\gamma \end{Bmatrix} A_{;\beta}$$

is symmetric in α and γ.

If, in particular, ϕ is continuous, or more generally if its jump is independent of surface coordinates, across σ, (2.4.12) reduces to

$$[\phi_{,ij}] = Cn_i n_j + a^{\alpha\beta} B_{;\alpha}(n_i x_{j\beta} + n_j x_{i\beta}) - Bb^{\alpha\beta} x_{i\alpha} x_{j\beta} \tag{2.4.14}$$

If we further assume that $\phi_{,i}$ are continuous across σ we obtain an important special case in which

$$[\phi_{,ij}] = Cn_i n_j \tag{2.4.15}$$

The higher-order geometrical conditions of compatibility can be obtained in exactly the same fashion. The results, however, are cumbersome and the formulas in the general case cease to be useful. If, however, the function ϕ and its spatial derivatives up to the order $n - 1$ are continuous across the surface σ, it can be shown that

$$[\phi_{,i_1 i_2 \cdots i_n}] = Cn_{i_1} n_{i_2} \cdots n_{i_n} \tag{2.4.16}$$

where

$$C \equiv [\phi_{,i_1 i_2 \cdots i_n}] n_{i_1} n_{i_2} \cdots n_{i_n} \tag{2.4.17}$$

2.5 KINEMATICAL CONDITIONS OF COMPATIBILITY

In this section we evaluate the jump of the time rates of tensors at the discontinuity surface σ. By taking the difference of (2.2.37) written on both sides of σ and using (2.4.3) we obtain the kinematical condition of compatibility of first order:

$$[\partial\phi/\partial t] = -u_n B + (\delta A/\delta t) \tag{2.5.1}$$

When $[\phi]$ is constant on σ, (2.5.1) is simplified to

$$[\partial\phi/\partial t] = -u_n B = -u_n[\phi_{,i}]n_i \tag{2.5.2}$$

If we replace ϕ in (2.4.4) and (2.5.1) by $\partial\phi/\partial t$, we obtain the relations

$$\begin{aligned}[\partial^2\phi/\partial x_i\,\partial t] &= B'n_i + a^{\alpha\beta}A'_{;\alpha}x_{i\beta}\\ [\partial^2\phi/\partial t^2] &= -u_n B' + (\delta A'/\delta t)\end{aligned} \tag{2.5.3}$$

where

$$A' \equiv [\partial\phi/\partial t], \qquad B' \equiv [\partial^2\phi/\partial x_i\,\partial t]n_i \tag{2.5.4}$$

We see from (2.5.1) that A' is already expressed in terms of A and B. Similarly, we can relate B' to the quantities appearing in the geometrical conditions of compatibility. In fact, from $(2.4.3)_2$ we have

$$\frac{\delta B}{\delta t} = \frac{\delta[\phi_{,i}]}{\delta t}n_i + [\phi_{,i}]\frac{\delta n_i}{\delta t} = [\partial\phi_{,i}/\partial t]n_i + u_n[\phi_{,ij}]n_i n_j - [\phi_{,i}]a^{\alpha\beta}x_{i\alpha}u_{n,\beta}$$

where on the extreme right we used (2.2.43). Employing (2.4.4), this equation is simplified to

$$\delta B/\delta t = [\partial^2\phi/\partial x_i\,\partial t]n_i + u_n[\phi_{,ij}]n_i n_j - a^{\alpha\beta}A_{;\alpha}u_{n,\beta}$$

or

$$B' = -u_n C + (\delta B/\delta t) + a^{\alpha\beta}A_{;\alpha}u_{n,\beta} \tag{2.5.5}$$

By introducing (2.5.5) into (2.5.3), we derive

$$\begin{aligned}[\partial^2\phi/\partial x_i\,\partial t] &= [-u_n C + (\delta B/\delta t) + a^{\alpha\beta}A_{;\alpha}u_{n,\beta}]n_i + a^{\alpha\beta}A'_{;\alpha}x_{i\beta}\\ [\partial^2\phi/\partial t^2] &= [u_n C - (\delta B/\delta t) - a^{\alpha\beta}A_{;\alpha}u_{n,\beta}]u_n + (\delta A'/\delta t)\end{aligned} \tag{2.5.6}$$

where

$$A' = -u_n B + (\delta A/\delta t) \tag{2.5.7}$$

If ϕ is continuous across σ, then $A = 0$ and (2.5.6) reduces to

$$\begin{aligned}[\partial^2\phi/\partial x_i\,\partial t] &= [-u_n C + (\delta B/\delta t)]n_i - a^{\alpha\beta}(u_n B)_{;\alpha}x_{i\beta}\\ [\partial^2\phi/\partial t^2] &= u_n{}^2 C - 2u_n(\delta B/\delta t) - B(\delta u_n/\delta t)\end{aligned} \tag{2.5.8}$$

When, in addition, the first-order derivatives of ϕ are continuous across σ, these are simplified to

$$[\partial^2\phi/\partial x_i\,\partial t] = -u_n C n_i, \qquad [\partial^2\phi/\partial t^2] = u_n{}^2 C \tag{2.5.9}$$

Higher-order kinematical compatibility conditions are found by successive applications of the above procedure.

An account of the compatibility conditions on moving discontinuity surfaces was given by Thomas [1966].

2.6 SINGULAR SURFACES ASSOCIATED WITH THE MOTION OF A MEDIUM

So far our discussions on singular surfaces did not involve the motion of the material body. We now consider the propagation of singular surfaces in a medium set into motion.

The motion is described by a one-parameter family of mappings of a material point $\mathbf{X}$ into a spatial position $\mathbf{x}$ at time t, i.e.,

$$\mathbf{x} = \mathbf{x}(\mathbf{X}, t) \qquad \text{or} \qquad x_k = x_k(X_K, t) \tag{2.6.1}$$

where x_k and X_K are the rectangular coordinates of the spatial and material points, respectively. The mapping (2.6.1) is single valued and continuous. According to the axiom of *continuity, impenetrability, and indestructibility* of matter the inverse motion

$$\mathbf{X} = \mathbf{X}(\mathbf{x}, t) \qquad \text{or} \qquad X_K = X_K(x_k, t) \tag{2.6.2}$$

exists in the neighborhood of all points except possibly at countable numbers of singular surfaces, lines, and points which may be present in the body. This is guaranteed by the continuity of partial derivatives $x_{k,K}$ and

$$\det x_{k,K} \neq 0 \tag{2.6.3}$$

at points $\mathbf{X}$ not lying on the singular surfaces, lines, and points at all times (cf. Section 1.3).

A moving spatial surface $\sigma(t)$ may be given by representations of the form

$$f(\mathbf{x}, t) = 0 \qquad \text{or} \qquad \mathbf{x} = \mathbf{x}(p^\alpha, t) \tag{2.6.4}$$

The material description of the surface $\sigma(t)$ is provided by

$$f[\mathbf{x}(\mathbf{X}, t), t] = F(\mathbf{X}, t) = 0 \tag{2.6.5}$$

or using the inverse motion

$$\mathbf{X} = \mathbf{X}(\mathbf{x}, t) = \mathbf{X}[\mathbf{x}(p^\alpha, t), t] = \hat{\mathbf{X}}(p^\alpha, t) \tag{2.6.6}$$

Equation (2.6.5) or (2.6.6) represents in the material coordinates a moving surface $\Sigma(t)$, which is the image of the surface $\sigma(t)$. Thus the alternative forms (2.6.4) and (2.6.5) or (2.6.6.) represent the same moving surface with different interpretations. While the former is a surface in the spatial system under

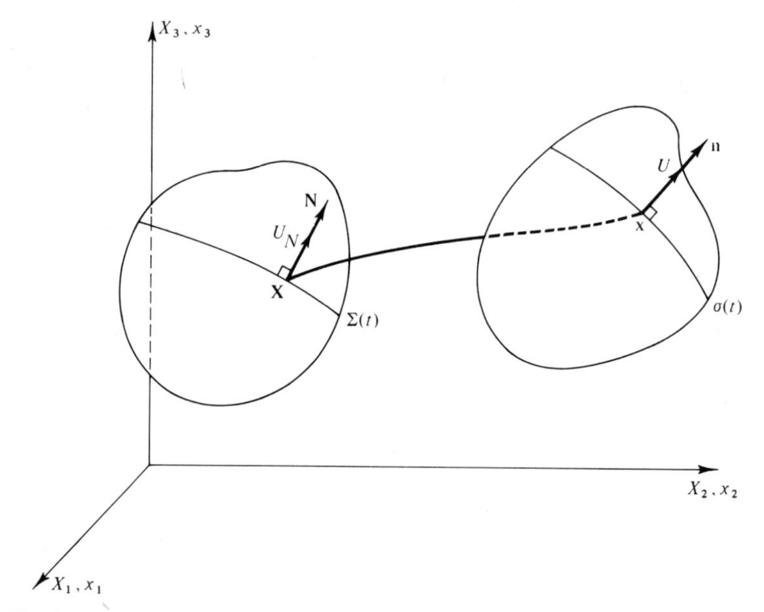

Fig. 2.6.1 Moving material singular surface $\Sigma(t)$, and spatial singular surface $\sigma(t)$.

observation, the latter is the locus of the initial position of material particles which are situated on the surface σ at time t (Fig. 2.6.1).

The first and second fundamental forms of the surface $\Sigma(t)$ (i.e., its intrinsic geometrical properties) are of no immediate concern to an observer of a singular surface in space. However, sometimes it may be advantageous to employ the material description in the formulation of a problem. Once the surface $\sigma(t)$ and the motion of the body are given, the geometrical properties of $\Sigma(t)$ are obtained. The base vectors $\mathbf{A}_\alpha$ on $\Sigma(t)$ are given by

$$\mathbf{A}_\alpha(\mathbf{p}, t) \equiv \partial \mathbf{X}/\partial p^\alpha = \hat{X}_{k\alpha} \mathbf{I}_K \tag{2.6.7}$$

where $\mathbf{I}_K$ are the base vectors in the rectangular material coordinates X_K. For simplicity, henceforth we drop the caret from $\hat{X}_K(p^\alpha, t)$. The arc length dS on $\Sigma(t)$ is calculated by

$$dS^2 = A_{\alpha\beta}\, dp^\alpha\, dp^\beta \tag{2.6.8}$$

where

$$A_{\alpha\beta}(\mathbf{p}, t) = X_{K\alpha} X_{K\beta} \tag{2.6.9}$$

is the metric tensor on $\Sigma(t)$. On substituting

$$X_{K\alpha} = X_{K,k}\, x_{k\alpha} \tag{2.6.10}$$

into (2.6.9), we obtain

(2.6.11) $$A_{\alpha\beta} = c_{kl} x_{k\alpha} x_{l\beta}$$

where c_{kl} is the Cauchy deformation tensor.

The unit normals **n** to the surface $\sigma(t)$ and **N** to $\Sigma(t)$ are given by

$$n_k = f_{,k}(f_{,l} f_{,l})^{-1/2}, \qquad N_K = F_{,K}(F_{,L} F_{,L})^{-1/2}$$

By virtue of (2.6.5), we have

$$f_{,k} = F_{,K} X_{K,k}, \qquad F_{,K} = f_{,k} x_{k,K}$$

whence follow the relations between **N** and **n**:

(2.6.12) $$N_K = C_n^{-1} n_k x_{k,K}, \qquad n_k = C_n N_K X_{K,k}$$

where

(2.6.13) $$C_n^{\ 2} \equiv \overset{-1}{c}_{kl} n_k n_l$$

and $\overset{-1}{c}_{kl}$ is the Finger deformation tensor (see Section 1.4).

To obtain the expression of the second fundamental tensor **B**, the coefficient of the second fundamental form B, i.e.,

(2.6.14) $$B = B_{\alpha\beta}\, dp^\alpha\, dp^\beta, \qquad B_{\alpha\beta} \equiv -N_{K;\alpha} X_{K\beta} = N_K X_{K\beta;\alpha}$$

we employ $(2.6.12)_1$ and (2.6.10). Hence

$$\begin{aligned} B_{\alpha\beta} &= C_n^{-1} n_k x_{k,K} X_{K\beta;\alpha} = C_n^{-1} n_k[(x_{k,K} X_{K\beta})_{;\alpha} - x_{k,KL} X_{K\alpha} X_{L\beta}] \\ &= C_n^{-1} n_k(x_{k;\beta\alpha} - x_{k,KL} X_{K\alpha} X_{L\beta}) \end{aligned}$$

On using (2.2.29), this gives

(2.6.15) $$B_{\alpha\beta} = C_n^{-1}(b_{\alpha\beta} - x_{r,KL} X_{K,k} X_{L,l} x_{k\alpha} x_{l\beta} n_r)$$

The normal velocity of Σ is called the *speed of propagation* and is calculated by

(2.6.16) $$U_N = -(\partial F/\partial t)/|\nabla F|$$

where ∇F denotes the gradient of F with components $F_{,K}$. Since

$$\dot{F} = \partial F/\partial t = \dot{f} = (\partial f/\partial t) + f_{,k} v_k$$

where **v** is the velocity of the material particles, it follows from (2.2.33) that

(2.6.17) $$\dot{F} = -|\nabla f|(u_n - v_n)$$

where u_n is the speed of displacement of the surface $\sigma(t)$, $v_n = \mathbf{v} \cdot \mathbf{n}$ on σ, and ∇f is the gradient of f with components $f_{,k}$. The quantity

(2.6.18) $$U \equiv u_n - v_n$$

is called the *local speed of propagation*. It is a measure of the normal speed of the surface $\sigma(t)$ with respect to the particles instantaneously comprising it. From (2.6.17), (2.6.18), (2.6.16), and (2.6.12), it follows that

$$U_N = U|\nabla f|/|\nabla F| = C_n^{-1}U \tag{2.6.19}$$

A singular surface is said to be a *propagating singular surface* or a *wave* if U_N (or U) is different from zero.

It should be remarked that the foregoing results fail to be valid for singular surfaces on which the velocity of the medium suffers a discontinuity, for example, in the case of shock waves. Also one can envision singular surfaces on which the inverse motion is not single valued, as in the case of *slip surfaces*. In such cases the fundamental equations should be modified accordingly. For an account on this topic the reader is referred to Truesdell and Toupin [1960] and Hill [1961].

2.7 DYNAMICAL CONDITIONS OF COMPATIBILITY

When a singular surface involves field variables that are affected by the motion and deformation of the medium, the geometrical and kinematical compatibility conditions obtained in Sections 2.4 and 2.5 should be supplemented by restrictions originating from the local balance equations. These conditions are called the *dynamical conditions of compatibility*. The dynamical conditions of compatibility are due to the local conservation of mass, balance of linear and angular momenta, balance of energy, and the local Clausius–Duhem inequality (cf. Section 1.11):

$$[\rho U] = 0 \tag{2.7.1}$$

$$[t_{kl}]n_k = -[\rho v_k U] \tag{2.7.2}$$

$$[\rho(\varepsilon + \tfrac{1}{2}v^2)U] = -[t_{kl}v_l]n_k \tag{2.7.3}$$

$$[\rho U\eta] \leq 0 \tag{2.7.4}$$

on $\sigma(t)$. In these expressions ρ is the density of the material at $\mathbf{x}$ at time t, t_{kl} the symmetric Cauchy stress, ε the internal energy density, η the entropy density, and $v^2 = v_k v_k$.

The material forms of the dynamical conditions of compatibility have the corresponding forms (cf. Section 1.12)

$$[T_{Kk}]N_K = -\rho_0 U_N[v_k] \tag{2.7.5}$$

$$[T_{Kk}v_k]N_K = -\rho_0 U_N[\varepsilon + \tfrac{1}{2}v^2] \tag{2.7.6}$$

$$\rho_0 U_N[\eta] \leq 0 \tag{2.7.7}$$

where ρ_0 is the density of the medium in the reference state and T_{Kk} is the Piola–Kirchhoff stress tensor. In these equations the effect of heat conduction is disregarded. For the jump conditions involving heat conduction see Sections 1.11–1.12 or the resume in Section 1.19.

2.8 CLASSIFICATION OF SINGULAR SURFACES ASSOCIATED WITH MOTION

Singular surfaces, in the theory of waves in elastic solids, are classified according to the order of the discontinuity suffered by the motion or some of its derivatives. To this end we take the quantity relative to which the singular surface is defined as

$$\phi \equiv x_k(\mathbf{X}, t) \tag{2.8.1}$$

Then the definition for the order of the singular surface in Section 2.3 becomes applicable to (2.8.1). Since fracture of the body is excluded from our consideration, the motion on a surface is assumed to be continuous. Therefore, a singular surface of order zero is assumed not to exist, and on every singular surface the relation $[\mathbf{x}] = \mathbf{0}$ is supposed to hold.

On a singular surface of order 1 the deformation gradient and the velocity of the medium may suffer finite discontinuity. Such a propagating singular surface will be called a *shock wave*.[1]

On a singular surface of order 2, the deformation gradients and the velocity of the medium will be continuous, while the second gradients of motion and the acceleration of material particles may suffer jumps. Such a propagating singular surface will be called an *acceleration wave*.

Higher-order singular surfaces are similarly defined. In this chapter we confine our attention to the first- and second-order waves.

2.9 SHOCK WAVES

A complete study of the propagation of shock waves in elastic solids is presently lacking. Since the early investigations of Jouguet [1920a,b,c,d, 1921], a considerable time has elapsed for the reconsideration of these problems in the context of modern continuum mechanics. The need for the explicit constitutive relations, which are discontinuous on the shock surface in general, and

[1] Truesdell and Toupin [1960] preserve the term "shock wave" for singular surfaces on which only the normal components of the material velocity suffer jumps. They use the term "*vortex sheets*" for tangential velocity jumps. We prefer to consider a shock wave as any singular surface across which the material velocity may be discontinuous in any direction.

the necessity of strong thermodynamic restrictions for a physically admissible theory have impeded the development of a satisfactory theory of shock waves in elastic materials. These obstacles can be avoided to some extent by considering special cases such as plane shock waves, special materials, and incompressible solids (cf. Dewey [1959], Verma [1964], Bland [1964a,b, 1965a], Coleman *et al.* [1965], Chu [1967]).

Let $\Sigma(t)$ and $\sigma(t)$ be, respectively, the material and spatial descriptions of a propagating shock surface (Fig. 2.9.1). We suppose that the motion and its

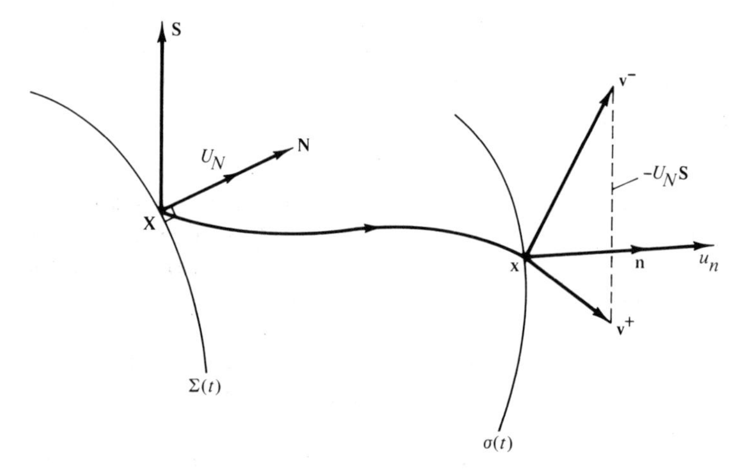

Fig. 2.9.1 Moving material and spatial shock surfaces.

inverse are continuous and single valued on the shock surface, i.e., $\sigma \leftrightarrow \Sigma$ is a mapping one-to-one onto. The velocity of the medium will have a finite jump across the shock surface $\sigma(t)$. Clearly the velocity jump entails finite jumps in the deformation gradients [cf. (2.5.2)]. For simplicity, we formulate the problem in material coordinates. We take $\phi \equiv x_k$ such that $[\mathbf{x}] = \mathbf{0}$ on the shock surface. From (2.4.4) and (2.5.1) we have

(2.9.1) $$[x_{k,K}] = S_k N_K, \qquad [v_k] = -U_N S_k$$

with

(2.9.2) $$A = 0, \qquad B \equiv S_k = [x_{k,K}] N_K$$

The jumps in the velocity and the deformation gradient are related by

$$[v_k] N_K + U_N [x_{k,K}] = 0$$

The vector S_k, defined on the material surface Σ, is in the direction of the velocity jump and may be considered as a measure of the singularity.

Recalling the definitions in Section 2.6, we see that the local speed of propagation U is discontinuous across $\sigma(t)$:

$$U^+ = u_n - v_n^+, \qquad U^- = u_n - v_n^-$$

Hence

$$[U] = -[v_n] = -[\mathbf{v}] \cdot \mathbf{n} \tag{2.9.3}$$

From (2.6.19), we have

$$U^+ = C_n^+ U_N, \qquad U^- = C_n^- U_N$$

where

$$C_n^+ \equiv \left(\overset{+}{c}{}^{-1}_{kl} n_k n_l \right)^{1/2}, \qquad C_n^- \equiv \left(\overset{-}{c}{}^{-1}_{kl} n_k n_l \right)^{1/2}$$

The assumptions on the equations of motion secure a unique normal $\mathbf{N}$ to the surface Σ. Therefore, we can write

$$N_K = (1/C_n^+) n_k \overset{+}{x}_{k,K} = (1/C_n^-) n_k \overset{-}{x}_{k,K} \tag{2.9.4}$$

or

$$N_K = (U_N/U^+) n_k \overset{+}{x}_{k,K} = (U_N/U^-) n_k \overset{-}{x}_{k,K} \tag{2.9.5}$$

The foregoing relations hold if we consider only (2.9.1). In fact, if we use (2.9.1) and (2.9.3) in (2.9.5) we obtain

$$N_K = \frac{U_N}{U^+} n_k \overset{+}{x}_{k,K} = \frac{U_N n_k(\overset{-}{x}_{k,K} + S_k N_K)}{U^- + U_N S_k n_k}$$

which also implies that

$$N_K = (U_N/U^-) n_k \overset{-}{x}_{k,K}$$

Thus we can suppress (+) or (−) superscripts in (2.9.4) or (2.9.5) without any ambiguity. In addition, the following relations should be satisfied on the surface $\Sigma(t)$:

$$[(1/U) x_{k,K}] n_k = [(1/C_n) x_{k,K}] n_k = 0 \tag{2.9.6}$$

The jump vector S_k cannot be chosen arbitrarily. For a nonheat-conducting material it should satisfy the dynamical conditions of compatibility (2.7.5)–(2.7.7) on the surface $\Sigma(t)$.

The constitutive equations for hyperelastic materials may be written as [cf. (1.13.25) and $(1.13.23)_4$]

$$T_{Kk} = \partial\Sigma/\partial x_{k,K}, \qquad \theta = (1/\rho_0)\, \partial\Sigma/\partial\eta \tag{2.9.7}$$

where $\Sigma(x_{k,K}, \eta) \equiv \rho_0 \varepsilon$ and θ are, respectively, the stress potential and the absolute temperature. In terms of the stored energy function, the dynamical conditions of compatibility (2.7.5)–(2.7.7) read

$$\begin{aligned} &[\partial\Sigma/\partial x_{k,K}]N_K = -\rho_0 U_N[v_k] \\ (2.9.8)\qquad &[(\partial\Sigma/\partial x_{k,K})v_k]N_K = -U_N[\Sigma + \tfrac{1}{2}\rho_0 v^2] \\ &\rho_0 U_N[\eta] \le 0 \end{aligned}$$

The last condition implies that, for thermodynamically admissible processes, the entropy density should increase after the passage of the shock wave. When $[\eta] = 0$ the wave is called *isentropic*. Since both Σ and $x_{k,K}$ are discontinuous across the wave, one cannot expect η to be continuous. Therefore, shock waves are not isentropic in general. However, in Section 2.10 we shall prove that weak shock waves can be considered *approximately isentropic*.

In view of equations (2.9.1) we have five unknowns S_k, U_N, and $[\eta]$ to determine the state behind the shock in terms of state ahead of the shock, which is presumed to be known. Since the last relation in (2.9.8) is generally an inequality, the set (2.9.8) provides only four equations. Therefore, the solution contains one arbitrary parameter, which can be taken as the measure of the strength of the shock. However, condition $(2.9.8)_3$ imposes some restrictions on the arbitrariness of this parameter for thermodynamically admissible shocks.

Introducing (2.9.1) into $(2.7.5)_1$ we have

$$(2.9.9)\qquad [T_{Kk}]N_K = \rho_0 U_N{}^2 S_k$$

Multiplying both sides of this expression by S_k and using $(2.9.1)_1$ we obtain, for the speed of propagation U_N,

$$(2.9.10)\qquad \rho_0 U_N{}^2 = S^{-2}[T_{Kk}][x_{k,K}] = [T_{Kk}][x_{k,K}]/[x_{l,L}][x_{l,L}]$$

with $S^2 = S_k S_k$. Since S_k should be a real vector for the existence of a shock, $S^2 > 0$, then the condition for the *propagation* of a shock wave $(U_N{}^2 > 0)$ is

$$[T_{Kk}][x_{k,K}] > 0$$

If we assume that the stored energy function Σ is convex, i.e.,[1]

$$\begin{aligned} \Sigma(\overset{+}{x}_{k,K}, \eta^+) - \Sigma(\overset{-}{x}_{k,K}, \eta^-) &> (\overset{+}{x}_{k,K} - \overset{-}{x}_{k,K})(\partial\Sigma/\partial x_{k,K})^- + (\eta^+ - \eta^-)(\partial\Sigma/\partial\eta)^- \\ \Sigma(\overset{-}{x}_{k,K}, \eta^-) - \Sigma(\overset{+}{x}_{k,K}, \eta^+) &> (\overset{-}{x}_{k,K} - \overset{+}{x}_{k,K})(\partial\Sigma/\partial x_{k,K})^+ + (\eta^- - \eta^+)(\partial\Sigma/\partial\eta)^+ \end{aligned}$$

[1] Strictly speaking, these inequalities hold, without any apparent contradiction, if $\overset{+}{x}_{k,K}$ and $\overset{-}{x}_{k,K}$ differ by a pure stretch. It can be shown that this is the case if **S** is in the direction of **n**, i.e., in normal shocks. However, the above restriction on the deformation gradients is only a sufficient condition and it cannot be proved that it is also necessary (cf. Truesdell and Noll [1965, Section 52]). Hence the inequalities will be thought here as applicable to a general shock wave unless they lead to an intuitively meaningless consequence [cf. Section 1.18 (iii)].

or

$$[\Sigma] > [x_{k,K}]T_{Kk}^{-} + \rho_0[\eta]\theta^{-}$$
$$-[\Sigma] > -[x_{k,K}]T_{Kk}^{+} - \rho_0[\eta]\theta^{+}$$

On adding these two inequalities we get

$$[T_{Kk}][x_{k,K}] + \rho_0[\eta][\theta] > 0 \tag{2.9.11}$$

From (2.9.11) it is not possible, in general, to decide whether U_N^2 is positive or not. In isentropic approximation $[\eta] = 0$, and in this case (2.9.11) provides the necessary information. In this case the first set of equations (2.9.8) determines the desired quantities behind the shock within an arbitrary parameter, and the second equation of (2.9.8) provides a means to calculate the internal energy and, consequently, the temperature behind the shock.

Equation $(2.9.8)_2$ can be put in a simpler and more meaningful form. Using the identity

$$[ab] = a^{+}[b] + b^{+}[a] - [a][b] \tag{2.9.12}$$

we can write $(2.9.8)_2$ as

$$T_{Kk}^{+}[v_k]N_K + v_k^{+}[T_{Kk}]N_K - [T_{Kk}][v_k]N_K$$
$$= -U_N[\Sigma] - \rho_0 U_N v_k^{+}[v_k] + \tfrac{1}{2}\rho_0 U_N[v_k][v_k]$$

or, making use of the relations (2.9.1),

$$T_{Kk}^{+}[x_{k,K}] = [\Sigma] + \tfrac{1}{2}\rho_0 U_N^2 S^2$$

and introducing $\rho_0 U_N^2 S^2$ from (2.9.10), we find

$$\tfrac{1}{2}(T_{Kk}^{+} + T_{Kk}^{-})[x_{k,K}] = [\Sigma] \tag{2.9.13}$$

which can be transformed into

$$\{(\partial\Sigma/\partial x_{k,K})^{+} + (\partial\Sigma/\partial x_{k,K})^{-}\}[x_{k,K}] = 2[\Sigma] \tag{2.9.14}$$

These equations are analogous to the Hugoniot relations in the theory of shock waves in gas dynamics. They help to determine the entropy increase behind the shock.

The foregoing equations can be simplified to some extent if a shock wave propagates into a medium at rest and unstrained in the reference state. Introducing the displacement gradients $u_{k,K}$ and noticing that $v_k^{+} = u_{k,K}^{+} = T_{Kk}^{+} = \Sigma^{+} = 0$, we have

$$T_{Kk}N_K = -\rho_0 U_N v_k \quad \text{or} \quad (\partial\Sigma/\partial u_{k,K})N_K = -\rho_0 U_N v_k \tag{2.9.15}$$

$$(\partial\Sigma/\partial u_{k,K})u_{k,K} = 2\Sigma \tag{2.9.16}$$

and

$$\rho_0 U_N{}^2 = S^{-2} T_{Kk} u_{k,K} = 2S^{-2}\Sigma \tag{2.9.17}$$

with

$$S_k \equiv -u_{k,K} N_K$$

where (−) superscripts are suspended without any ambiguity. The entropy density is constant ahead of the shock, and for an isentropic shock wave this constant gives the entropy density behind the shock. Therefore, for isentropic waves propagating in such a medium Σ may be thought of as independent of the entropy density; hence (2.9.16) indicates that Σ is a quadratic homogeneous function of the displacement gradient. Furthermore, following a common practice, if we assume that Σ is a polynomial function of its arguments, we conclude that Σ is a quadratic polynomial in displacement gradients; thus it is none other than the strain energy function corresponding to the linear elastic solids. *Therefore, only in the case of the linear theory of elasticity can an isentropic shock wave be realized.* For nonlinear elasticity, the isentropy assumption is merely an approximation which is not even valid for weak shocks.

To illustrate the foregoing analysis, consider a linear elastic solid. If $\mathbf{u}$ is an infinitesimal displacement vector, then the stress tensor may be approximated by (cf. Sections 1.12 and 1.15)

$$T_{Kk} = \{\lambda u_{m,m}\,\delta_{kl} + \mu(u_{k,l} + u_{l,k})\}\,\delta_{Kl}$$

where λ and μ are Lamé's constants and $u_{k,l} = \delta_{Kl} u_{k,K}$. From (2.9.10) it now follows that

$$\rho_0 U_N{}^2 = [(\lambda + \mu)(S_k n_k)^2 + \mu S^2]/S^2$$

where we wrote

$$n_k = N_K\,\delta_{Kk}$$

If we also write

$$\mathbf{S} = S\mathbf{v}$$

where $\mathbf{v}$ is the unit vector in the direction of the velocity jump, then we get

$$U_N{}^2 = \rho_0^{-1}[(\lambda + \mu)(\mathbf{v}\cdot\mathbf{n})^2 + \mu] \tag{2.9.18}$$

which indicates that purely longitudinal and purely transverse shock waves are realizable in linear solids.[1]

For *longitudinal waves* or *normal shock waves* $\mathbf{v} = \mathbf{n}$; therefore,

$$U_1{}^2 = (\lambda + 2\mu)/\rho_0$$

[1] In fact, one can easily prove using (2.9.9) that they are either longitudinal or transverse.

while for *transverse waves* or *transverse shock waves* $\mathbf{v} \cdot \mathbf{n} = 0$ and

$$U_2^2 = \mu/\rho_0$$

As is well known, these are the speeds of propagation of irrotational and equivoluminal waves in a linear elastic solid.

We conclude this section with the calculation of the jump in the density of the medium across a shock wave. From (2.7.1) it follows that

$$\rho^+/\rho^- = U^-/U^+$$

on the shock surface $\sigma(t)$, or we can write

$$[\rho]/\rho^- = -[U]/U^+ = [v_k]n_k/C_n^+ U_N = -S_k n_k/C_n^+ \tag{2.9.19}$$

If $\mathbf{S} \cdot \mathbf{n} > 0$ then $\rho^- > \rho^+$, and if $\mathbf{S} \cdot \mathbf{n} < 0$ then $\rho^- < \rho^+$. Thus, if we define a *shock strength* by

$$\delta = (\rho^-/\rho^+) - 1$$

we see that $\delta > 0$ if $\mathbf{S} \cdot \mathbf{n} > 0$, and $\delta < 0$ if $\mathbf{S} \cdot \mathbf{n} < 0$. The former case corresponds to a condensation wave, whereas the latter corresponds to a rarefaction wave. From (2.9.19) we have

$$\delta = \frac{1}{C_n^+} \frac{\mathbf{S} \cdot \mathbf{n}}{1 - (\mathbf{S} \cdot \mathbf{n}/C_n^+)} \tag{2.9.20}$$

If the elastic material is incompressible, then $\rho^+ = \rho^-$; hence $[\rho] = 0$ and

$$\mathbf{S} \cdot \mathbf{n} = 0 \tag{2.9.21}$$

Therefore, incompressible elastic materials may transmit only transverse shock waves.

2.10 SHOCK WAVES IN INCOMPRESSIBLE MATERIALS[1]

In incompressible materials the mass density remains unchanged during the motion. Thus $[\rho] = 0$ on shock surfaces. This implies that [cf. (2.7.1)] $[U] = 0$; hence from (2.9.3)

$$[v_n] = 0 \tag{2.10.1}$$

Thus in incompressible materials the velocity may suffer only a tangential jump across the shock surface. From (2.9.6) it now follows that

$$n_k[x_{k,K}] = 0 \tag{2.10.2}$$

[1] See Chu [1967].

Therefore, there is no jump in the deformation gradient in the direction normal to the surface $\sigma(t)$. The continuity of the local speed of propagation U requires that (cf. 2.6.19)[1]

$$C_n^2 = \overset{-1}{c}_{kl} n_k n_l = B_{kl} n_k n_l$$

is also continuous on the wave front. We now define a spatial vector **a**, called the *amplitude vector*, perpendicular to **n**, by

$$\mathbf{a} = \mathbf{S}/C_n \tag{2.10.3}$$

On using (2.9.4) and (2.10.2) the compatibility conditions (2.9.1) take the form

$$[x_{k,K}] = C_n a_k N_K = a_k n_l x_{l,K}, \qquad [v_k] = -Ua_k \tag{2.10.4}$$

The dynamical compatibility conditions (2.7.2)–(2.7.4) are simplified to

$$[t_{kl}]n_l = -\rho U[v_k], \qquad [t_{kl} v_l]n_k = -U[\Sigma + \tfrac{1}{2}\rho v^2], \qquad \rho U[\eta] \leq 0 \tag{2.10.5}$$

where $\Sigma \equiv \rho\varepsilon$. We define three mutually orthogonal unit vectors **l**, **m**, **n** on the shock surface with **m** being in the direction of the transverse velocity jump [**v**], and **l** is given by

$$\mathbf{l} = \mathbf{m} \times \mathbf{n}$$

The vectors (**m**, **n**) determine the plane of polarization for the amplitude at a point **x** on the shock surface $\sigma(t)$. Since **a** is in the direction of **m** we have $\mathbf{l} \cdot \mathbf{a} = 0$, and on multiplying $(2.10.4)_1$ by l_k we obtain

$$l_k[x_{k,K}] = 0 \tag{2.10.6}$$

Therefore, the displacement gradient has no component in the spatial direction **l**.

Let **L**, **M**, **N** be three mutually perpendicular unit vectors on the material shock surface $\Sigma(t)$ at a point **X**, which at time t occupies the spatial point **x** (Fig. 2.10.1). We take **L** and **M** to lie in the shock surface. Because of the definition of the set **L**, **M**, **N**, the components of the jump in the deformation gradients in **L** and **M** directions vanish [cf. (2.9.1)]:

$$M_K[x_{k,K}] = L_K[x_{k,K}] = 0 \tag{2.10.7}$$

Now consider a local rectangular material frame (**L**, **M**, **N**) and a spatial frame (**l**, **m**, **n**) directed, respectively, along **L**, **M**, **N** and **l**, **m**, **n**. In view of (2.10.2), (2.10.6), and (2.10.7), the only nonvanishing components of the jump in the deformation gradients in these systems are

$$[x_{m,N}] = m_k N_K[x_{k,K}] \tag{2.10.8}$$

[1] Some authors use B_{kl} in place of $\overset{-1}{c}_{kl}$. For notational simplicity in this section we employ B_{kl}.

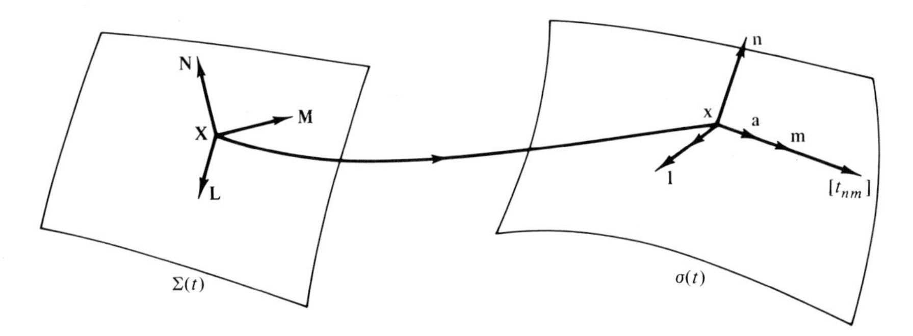

Fig. 2.10.1 Orthogonal triads on material and spatial shock surfaces.

We also see from (2.10.5) that[1]

(2.10.9)

$$[t_{nn}] = [t_{ij}]n_i n_j = 0, \qquad [t_{nl}] = [t_{ij}]n_i l_j = 0, \qquad [t_{nm}] = [t_{ij}]n_i m_j \neq 0$$

Therefore, *in incompressible materials during shock the normal stress t_{nn} and the shear stress t_{nl} remain continuous, but the shear stress in the direction of the velocity jump suffers a discontinuity*. The jump in t_{nm} can be evaluated from (2.10.5) as

$$[t_{nm}] = \rho U^2 a_m \tag{2.10.10}$$

Since **a** is in the direction of **m**, a_m is actually the magnitude of the amplitude of **a**. On the other hand, from (2.10.3) and (2.10.8) we can write

$$a_m = (1/C_n)[x_{m,N}]$$

and using $(2.6.12)_1$, we obtain, for this particular choice of coordinate axes

$$C_n = x_{n,N}, \qquad x_{n,M} = x_{n,L} = 0$$

Therefore, (2.10.10) yields the local speed of propagation

$$\rho U^2 = [t_{nm}]x_{n,N}/[x_{m,N}] \tag{2.10.11}$$

Furthermore, it can be seen that the jumps in the components of Finger's strain tensor in the (**l**, **m**, **n**) system are given by

$$\begin{aligned} &[B_{nn}] = [B_{nl}] = [B_{ll}] = 0, \qquad && [B_{mn}] = [x_{m,N}]x_{n,N} \\ &[B_{ml}] = [x_{m,N}]x_{l,N}, && [B_{mm}] = [(x_{m,N})^2] \end{aligned} \tag{2.10.12}$$

[1] We suspend the summation convention on indices l, m, n.

By use of

(2.10.13) $$B_{nn} = (x_{n,N})^2 = C_n{}^2$$

we may express (2.10.11) as

(2.10.14) $$\rho U^2 = B_{nn}[t_{mn}]/[B_{mn}]$$

Repeating the calculations leading to (2.9.13) we find that

$$\tfrac{1}{2}(t_{mn}^{+} + t_{mn}^{-})[x_{m,N}] = x_{n,N}[\Sigma]$$

or using $(2.10.12)_2$ and (2.10.13)

(2.10.15) $$\tfrac{1}{2}(t_{mn}^{+} + t_{mn}^{-})[B_{mn}] = B_{nn}[\Sigma]$$

In an incompressible elastic material the constitutive equations for the stress tensor and the temperature are given by

(2.10.16) $$t_{kl} = -p\,\delta_{kl} + x_{k,K}\frac{\partial\Sigma}{\partial x_{l,K}}, \qquad \det(x_{k,K}) = 1, \qquad \theta = \frac{1}{\rho}\frac{\partial\Sigma}{\partial\eta}$$

where p is an unknown pressure function. The jump in p can be determined by introducing $(2.10.16)_1$ into $(2.10.9)_1$:

(2.10.17) $$[p] = [x_{k,K}\,\partial\Sigma/\partial x_{l,K}]n_k n_l = x_{n,N}[\partial\Sigma/\partial x_{n,N}]$$

The second equation of (2.10.9) determines the plane of polarization since $\mathbf{l}$ is perpendicular to both $\mathbf{n}$ and the vector $[t_{kl}]n_k$. From (2.10.16) it thus follows that

(2.10.18) $$[t_{mn}] = [x_{k,K}\,\partial\Sigma/\partial x_{l,K}]m_k n_l$$

Elimination of $[\eta]$ between (2.10.18) and (2.10.15) enables us to express $[t_{mn}]$ in terms of $[x_{m,N}]$ and the existing state in front of the shock. Introducing this into (2.10.11) we can determine the propagation velocity in terms of the jump $[x_{m,N}]$, which may be taken as a measure of shock strength.

An alternate formulation, which may be of use in some cases, can be developed by changing the fundamental constitutive variable from the pair $(x_{k,K}, \eta)$ to the pair $(x_{k,K}, \theta)$. The stored energy is now determined in terms of the free energy $\psi = \Sigma/\rho$, and the constitutive equations read (cf. Section 1.13)

(2.10.19)
$$t_{kl} = -p\,\delta_{kl} + \frac{\partial\Sigma}{\partial x_{k,K}}x_{l,K}, \qquad \varepsilon = \frac{1}{\rho}\left(\Sigma - \theta\frac{\partial\Sigma}{\partial\theta}\right), \qquad \eta = -\frac{1}{\rho}\frac{\partial\Sigma}{\partial\theta}$$

Using constitutive equations of this type, Chu [1967] gave explicit results concerning the shock propagation in an *ideal rubber*. According to Flory

[1953], the entropy and internal energy densities in an ideal rubber are given by

$$\eta = \eta_0 - \tfrac{1}{2}Nk(I-3) + \kappa \log(\theta/\theta_0), \qquad \varepsilon = \varepsilon_0 + \kappa(\theta - \theta_0) \tag{2.10.20}$$

where $I \equiv \operatorname{tr} \mathbf{B}$, N is the number of effective molecular chains (monomers) per unit mass, k Boltzmann's constant, κ the specific heat under constant deformation, and θ_0 the temperature of the natural state. The definition of the free energy $\psi = \varepsilon - \theta\eta$ yields

$$\begin{aligned}\psi = \Sigma/\rho = \varepsilon_0 - \theta_0\eta_0 &+ (\kappa - \eta_0)(\theta - \theta_0)\\ &- \kappa\theta\log(\theta/\theta_0) + \tfrac{1}{2}Nk\theta(I-3)\end{aligned} \tag{2.10.21}$$

Employing expression (2.10.21) in the constitutive equations (1.14.26) for an isotropic incompressible material, we obtain

$$t_{kl} = -p\,\delta_{kl} + \rho Nk\theta B_{kl} \tag{2.10.22}$$

for the components of the stress tensor. Using $(2.10.20)_2$ and (2.10.22) in (2.10.15) we have

$$\tfrac{1}{2}\rho Nk(\theta^+ B^+_{mn} + \theta^- B^-_{mn})[B_{mn}] = \rho\kappa B_{nn}[\theta]$$

from which the temperature jump follows:

$$[\theta] = \theta^+\gamma(B^+_{mn} + B^-_{mn})[B_{mn}](B_{nn})^{-1}/\{1 + \gamma B^-_{mn}[B_{mn}](B_{nn})^{-1}\} \tag{2.10.23}$$

where the dimensionless ratio

$$\gamma \equiv Nk/2\kappa \tag{2.10.24}$$

is usually rather small (of the order of 10^{-2} for natural vulcanized rubber). Thus we may approximately write (2.10.23) as

$$[\theta] \simeq \gamma\theta^+ \frac{B^+_{mn} + B^-_{mn}}{B_{nn}}[B_{mn}] = \gamma\theta^+ \frac{2B^+_{mn} - [B_{mn}]}{B_{nn}}[B_{mn}] = \gamma\theta^+[(B_{mn})^2]/B_{nn}$$

If the shock is weak, i.e., if $[B_{mn}]$ is small, we further simplify the above equation to obtain

$$[\theta] \simeq 2\gamma\theta^+ B^+_{mn}[B_{mn}]/B_{nn}$$

The jump in the stress component may be obtained from the relation

$$t_{mn} = \rho Nk\theta B_{mn}$$

which upon utilizing (2.10.23) gives

$$[t_{mn}] = t^+_{mn}\frac{\{1 - (B^-_{mn}/B^+_{mn}) + 2\gamma B^-_{mn}[B_{mn}](B_{nn})^{-1}\}}{1 + \gamma(B^-_{mn}/B_{nn})[B_{mn}]} \tag{2.10.25}$$

For small γ we can write, approximately,

$$[t_{mn}] = t^+_{mn}\{1 - (B^-_{mn}/B^+_{mn}) + \gamma(B^-_{mn}/B^+_{mn})[(B_{mn})^2](B_{nn})^{-1}\}$$

and for weak waves

$$[t_{mn}] \simeq (t^+_{mn}/B^+_{mn})\{1 + 2\gamma(B^+_{mn})^2(B_{nn})^{-1}\}[B_{mn}]$$

The local speed of propagation can now be evaluated from (2.10.14):

$$U^2 = Nk\theta^+ B_{nn}\{1 + 2\gamma B^+_{mn} B^-_{mn}(B_{nn})^{-1}\}/\{1 + \gamma B^-_{mn}[B_{mn}](B_{nn})^{-1}\} \tag{2.10.26}$$

For small γ we get

$$U = (Nk\theta^+ B_{nn})^{1/2}\{1 + \tfrac{1}{2}\gamma\, B^-_{mn}(B_{nn})^{-1}(B^+_{mn} + B^-_{mn})\}$$

and for weak waves

$$U = (Nk\theta^+ B_{nn})^{1/2}\{1 + \gamma(B^+_{mn}/B_{nn})(B^+_{mn} - \tfrac{3}{2}[B_{mn}])\}$$

By use of $(2.10.4)_2$ and $(2.10.12)_2$, we calculate the jump in the material velocity across the shock wave:

$$[v_m] = -Ua_m = -U[B_{mn}]/B_{nn} \tag{2.10.27}$$

For small γ, (2.10.27) may be approximated by

$$[v_m] \simeq -(Nk\theta^+/B_{nn})^{1/2}\{1 + \tfrac{1}{2}\gamma(B^-_{mn}/B_{nn})(B^+_{mn} + B^-_{mn})\}[B_{mn}]$$

and for weak waves this further reduces to

$$[v_m] \simeq -(Nk\theta^+/B_{nn})^{1/2}\{1 + \gamma(B^+_{mn})^2(B_{nn})^{-1}\}[B_{mn}]$$

Introduction of (2.10.22) into $(2.10.9)_1$ determines the pressure jump as

$$[p] = \rho Nk B_{nn}[\theta]$$

Substitution from (2.10.23) into this relation gives the dependence of $[p]$ on the jump $[B_{mn}]$. Furthermore, $(2.10.9)_2$ is satisfied only if [cf. $(2.10.12)_1$] $[\theta]B_{nl} = 0$, or

$$l_i n_j B_{ij} = 0 \tag{2.10.28}$$

Therefore, the vector **l** is perpendicular to the vector **Bn**. Thus, the plane of polarization is the plane containing the vectors **n** and **Bn**.

Finally, we determine the jump in the entropy and discuss the physical admissibility of the solution. From $(2.10.20)_1$ it follows that

$$[\eta] = -\tfrac{1}{2}Nk[I] + \kappa \log(\theta^+/\theta^-)$$

The second term on the right-hand side of this expression can be obtained from (2.10.23) and

$$[I] = [B_{mm}] + [B_{nn}] + [B_{ll}] = [(B_{mn})^2]/B_{nn}$$
$$= (B^+_{mn} + B^-_{mn})[B_{mn}]/B_{nn}$$

Consequently,

(2.10.29) $$[\eta] = -\tfrac{1}{2}Nk(B^+_{mn} + B^-_{mn})(B_{nn})^{-1}[B_{mn}] + \kappa \log\{1 + \gamma B^-_{mn}[B_{mn}](B_{nn})^{-1}\}\{1 - \gamma B^+_{mn}[B_{mn}](B_{nn})^{-1}\}^{-1}$$

For weak shock waves we have, approximately,

$$[\eta] = [\gamma^2\kappa/(B_{nn})^{-2}]B^+_{mn}[B_{mn}]^3\{1 + \tfrac{2}{3}\gamma(B^+_{mn})^2(B_{nn})^{-1}\} + \cdots$$

Hence the entropy jump across a weak shock wave varies with the third power of the shock strength. In the following section we shall see that this result is true in all hyperelastic materials.

Since $[\eta]$ should be nonpositive for physically admissible shocks, we conclude from the above expression that

$$B^+_{mn}[B_{mn}] \le 0 \qquad \text{or} \qquad B^-_{mn}\, B^+_{mn} \ge (B^+_{mn})^2$$

that is, we must have $B^-_{mn} > B^+_{mn}$ if $B^+_{mn} > 0$, and $B^-_{mn} < B^+_{mn}$ if $B^+_{mn} < 0$. Therefore, in physically admissible shocks, the wave will tend to increase the existing state of strain in the material after its passage. It should be remarked immediately that this is proved here only for weak shock waves.

2.11 WEAK SHOCK WAVES

If the shock strength is small, it is reasonable to assume that the magnitudes of jumps across the shock front are also small. Therefore, the strain energy function and its derivatives with respect to its arguments evaluated on the negative side of the shock surface may be expressible as Taylor's expansion about the values taken on the positive side of the surface. Using this expansion technique, we shall now show that an approximate explicit solution can be found to the shock equations (Şuhubi [1972]).

The equations to be solved are [cf. (2.9.8)$_1$ and (2.9.14)]

(2.11.1)

$$[\partial\Sigma/\partial x_{k,K}]N_K = \rho_0 U_N{}^2 S_k, \qquad \{(\partial\Sigma/\partial x_{k,K})^+ + (\partial\Sigma/\partial x_{k,K})^-\}[x_{k,K}] = 2[\Sigma]$$

Now we write

$$x^-_{k,K} = x^+_{k,K} - [x_{k,K}], \qquad \eta^- = \eta^+ - [\eta]$$

and expand Σ^- and $(\partial\Sigma/\partial x_{k,K})^-$ about $x_{k,K}^+$ and η^+ to obtain

(2.11.2)

$$\Sigma^- = \Sigma^+ - T_{Kk}^+[x_{k,K}] + \tfrac{1}{2}A_{KkLl}^+[x_{k,K}][x_{l,L}] - \tfrac{1}{6}A_{KkLlMm}^+[x_{k,K}][x_{l,L}][x_{m,M}] - \rho_0\theta^+[\eta] + \rho_0\theta^+B_{Kk}^+[\eta][x_{k,K}] - \cdots$$

$$(\partial\Sigma/\partial x_{k,K})^- = T_{Kk}^+ - A_{KkLl}^+[x_{l,L}] + \tfrac{1}{2}A_{KkLlMm}^+[x_{l,L}][x_{m,M}] + \cdots - \rho_0\theta^+B_{Kk}^+[\eta] + \cdots$$

where we defined

(2.11.3)

$$A_{KkLl}^+ \equiv (\partial^2\Sigma/\partial x_{k,K}\,\partial x_{l,L})^+, \qquad A_{KkLlMm}^+ \equiv (\partial^3\Sigma/\partial x_{k,K}\,\partial x_{l,L}\,\partial x_{m,M})^+$$
$$\rho_0\theta^+B_{Kk}^+ \equiv (\partial^2\Sigma/\partial x_{k,K}\,\partial\eta)^+$$

If we introduce (2.11.2) into the second equation of (2.11.1), we obtain the entropy jump:

$$2\rho_0\theta^+(1 - \tfrac{1}{2}B_{Kk}^+[x_{k,K}])[\eta] = \tfrac{1}{6}A_{KkLlMm}^+[x_{k,K}][x_{l,L}][x_{m,M}] + \cdots$$

Using (2.9.1) in this expression and defining a unit vector $\mathbf{v}$ in the direction of the velocity jump by $\mathbf{S} = S\mathbf{v}$, where S is the magnitude of the discontinuity vector $\mathbf{S}$, we conclude that

(2.11.4)
$$[\eta] = (S^3/12\rho_0\theta^+)Q_{klm}^+ v_k v_l v_m + 0(S^4)$$

where

(2.11.5)
$$Q_{klm}^+ \equiv A_{KkLlMm}^+ N_K N_L N_M$$

Equation (2.11.4) proves that *the entropy jump across the shock is of the order of the third power of the shock strength*, and this justifies the isentropic approximation in the case of weak shock waves.[1]

Within the same order of approximation the temperature jump may be evaluated by

(2.11.6)

$$[\theta] = \frac{1}{\rho_0}\left[\frac{\partial\Sigma}{\partial\eta}\right] = \frac{1}{\rho_0}\left(\frac{\partial^2\Sigma}{\partial x_{k,K}\,\partial\eta}\right)^+[x_{k,K}] + \frac{1}{\rho_0}\left(\frac{\partial^2\Sigma}{\partial\eta^2}\right)^+[\eta] - \frac{1}{2}\left(\frac{\partial^3\Sigma}{\partial x_{k,K}\,\partial x_{l,L}\,\partial x_{m,M}}\right)^+[x_{k,K}][x_{l,L}] + \cdots$$
$$= \theta^+ S Q_k{}^+ v_k + 0(S^2)$$

where

(2.11.7)
$$Q_k{}^+ \equiv B_{Kk}^+ N_K$$

[1] This result is the generalization of the earlier findings obtained for special cases (see Bland [1964] and Chu [1967]).

According to (2.11.4), the physical admissibility of the solution requires that

$$Q^{+}_{klm} v_k v_l v_m \leq 0 \tag{2.11.8}$$

which imposes restrictions on the direction of the velocity jump.

The speed of propagation may now be obtained by substituting $[\partial\Sigma/\partial x_{k,K}]$ from $(2.11.2)_2$ into $(2.11.1)_1$ and keeping only terms up to second order. Hence

$$(Q^{+}_{kl} - \rho_0 U_N \delta_{kl} - \tfrac{1}{2} Q^{+}_{klm} S v_m) v_l = 0 \tag{2.11.9}$$

where

$$Q^{+}_{kl} \equiv A^{+}_{KkLl} N_K N_L \tag{2.11.10}$$

It can be seen that Q^{+}_{kl} is symmetric in indices k and l. The speed of propagation is determined from (2.11.9) as the roots of the following determinantal equation:

$$\det(Q^{+}_{kl} - \tfrac{1}{2} Q^{+}_{klm} v_m S - \rho_0 U_N{}^2 \delta_{kl}) = 0 \tag{2.11.11}$$

The nontrivial solutions of (2.11.9) give the directions for velocity jumps associated with a specific speed of propagation. We seek a linear solution for small S. First we note that if the unit vectors $\mathbf{v}$ are known the corresponding wave speed may be obtained by multiplying (2.11.9) by v_k:

$$\rho_0 U_N{}^2 = Q^{+}_{kl} v_k v_l - \tfrac{1}{2} Q^{+}_{klm} S v_k v_l v_m \tag{2.11.12}$$

If we further assume that Q^{+}_{kl} is a positive definite matrix, then inequality (2.11.8) assures the existence of real speeds of propagations for physically admissible shocks and indicates that the speed of propagation increases with the magnitude of the velocity discontinuity.

The speed of propagation $\bar{U}_N$ and associated unit vector $\bar{\mathbf{v}}$ corresponding to the vanishing shock strength are obtained from the system of equations

$$(Q^{+}_{kl} - \rho_0 \bar{U}_N{}^2 \delta_{kl}) \bar{v}_l = 0 \tag{2.11.13}$$

with

$$\det(Q^{+}_{kl} - \rho_0 \bar{U}_N{}^2 \delta_{kl}) = 0 \tag{2.11.14}$$

The roots $\rho_0 \bar{U}_N{}^2$ may be written in the proper unit vectors $\bar{\mathbf{v}}$ as

$$\rho_0 \bar{U}_N{}^2 = Q^{+}_{kl} \bar{v}_k \bar{v}_l \tag{2.11.15}$$

In Section 2.12 we shall see that the speed $\bar{U}_N$ corresponding to a zero jump across the singular surface is actually the speed of propagation of an acceleration wave. $\bar{U}_N$ and $\bar{\mathbf{v}}$ are completely determined from the existing state ahead of the shock.

Now suppose that the proper vector $\mathbf{v}$ be written approximately as

$$\mathbf{v} = \bar{\mathbf{v}} + S\boldsymbol{\mu} \tag{2.11.16}$$

where $\boldsymbol{\mu}$ is an unknown unit vector. Since $\mathbf{v}$ and $\bar{\mathbf{v}}$ are both unit vectors, $\boldsymbol{\mu}$ should be perpendicular to $\bar{\mathbf{v}}$ for a linear approximation in S:

$$\bar{\mathbf{v}} \cdot \boldsymbol{\mu} = 0 \tag{2.11.17}$$

Introducing (2.11.16) into (2.11.12), using (2.11.15), and keeping terms linear in S only, we have

$$\rho_0 U_N{}^2 = \rho_0 \bar{U}_N{}^2 + (2Q^+_{kl}\bar{v}_k \mu_l - \tfrac{1}{2}Q^+_{klm}\bar{v}_k\bar{v}_l\bar{v}_m)S \tag{2.11.18}$$

Substitution of (2.11.18) into (2.11.9) results in

$$\begin{aligned}(Q^+_{kl} - \rho_0 \bar{U}_N{}^2\delta_{kl})\bar{v}_l + S(Q^+_{kl}\mu_l - \rho_0\bar{U}_N{}^2\mu_k - 2Q^+_{mn}\bar{v}_m\bar{v}_k\mu_n \\ - \tfrac{1}{2}Q^+_{klm}\bar{v}_l\bar{v}_m + \tfrac{1}{2}Q^+_{mnp}\bar{v}_m\bar{v}_n\bar{v}_p\bar{v}_k) = 0\end{aligned} \tag{2.11.19}$$

Because of (2.11.13), the first term of this equation vanishes. The remaining part then determines the components of $\boldsymbol{\mu}$. The direction of $\boldsymbol{\mu}$ can be most easily found by multiplying the remaining part of (2.11.19) by $\bar{v}_k$ and considering (2.11.17). This gives

$$Q^+_{kl}\bar{v}_k\mu_l = 0$$

which implies that $\boldsymbol{\mu}$ is perpendicular to a vector $\boldsymbol{\lambda}$ defined by

$$\lambda_k = Q^+_{kl}\bar{v}_l \tag{2.11.20}$$

Therefore, $\boldsymbol{\mu}$ should be orthogonal to a plane determined by the vectors $\bar{\mathbf{v}}$ and $\boldsymbol{\lambda}$.

Finally, the physical admissibility of the solution restricts the sense of $\boldsymbol{\mu}$. From (2.11.8) follows the condition

$$Q^+_{klm}\bar{v}_k\bar{v}_l\bar{v}_m + 3Q^+_{klm}\bar{v}_k\bar{v}_l\mu_m S \leq 0 \tag{2.11.21}$$

The direction of the velocity jump is determined by (2.11.16). However, for a linear approximation we have the relations

$$\mathbf{S} = S\bar{\mathbf{v}}, \qquad [\mathbf{v}] = -\bar{U}_N\bar{\mathbf{v}}S$$

As an example, consider a plane shock wave propagating in the direction of the positive X_1-axis.[1] We consider a deformation field in the form:

$$x_k = x_k(X_1, t)$$

The displacement vector $\mathbf{u}$ has the form

$$u_k = u_k(X_1, t) \tag{2.11.22}$$

[1] For a similar analysis, see Bland [1964a].

The only nonvanishing displacement gradients are

$$p_k \equiv \partial u_k/\partial X_1 \tag{2.11.23}$$

The shock conditions (2.9.8) take the form

$$[\partial\Sigma/\partial p_k] = \rho_0 U_N{}^2 S_k, \qquad \{(\partial\Sigma/\partial p_k)^+ + (\partial\Sigma/\partial p_k)^-\}[p_k] = 2[\Sigma], \qquad \rho_0 U_N[\eta] \leq 0$$

with $\mathbf{N} = \{1, 0, 0\}$, $\Sigma = \Sigma(p_k, \eta)$, and

$$S_k = [p_k] = S v_k \tag{2.11.24}$$

The components of the Lagrangian strain tensor corresponding to deformation (2.11.22) are

$$E_{11} = p_1 + \tfrac{1}{2}p_k p_k, \qquad E_{12} = \tfrac{1}{2}p_2, \qquad E_{13} = \tfrac{1}{2}p_3, \qquad E_{22} = E_{33} = E_{23} = 0$$

with invariants

$$I_E = p_1 + \tfrac{1}{2}p_k p_k, \qquad II_E = -\tfrac{1}{4}(p_2{}^2 + p_3{}^2), \qquad III_E = 0 \tag{2.11.25}$$

We now consider a nonlinear isotropic elastic solid having an internal energy function of third degree in strains and entropy density (cf. Section 1.16):

$$\begin{aligned}\Sigma = {} & \tfrac{1}{2}(\lambda + 2\mu)I_E{}^2 - 2\mu II_E + lI_E{}^3 + mI_E II_E + nIII_E \\ & + \theta_0 \eta + \alpha\eta^2 + \beta\eta I_E + \gamma\eta^3 + \delta\eta I_E{}^2 + \varepsilon\eta^2 I_E + \nu\eta II_E\end{aligned}$$

In terms of displacement gradients and the entropy density up to third degree, this reads

$$\begin{aligned}\Sigma = {} & \tfrac{1}{2}(\lambda + 2\mu)p_1{}^2 + \tfrac{1}{2}\mu(p_2{}^2 + p_3{}^2) + \tfrac{1}{2}(\lambda + 2\mu + 2l)p_1{}^3 \\ & + \tfrac{1}{2}(\lambda + 2\mu - \tfrac{1}{2}m)p_1(p_2{}^2 + p_3{}^2) + \theta_0\eta + \alpha\eta^2 \\ & + \beta\eta[p_1 + \tfrac{1}{2}(p_1{}^2 + p_2{}^2 + p_3{}^2)] + \gamma\eta^3 + \delta\eta p_1{}^2 + \varepsilon\eta^2 p_1 - \tfrac{1}{4}\nu(p_2{}^2 + p_3{}^2)\eta\end{aligned} \tag{2.11.26}$$

where θ_0 is the temperature at the natural state.

One can readily establish that (2.11.5) and (2.11.10) are transformed into

$$Q_{kl}^+ = (\partial^2\Sigma/\partial p_k\,\partial p_l)^+, \qquad Q_{klm}^+ = (\partial^3\Sigma/\partial p_k\,\partial p_l\,\partial p_m)^+ \tag{2.11.27}$$

from which we calculate

$$\begin{aligned}Q_{11}^+ &= \lambda + 2\mu + 3(\lambda + 2\mu + 2l)p_1{}^+ + (\beta + 2\delta)\eta^+ \\ Q_{12}^+ &= Q_{21}^+ = (\lambda + 2\mu - \tfrac{1}{2}m)p_2{}^+ \\ Q_{13}^+ &= Q_{31}^+ = (\lambda + 2\mu - \tfrac{1}{2}m)p_3{}^+ \\ Q_{22}^+ &= Q_{33}^+ = \mu + (\lambda + 2\mu - \tfrac{1}{2}m)p_1{}^+ + (\beta - \tfrac{1}{2}\nu)\eta^+ \\ Q_{23}^+ &= Q_{32}^+ = 0\end{aligned} \tag{2.11.28}$$

and

$$Q^+_{111} = 3(\lambda + 2\mu + 2l)$$
(2.11.29)
$$Q^+_{122} = Q^+_{221} = Q^+_{212} = Q^+_{133} = Q^+_{331} = Q^+_{313} = \lambda + 2\mu - \tfrac{1}{2}m$$

all other components of Q^+_{klm} being equal to zero. Substituting (2.11.28) and (2.11.29) into (2.11.4), we compute the entropy jump across the shock:

$$[\eta] = (1/12\rho_0\,\theta^+)\{3(\lambda + 2\mu + 2l)v_1{}^3 + 3(\lambda + 2\mu - \tfrac{1}{2}m)v_1(v_2{}^2 + v_3{}^2)\}S^3 + 0(S^4)$$

or, using the relation $v_k v_k = 1$,

(2.11.30) $$[\eta] = (1/2\rho_0\,\theta^+)\{\tfrac{1}{2}\lambda + \mu - \tfrac{1}{4}m + (l + \tfrac{1}{4}m)v_1{}^2\}v_1 S^3 + \cdots$$

In the same way, from (2.11.9) it follows that

(2.11.31)
$$\{\lambda + 2\mu + 3(\lambda + 2\mu + 2l)p_1{}^+ + (\beta + 2\delta)\eta^+ - \tfrac{3}{2}(\lambda + 2\mu + 2l)v_1 S - \rho_0\, U_N{}^2\}v_1 + (\lambda + 2\mu - \tfrac{1}{2}m)\{(p_2{}^+ - \tfrac{1}{2}Sv_2)v_2 + (p_3{}^+ - \tfrac{1}{2}Sv_3)v_3\} = 0$$
$$(\lambda + 2\mu - \tfrac{1}{2}m)(p_2{}^+ - \tfrac{1}{2}Sv_2)v_1 + \{\mu + (\lambda + 2\mu - \tfrac{1}{2}m)p_1{}^+ + (\beta - \tfrac{1}{2}\nu)\eta^+ - \tfrac{1}{2}(\lambda + 2\mu - \tfrac{1}{2}m)Sv_1 - \rho_0\, U_N{}^2\}\, v_2 = 0$$
$$(\lambda + 2\mu - \tfrac{1}{2}m)(p_3{}^+ - \tfrac{1}{2}Sv_3)v_1 + \{\mu + (\lambda + 2\mu - \tfrac{1}{2}m)p_1{}^+ + (\beta - \tfrac{1}{2}\nu)\eta^+ - \tfrac{1}{2}(\lambda + 2\mu - \tfrac{1}{2}m)Sv_1 - \rho_0\, U_N{}^2\}\, v_3 = 0$$

One can verify that (2.11.31) admits a purely longitudinal, or dilatational shock wave. Assume that $p_2 = p_3 = 0$; then $v_1{}^2 = 1$, $v_2 = v_3 = 0$. Thus the speed of propagation follows from $(2.11.31)_1$:

(2.11.32) $$\rho_0\, U_N{}^2 = \lambda + 2\mu + 3(\lambda + 2\mu + 2l)p_1{}^+ + (\beta + 2\delta)\eta^+ - \tfrac{3}{2}(\lambda + 2\mu + 2l)v_1 S$$

From (2.11.30) we have for the entropy jump

(2.11.33) $$[\eta] = (1/4\rho_0\,\theta^+)(\lambda + 2\mu + 2l)S^3 v_1$$

In these expressions v_1 may be $+1$ or -1. The thermodynamic admissibility requires that $[\eta] \le 0$. Thus (2.11.33) implies that if $\lambda + 2\mu + 2l > 0$, then $v_1 = -1$. Consequently $[p_1] = -S$ or $p_1{}^- > p_1{}^+$, namely, expansion shocks are permissible only in this case. If $\lambda + 2\mu + 2l < 0$, then $v_1 = 1$, $[p_1] = S$, or $p_1{}^+ > p_1{}^-$, and only compressive shocks become physically realistic. Nevertheless, for physically permissible shocks, we can write

$$\rho_0\, U_N{}^2 = \lambda + 2\mu + 3(\lambda + 2\mu + 2l)p_1{}^+ + (\beta + 2\delta)\eta^+ + \tfrac{3}{2}|\lambda + 2\mu + 2l|\,|[p_1]|$$

reflecting the property that the shock velocity increases with the magnitude of the discontinuity.

For transverse shocks (if they exist) we must have $v_1 = 0$. From $(2.11.31)_{2,3}$ the speed of propagation follows:

$$\rho_0 U_N^2 = \mu + (\lambda + 2\mu - \tfrac{1}{2}m)p_1^+ + (\beta - \tfrac{1}{2}v)\eta^+$$

The first of (2.11.31) yields approximately

$$[p_2] = -(p_3^+/p_2^+)[p_3]$$

If a transverse shock propagates into a medium at rest in its natural state, then $p_k^+ = 0$ and $(2.11.31)_1$ yields $S = 0$, which means that no propagation can occur in such media.

As another special case we may consider a shock wave propagating in a material at rest and producing a plane deformation in the (1,2)-plane. When $v_3 = 0$ and $p_k^+ = \eta^+ = 0$, we obtain from the first equation of (2.11.31)

$$v_1 = \frac{1}{2}\frac{\lambda + 2\mu - \frac{1}{2}m}{\lambda + \mu} v_2^2 S \tag{2.11.34}$$

with the propagation velocity

$$\rho_0 U_N^2 = \mu - S(\lambda + 2\mu - \tfrac{1}{2}m)v_1$$

obtained from the second equation of this set. Using (2.11.34), we get

$$\rho_0 U_N^2 \simeq \mu \tag{2.11.35}$$

We may also write from (2.11.34)

$$p_1^- = -\frac{\lambda + 2\mu - \frac{1}{2}m}{2(\lambda + \mu)}(p_2^-)^2 \tag{2.11.36}$$

2.12 ACCELERATION WAVES

Acceleration waves, more commonly known as sound waves, in continuum physics were defined in Section 2.8. Since the discontinuities are in the second derivatives of the motion itself, acceleration waves are simpler to treat mathematically than shock waves and have thus been studied extensively. For an early account on the subject, we refer the reader to Truesdell [1961] and Hill [1962].

The study of acceleration waves supplies valuable information about the dynamic response of elastic solids. Although it is difficult to establish a direct relationship between the acceleration fronts and the finite wave motions of solids, there is no doubt that the results obtained from the propagation conditions of a singular surface are relevant. In the case of linear elastic solids it has, however, been shown that there is a close correspondence between the acceleration fronts and very high frequency plane waves (cf. Section 4.3).

Consider a surface $\sigma(t)$ propagating into a medium with a given motion $\mathbf{x} = \mathbf{x}(\mathbf{X}, t)$. The surface $\sigma(t)$ is the acceleration wave if the motion, deformation gradients, and velocity of the material are continuous, and the acceleration and second deformation gradients of the material are discontinuous across it. Let $\Sigma(t)$ be the material image of the surface $\sigma(t)$. Then the jumps suffered by the field quantities satisfy the following compatibility conditions on Σ:

$$[x_{k,KL}] = S_k N_K N_L, \qquad [\dot{x}_{k,K}] = -U_N S_k N_K, \qquad [\ddot{x}_k] = U_N{}^2 S_k \tag{2.12.1}$$

where the vector $\mathbf{S}$ defined by the relation

$$S_k \equiv [x_{k,KL}] N_K N_L \tag{2.12.2}$$

is in the direction of the acceleration jump and is a measure of the discontinuity with dimension L^{-1}. These relations are obtained by replacing ϕ in (2.4.15) and (2.5.9) by x_k. Since the velocity of particles is to be continuous across the wave, the local speed of propagation U is also continuous across the acceleration front. Thus from (2.7.1) it follows that the density is continuous, i.e.,

$$[\rho] = 0 \tag{2.12.3}$$

Condition (2.7.2) reduces to

$$[t_{kl}] n_l = 0 \tag{2.12.4}$$

which implies that the stress vector is continuous across the wave. Using (2.12.4) in (2.7.3), we find that the internal energy density has no jump on $\sigma(t)$:

$$[\varepsilon] = 0 \tag{2.12.5}$$

In elastic materials both (2.12.4) and (2.12.5) are automatically satisfied, also implying that $[\eta] = 0$. Therefore, acceleration waves in non-heat conducting elastic solids are necessarily isentropic.

The material expressions (2.12.1) can be transformed into the following spatial forms by recalling (2.6.12) and (2.6.19):

(2.12.6)

$$[x_{k,KL}] = a_k x_{m,K} x_{n,L} n_m n_n, \qquad [\dot{x}_{k,K}] = -U a_k x_{l,K} n_l, \qquad [\ddot{x}_k] = U^2 a_k$$

where the vector $\mathbf{a}$ is related to $\mathbf{S}$ by

$$\mathbf{a} = \mathbf{S}/C_n{}^2 \tag{2.12.7}$$

and $C_n{}^2$ is given by (2.6.13). The vector **a** is in the direction of the acceleration discontinuity and is a measure of the strength of the discontinuity. Therefore, it will be called the *amplitude* of the singularity. If **a** is parallel to the unit normal **n** of the wave surface $\sigma(t)$, we have a *longitudinal* wave; if it is perpendicular to **n**, we have a *transverse* wave. Later we shall see that an acceleration wave is, in general, neither longitudinal nor transverse.

Equation $(2.12.6)_2$ may be written

$$[\dot{x}_{k,l}] = -Ua_k n_l$$

Contracting indices,

$$[\operatorname{div} \mathbf{v}] = -U\mathbf{a} \cdot \mathbf{n}$$

Thus div **v** is continuous across the wave front $\sigma(t)$ *for transverse* waves.

From the continuity equation

$$\dot{\rho} = -\rho \operatorname{div} \mathbf{v}$$

it follows that

$$[\dot{\rho}] = -\rho[\operatorname{div} \mathbf{v}] = \rho U a_k n_k \tag{2.12.8}$$

Therefore, for transverse waves the material time rate of the density is continuous across $\sigma(t)$. For longitudinal waves, if we write $\mathbf{a} = a\mathbf{n}$, we get

$$[\dot{\rho}] = \rho U a$$

When $a > 0$, $\dot{\rho}^- < \dot{\rho}^+$ and the wave is called *expansive* since the density tends to decrease after the passage of the wave; when $a < 0$, $\dot{\rho}^- > \dot{\rho}^+$ and the wave is called *compressive*. In a general wave, the wave will be said to be expansive if $\mathbf{a} \cdot \mathbf{n} > 0$ and compressive if $\mathbf{a} \cdot \mathbf{n} < 0$.

The dynamical conditions of compatibility (2.7.1)–(2.7.7) are satisfied identically. Therefore, the propagation of acceleration waves will be governed by relations involving derivatives of the field quantity entering into the dynamical conditions of compatibility. To find these relations, we consider differential equations of motion of the body in the reference state (cf. Section 1.12):

$$T_{Kk,K} + \rho_0 f_k = \rho_0 \ddot{x}_k \tag{2.12.9}$$

Introducing the elastic constitutive relations

$$T_{Kk} = \partial\Sigma/\partial x_{k,K}$$

into (2.12.9), we have

$$A_{KkLl} x_{l,KL} + \rho_0 f_k = \rho_0 \ddot{x}_k$$

where

$$A_{KkLl}(x_{m,M}) = \partial^2\Sigma/\partial x_{k,K}\,\partial x_{l,L}$$

are called *elasticities* of the material (cf. Section 1.17) and may be assumed to be symmetrical in indices K and L without any loss in generality. Clearly elasticities are only functions of the deformation gradients. The field equations (2.12.9) must hold on each side of the singular surface. If we assume that the body force is continuous across the singular surface and recall that A_{KkLl} are continuous on the wave front, we have on $\Sigma(t)$

(2.12.10) $$A_{KkLl}[x_{l,KL}] = \rho_0[\ddot{x}_k]$$

Substitution of (2.12.1) into (2.12.10) leads to

(2.12.11) $$(\mathcal{Q}_{kl} - \rho_0\, U_N^{\ 2}\, \delta_{kl})S_l = 0$$

where $\mathcal{Q}_{kl}$ is a symmetric tensor defined by

(2.12.12) $$\mathcal{Q}_{kl}(\mathbf{N}) \equiv A_{KkLl}\, N_K\, N_L$$

The speed of propagation U_N is therefore obtained as the roots of the equation

(2.12.13) $$\det[\mathcal{Q}(\mathbf{N}) - \rho_0\, U_N^{\ 2}\mathbf{I}] = 0$$

Equations (2.12.11) and (2.12.13) show that $\rho_0\, U_N^{\ 2}$ is a proper value and $\mathbf{S}$ a proper vector of the tensor $\mathcal{Q}$. The spatial form of (2.12.11) may be obtained by using (2.6.12), (2.6.19), (2.12.7), and the relation $\rho_0 = j\rho$. Hence[1]

(2.12.14) $$[Q_{kl}(\mathbf{n}) - \rho U^2\, \delta_{kl}]a_l = 0$$

where the symmetric tensor

$$Q_{kl}(\mathbf{n}) = j^{-1}A_{KkLl}\, x_{m,K}\, x_{n,L}\, n_m\, n_n = B_{mknl}\, n_m\, n_n$$

or in short

(2.12.15) $$\mathbf{Q}(\mathbf{n}) = j^{-1}C_n^{\ 2}\mathcal{Q}(\mathbf{N})$$

is a function of deformation gradients and the propagation direction $\mathbf{n}$ of the wave front in the strained material. It is called the *acoustical tensor*.

The local speed of propagation satisfies the equation

(2.12.16) $$\det[\mathbf{Q}(\mathbf{n}) - \rho U^2\mathbf{I}] = 0$$

or equivalently the cubic equation

(2.12.17) $$-(\rho U^2)^3 + I_Q(\rho U^2)^2 - II_Q\,\rho U^2 + III_Q = 0$$

[1] Truesdell [1961] calls this result the Fresnel–Hadamard theorem.

where I_Q, II_Q, and III_Q are invariants of the acoustical tensor $\mathbf{Q}$. Equations (2.12.16) and (2.12.14) indicate that ρU^2 are proper values and $\mathbf{a}$ proper vectors of the acoustical tensor.

The proper unit vectors $\mathbf{v}$ of the acoustical tensor are called the *acoustical axes* for waves traveling in a given direction $\mathbf{n}$ in the deformed body, and they determine the direction of the singularity $\mathbf{a}$. Since $\mathbf{Q}$ is a symmetric tensor, all of its proper numbers are real, and consequently, if they are distinct, the acoustical axes form an orthogonal triad at a point on the wave surface. However, the speeds of propagation associated with the acoustical axes should be real for physically significant waves. This requires that the proper numbers of the acoustical tensor be positive. In other words, in order for a *real* wave to propagate in every direction in a deformed elastic material the acoustical tensor must be positive definite. It can be shown that the necessary and sufficient condition for this is the *strong ellipticity* of the elasticities[1] of the material, that is,

$$A_{KkLl} m_k m_l M_K M_L > 0 \tag{2.12.18}$$

for arbitrary vectors $\mathbf{m}$ and $\mathbf{M}$.

If two proper numbers coincide then one acoustical axis is determined, and any element of the set of directions perpendicular to this axis constitutes an acoustical axis, and a single wave speed is associated with the whole set. If all the proper numbers coincide then there is only one wave speed and any direction is an acoustical axis.

If, by chance, $\mathbf{a}$ is a proper vector of $\mathbf{Q}(\mathbf{n})$ then it follows from (2.12.14) that the corresponding wave speed is

$$\rho U^2 = Q_{kl} a_k a_l / a_m a_m = Q_{kl} v_k v_l \tag{2.12.19}$$

where $\mathbf{v}$ is the unit proper vector. We may thus define

$$\mathbf{a} = a\mathbf{v}$$

and choose the sense of $\mathbf{v}$ in such a way that we have $\mathbf{v} \cdot \mathbf{n} \geq 0$. Therefore, the magnitude of the amplitude vector $\mathbf{a}$ is $|a|$, and $a > 0$ corresponds to an expansive wave while $a < 0$ corresponds to a compressive wave. Equation (2.12.19) emphasizes once again the fact that the positive-definiteness of the acoustical tensor secures the existence of a real speed of propagation.

Finally, from (2.12.15) we see that all of the foregoing results remain valid if we replace $\mathbf{n}$ by $-\mathbf{n}$, since $\mathbf{Q}$ is an even function of the components of the propagation direction. Moreover, (2.12.14) indicates that if $\mathbf{a}$ is a proper vector of $\mathbf{Q}$ so is $k\mathbf{a}$, where k is a nonzero real number.

[1] It has been proved by Truesdell [1966] that the condition of strong ellipticity guarantees the existence of at least one direction in which a *longitudinal* discontinuity may propagate.

2.13 PROPAGATION OF ACCELERATION WAVES IN ISOTROPIC MATERIALS

In this section we discuss the propagation of acceleration waves in isotropic elastic materials. In this case the elasticities **A** are given by (cf. Section 1.17)

$$A_{KkLl} = 2j(X_{[K,|m|} X_{L],l} t_{mk} + X_{K,m} x_{n,L} F_{kmln}) \tag{2.13.1}$$

where

(2.13.2)

$$\begin{aligned} F_{klmn} &= \frac{\partial t_{kl}}{\partial \overset{-1}{c}_{mn}} \\ &= \frac{h_1}{2}(\delta_{km}\delta_{ln} + \delta_{lm}\delta_{kn}) + \frac{h_2}{2}\left(\delta_{km}\overset{-1}{c}_{ln} + \delta_{kn}\overset{-1}{c}_{lm} + \delta_{lm}\overset{-1}{c}_{kn} + \delta_{ln}\overset{-1}{c}_{km}\right) \\ &\quad + \delta_{kl}\left[\frac{\partial h_0}{\partial I}\delta_{mn} + \frac{\partial h_0}{\partial II}\left(I\,\delta_{mn} - \overset{-1}{c}_{mn}\right) + III\frac{\partial h_0}{\partial III}c_{mn}\right] \\ &\quad + \overset{-1}{c}_{kl}\left[\frac{\partial h_1}{\partial I}\delta_{mn} + \frac{\partial h_1}{\partial II}\left(I\,\delta_{mn} - \overset{-1}{c}_{mn}\right) + III\frac{\partial h_1}{\partial III}c_{mn}\right] \\ &\quad + \overset{-2}{c}_{kl}\left[\frac{\partial h_2}{\partial I}\delta_{mn} + \frac{\partial h_2}{\partial II}\left(I\,\delta_{mn} - \overset{-1}{c}_{mn}\right) + III\frac{\partial h_2}{\partial III}c_{mn}\right] \end{aligned}$$

The response functions h_0, h_1, and h_2 which appear in the stress–strain relations

$$\mathbf{t} = h_0\mathbf{I} + h_1\overset{-1}{\mathbf{c}} + h_2\overset{-2}{\mathbf{c}}$$

are given by (cf. Section 1.14)

$$h_0 = 2(III)^{1/2}\frac{\partial\Sigma}{\partial III}, \qquad h_1 = \frac{2}{(III)^{1/2}}\left(\frac{\partial\Sigma}{\partial I} + I\frac{\partial\Sigma}{\partial II}\right), \qquad h_2 = -\frac{2}{(III)^{1/2}}\frac{\partial\Sigma}{\partial II}$$

If we substitute (2.13.1) into (2.12.15) we obtain (Truesdell [1961])

$$Q_{kl} = 2F_{kmln}\overset{-1}{c}_{np}n_m n_p \tag{2.13.3}$$

Now substituting (2.13.2) into (2.13.3) and arranging the resulting relation, we get the following expression for the acoustical tensor:

$$\begin{aligned} \mathbf{Q}(\mathbf{n}) &= \left(h_1\mathbf{n}\cdot\overset{-1}{\mathbf{c}}\mathbf{n} + h_2\mathbf{n}\cdot\overset{-2}{\mathbf{c}}\mathbf{n}\right)\mathbf{I} + h_2\overset{-1}{\mathbf{c}}\mathbf{n}\otimes\overset{-1}{\mathbf{c}}\mathbf{n} + h_2\left(\mathbf{n}\cdot\overset{-1}{\mathbf{c}}\mathbf{n}\right)\overset{-1}{\mathbf{c}} \\ &\quad + h_1\overset{-1}{\mathbf{c}}\mathbf{n}\otimes\mathbf{n} + h_2\overset{-2}{\mathbf{c}}\mathbf{n}\otimes\mathbf{n} + 2\sum_{a=0}^{2}\left[\left(\frac{\partial h_a}{\partial I} + I\frac{\partial h_a}{\partial II}\right)\overset{-a}{\mathbf{c}}\mathbf{n}\otimes\overset{-1}{\mathbf{c}}\mathbf{n}\right. \\ &\quad \left. - \frac{\partial h_a}{\partial II}\overset{-a}{\mathbf{c}}\mathbf{n}\otimes\overset{-2}{\mathbf{c}}\mathbf{n} + III\frac{\partial h_a}{\partial III}\overset{-a}{\mathbf{c}}\mathbf{n}\otimes\mathbf{n}\right] \end{aligned} \tag{2.13.4}$$

where the symbol $\otimes$ denotes tensor multiplication, e.g.,

$$\left(\mathbf{a} \otimes \mathbf{b}\right)_{ij} = a_i b_j, \qquad \left(\overset{-1}{\mathbf{c}}\mathbf{n} \otimes \overset{-1}{\mathbf{c}}\mathbf{n}\right)_{ij} = \left(\overset{-1}{c}_{ik} n_k \overset{-1}{c}_{jl} n_l\right)$$

We transform this expression into a symmetric form

(2.13.4)′

$$\begin{aligned}
\mathbf{Q} = {} & \left(h_1 \mathbf{n} \cdot \overset{-1}{\mathbf{c}}\mathbf{n} + h_2 \mathbf{n} \cdot \overset{-2}{\mathbf{c}}\mathbf{n}\right)\mathbf{I} + 2III \frac{\partial h_0}{\partial III} \mathbf{n} \otimes \mathbf{n} \\
& + h_2(\mathbf{n} \cdot \overset{-1}{\mathbf{c}}\mathbf{n})\overset{-1}{\mathbf{c}} + 2\left(\frac{h_2}{2} + \frac{\partial h_1}{\partial I} + I \frac{\partial h_1}{\partial II}\right) \overset{-1}{\mathbf{c}}\mathbf{n} \otimes \overset{-1}{\mathbf{c}}\mathbf{n} \\
& - 2 \frac{\partial h_2}{\partial II} \overset{-2}{\mathbf{c}}\mathbf{n} \otimes \overset{-2}{\mathbf{c}}\mathbf{n} + 2\left(\frac{\partial h_0}{\partial I} + I \frac{\partial h_0}{\partial II}\right)\left(\mathbf{n} \otimes \overset{-1}{\mathbf{c}}\mathbf{n} + \overset{-1}{\mathbf{c}}\mathbf{n} \otimes \mathbf{n}\right) \\
& - 2 \frac{\partial h_0}{\partial II}\left(\mathbf{n} \otimes \overset{-2}{\mathbf{c}}\mathbf{n} + \overset{-2}{\mathbf{c}}\mathbf{n} \otimes \mathbf{n}\right) - 2 \frac{\partial h_1}{\partial II}\left(\overset{-1}{\mathbf{c}}\mathbf{n} \otimes \overset{-2}{\mathbf{c}}\mathbf{n} + \overset{-2}{\mathbf{c}}\mathbf{n} \otimes \overset{-1}{\mathbf{c}}\mathbf{n}\right)
\end{aligned}$$

In deriving the above expression we employed the compatibility conditions for the response functions h_0, h_1, and h_2, which follow from their expression given above by eliminating Σ, i.e.,

$$\frac{h_1}{2} + III \frac{\partial h_1}{\partial III} - \frac{\partial h_0}{\partial I} - I \frac{\partial h_0}{\partial II} = 0$$

(2.13.5)
$$\frac{h_2}{2} + III \frac{\partial h_2}{\partial III} + \frac{\partial h_0}{\partial II} = 0$$

$$\frac{\partial h_2}{\partial I} + I \frac{\partial h_2}{\partial II} + \frac{\partial h_1}{\partial II} = 0$$

For a given material, state of deformation, and wave shape, the symmetrical acoustical tensor is determined by (2.13.4). In contrast to those known in the propagation of infinitesimal waves in linear elastic solids, the proper vectors of $\mathbf{Q}$ are in general neither longitudinal nor transverse. We now show that such preferred directions do exist if the acoustical axes coincide with the principal directions of the deformation tensors of the medium. In this case then, a wave, called a *principal wave*, can propagate in the direction of one of the principal axes.

Let $\mathbf{n}_\alpha$ ($\alpha = 1, 2, 3,$) be the principal directions of the deformation tensor $\overset{-1}{\mathbf{c}}$, i.e.,

(2.13.6)
$$\overset{-1}{\mathbf{c}}\mathbf{n}_\alpha = \lambda_{\underline{\alpha}}{}^2 \mathbf{n}_{\underline{\alpha}}$$

where λ_α is the principal stretch in the αth principal direction. Suppose that at every point of the wave surface the propagation direction is parallel to one of the principal directions, say $\mathbf{n}_1$, of the deformation tensor at the spatial point of the body coinciding instantaneously with the surface point. Then substituting (2.13.6) and the relation

$$\overset{-2}{\mathbf{c}}\mathbf{n}_\alpha = \lambda_{\underline{\alpha}}{}^4 \mathbf{n}_\alpha$$

derived from it into (2.13.4), we get

(2.13.7)

$$\frac{1}{\lambda_1{}^2}\mathbf{Q}(\mathbf{n}_1) = (h_1 + \lambda_1{}^2 h_2)\mathbf{I} + h_2 \overset{-1}{\mathbf{c}} + 2\left\{\frac{h_1}{2} + \lambda_1{}^2 h_2 + \sum_{a=0}^{2} \lambda_1^{2a}\left[\frac{\partial h_a}{\partial I} + (\lambda_2{}^2 + \lambda_3{}^2)\frac{\partial h_a}{\partial II} + \lambda_2{}^2\lambda_3{}^2 \frac{\partial h_a}{\partial III}\right]\right\}\mathbf{n}_1 \otimes \mathbf{n}_1$$

It is now clear that the vector $\mathbf{Q}(\mathbf{n}_1)\mathbf{n}_\alpha$ is in the direction of the principal direction $\mathbf{n}_\alpha$. Therefore, $\mathbf{n}_\alpha$ are three proper vectors of the tensor $\mathbf{Q}$; that is, the principal axes of the deformation coincide with the acoustical axes of $\mathbf{Q}$ for waves propagating in one of the principal directions. Since the amplitude vector $\mathbf{a}$ should be in the direction of one of the proper vectors of the acoustical tensor, it will be in the direction of one of $\mathbf{n}_1$, $\mathbf{n}_2$, or $\mathbf{n}_3$, in this particular case. If $\mathbf{a}$ is parallel to $\mathbf{n}_1$, the direction of propagation, then the acceleration wave is said to be longitudinal; if it is parallel to either $\mathbf{n}_2$ or $\mathbf{n}_3$, it is perpendicular to the propagation direction; hence it is said to be transverse. Therefore, we can state that *every principal wave is either longitudinal or transverse.*

The proper vectors of the acoustical tensor thus determined, it is straightforward to calculate the corresponding wave speeds directly from (2.12.19). Replacing $\mathbf{v}$ by $\mathbf{n}_1$ in (2.12.19) and using (2.13.7), we obtain the *longitudinal wave speed* as

(2.13.8)

$$\rho U_{11}^2 = 2\lambda_1{}^2\left\{h_1 + 2\lambda_1{}^2 h_2 + \sum_{a=0}^{2} \lambda_1^{2a}\left[\frac{\partial h_a}{\partial I} + (\lambda_2{}^2 + \lambda_3{}^2)\frac{\partial h_a}{\partial II} + \lambda_2{}^2\lambda_3{}^2\frac{\partial h_a}{\partial III}\right]\right\}$$

Replacing $\mathbf{v}$ in (2.12.19) by $\mathbf{n}_2$ (or $\mathbf{n}_3$), we get two transverse wave speeds

(2.13.9)

$$\rho U_{12}^2 = \lambda_1{}^2[h_1 + (\lambda_1{}^2 + \lambda_2{}^2)h_2] = \frac{2\lambda_1}{\lambda_2\lambda_3}\left(\frac{\partial \Sigma}{\partial I} + \lambda_3{}^2\frac{\partial \Sigma}{\partial II}\right)$$

$$\rho U_{13}^2 = \lambda_1{}^2[h_1 + (\lambda_1{}^2 + \lambda_3{}^2)h_2] = \frac{2\lambda_1}{\lambda_2\lambda_3}\left(\frac{\partial \Sigma}{\partial I} + \lambda_2{}^2\frac{\partial \Sigma}{\partial II}\right)$$

where h_0, h_1, and h_2 are given functions of λ_1, λ_2, and λ_3. Therefore, three principal waves whose speeds are given by (2.13.8) and (2.13.9) are associated with a principal direction, say $\mathbf{n}_1$. Hence we deduce that there exist altogether nine principal waves in an isotropic compressible elastic medium—every three of which are associated with one of the principal directions. The corresponding wave speeds will be denoted $U_{\alpha\beta}$, where the indices designate the three principal directions, the first representing the principal direction coinciding with the direction of propagation. Thus the matrix of wave speed is

$$[U_{\alpha\beta}] = \begin{bmatrix} U_{11} & U_{12} & U_{13} \\ U_{21} & U_{22} & U_{23} \\ U_{31} & U_{32} & U_{33} \end{bmatrix}$$

The first row of this matrix is determined by (2.13.8) and (2.13.9), and the second and third rows are obtained from the first by suitably changing the indices of the principal stretches.

For transverse waves one can easily establish the relations

$$\begin{gathered} U_{12}^2/\lambda_1^2 = U_{21}^2/\lambda_2^2, \qquad U_{23}^2/\lambda_2^2 = U_{32}^2/\lambda_3^2, \qquad U_{31}^2/\lambda_3^2 = U_{13}^2/\lambda_1^2 \\ U_{23}^2 = [\lambda_2^2/\lambda_1^2(\lambda_2^2 - \lambda_3^2)][(\lambda_2^2 - \lambda_1^2)U_{12}^2 - (\lambda_3^2 - \lambda_1^2)U_{13}^2] \end{gathered} \tag{2.13.10}$$

Expressions (2.13.10) determine the speeds of all transverse waves in terms of those corresponding to two transverse waves propagating in a principal direction, and they are universal relations in the sense that they are true regardless of the form of the response functions for specific materials.

We can also show that the knowledge of the wave speed in terms of principal stretches determines the stress–strain relations for isotropic elastic solids. We can indeed derive from (2.13.9)

$$\begin{aligned} h_2 &= \rho(U_{12}^2 - U_{13}^2)/\lambda_1^2(\lambda_2^2 - \lambda_3^2) \\ h_1 &= \rho[(\lambda_1^2 + \lambda_2^2)U_{13}^2 - (\lambda_1^2 + \lambda_3^2)U_{12}^2]/\lambda_1^2(\lambda_2^2 - \lambda_3^2) \end{aligned} \tag{2.13.11}$$

From this result we see that the measurements of two transverse wave speeds in a principal direction determine uniquely the two response functions h_1 and h_2 as functions of λ_1, λ_2, and λ_3. From $(2.13.11)_1$ it follows that these two kinds of transverse waves have the same absolute speed of propagation when $h_2 = 0$ if all principal stretches are distinct. Otherwise the waves will have the same absolute speed if and only if the corresponding principal stretches are equal. In this last case, the amplitude of the transverse wave is an arbitrary vector perpendicular to the direction of propagation. Moreover, we see that $h_2 < 0$ if and only if $U_{12}^2 - U_{13}^2$ and $\lambda_2^2 - \lambda_3^2$ are of opposite sign [i.e., if transverse waves with amplitude parallel to the principal direction with lesser stretch travel faster than the others (Truesdell and Noll [1965, Section 74]).

On account of relations (2.13.10), the choice of any principal direction for the measurements of h_1 and h_2 is immaterial. However, measurements in other principal directions may help to verify relations (2.13.10) and consequently the elasticity of the material.

The determination of h_0 from the wave speeds is a little more involved. To this end, we consider (2.13.8) and set

(2.13.12)

$$\frac{\partial h_0}{\partial I} + (\lambda_2{}^2 + \lambda_3{}^2)\frac{\partial h_0}{\partial II} + \lambda_2{}^2\lambda_3{}^2\frac{\partial h_0}{\partial III} = \frac{\rho U_{11}^2}{2\lambda_1{}^2} - h_1 - 2\lambda_1 h_2 - \sum_{a=1}^{2}\lambda_1^{2a}\left[\frac{\partial h_a}{\partial I} + (\lambda_2{}^2 + \lambda_3{}^2)\frac{\partial h_a}{\partial II} + \lambda_2{}^2\lambda_3{}^2\frac{\partial h_a}{\partial III}\right] = A_1$$

Two similar expressions may be obtained from the relations for U_{22} and U_{33}:

$$\frac{\partial h_0}{\partial I} + (\lambda_1{}^2 + \lambda_3{}^2)\frac{\partial h_0}{\partial II} + \lambda_1{}^2\lambda_3{}^2\frac{\partial h_0}{\partial III} = A_2$$

(2.13.13)

$$\frac{\partial h_0}{\partial I} + (\lambda_1{}^2 + \lambda_2{}^2)\frac{\partial h_0}{\partial II} + \lambda_1{}^2\lambda_2{}^2\frac{\partial h_0}{\partial III} = A_3$$

If principal stretches are distinct, we can solve (2.13.12) and (2.13.13) for the derivatives of h_0:

(2.13.14)

$$\frac{\partial h_0}{\partial I} = -\frac{A_1\lambda_1{}^4(\lambda_2{}^2 - \lambda_3{}^2) + A_2\lambda_2{}^4(\lambda_3{}^2 - \lambda_1{}^2) + A_3\lambda_3{}^4(\lambda_1{}^2 - \lambda_2{}^2)}{(\lambda_1{}^2 - \lambda_2{}^2)(\lambda_2{}^2 - \lambda_3{}^2)(\lambda_3{}^2 - \lambda_1{}^2)} \equiv B_1(\lambda_1, \lambda_2, \lambda_3)$$

$$\frac{\partial h_0}{\partial II} = \frac{A_1\lambda_1{}^2(\lambda_2{}^2 - \lambda_3{}^2) + A_2\lambda_2{}^2(\lambda_3{}^2 - \lambda_1{}^2) + A_3\lambda_3{}^2(\lambda_1{}^2 - \lambda_2{}^2)}{(\lambda_1{}^2 - \lambda_2{}^2)(\lambda_2{}^2 - \lambda_3{}^2)(\lambda_3{}^2 - \lambda_1{}^2)} \equiv B_2(\lambda_1, \lambda_2, \lambda_3)$$

$$\frac{\partial h_0}{\partial III} = -\frac{A_1(\lambda_2{}^2 - \lambda_3{}^2) + A_2(\lambda_3{}^2 - \lambda_1{}^2) + A_3(\lambda_1{}^2 - \lambda_2{}^2)}{(\lambda_1{}^2 - \lambda_2{}^2)(\lambda_2{}^2 - \lambda_3{}^2)(\lambda_3{}^2 - \lambda_1{}^2)} \equiv B_3(\lambda_1, \lambda_2, \lambda_3)$$

If the material is elastic, the known functions B_1, B_2, and B_3 should satisfy the conditions of integrability

$$\partial B_1/\partial II = \partial B_2/\partial I, \qquad \partial B_1/\partial III = \partial B_3/\partial I, \qquad \partial B_2/\partial III = \partial B_3/\partial II$$

If these conditions are satisfied, then

(2.13.15)

$$h_0 = \int B_1\, dI + B_2\, dII + B_3\, dIII$$

The integration can be performed by remembering that (cf. Section 1.6)

(2.13.16)
$$I = \lambda_1{}^2 + \lambda_2{}^2 + \lambda_3{}^2, \qquad II = \lambda_1{}^2\lambda_2{}^2 + \lambda_1{}^2\lambda_3{}^2 + \lambda_2{}^2\lambda_3{}^2, \qquad III = \lambda_1{}^2\lambda_2{}^2\lambda_3{}^2$$

Since we consider here only hyperelastic materials, the evaluated response functions should also satisfy the compatibility conditions (2.13.5) that secure the existence of a strain energy function Σ. Therefore, requirement (2.13.5) must also be satisfied to assure the hyperelastic nature of the material.

Although the preceding suggested experiment sounds simple and reasonable in principle, the difficulties inherent in the wave-speed measurements, in general, make this rational method intractable for the determination of response functions.[1]

We now return to Eqs. (2.13.8) and (2.13.9) to put them into a more elegant and meaningful form in principal stresses. The principal stresses are given by

(2.13.17)
$$t_\alpha = h_0 + \lambda_\alpha{}^2 h_1 + \lambda_\alpha{}^4 h_2$$

From this, by using (2.13.16) we find that

$$\frac{\partial t_1}{\partial \lambda_1} = 2\lambda_1 h_1 + 4\lambda_1{}^3 h_2 + \sum_{a=0}^{2} \lambda_1^{2a} \frac{\partial h_a}{\partial \lambda_1}$$
$$= 2\lambda_1\left\{h_1 + 2\lambda_1{}^2 h_2 + \sum_{a=0}^{2} \lambda_1^{2a}\left[\frac{\partial h_a}{\partial I} + (\lambda_2{}^2 + \lambda_3{}^2)\frac{\partial h_a}{\partial II} + \lambda_2{}^2\lambda_3{}^2 \frac{\partial h_a}{\partial III}\right]\right\}$$

Comparing this with (2.13.8), we arrive at the simple relation for a principal wave

(2.13.18)
$$\rho U_{11}^2 = \lambda_1\, \partial t_1/\partial \lambda_1 = \partial t_1/\partial \log \lambda_1$$

In terms of the differences $t_1 - t_2$ and $t_1 - t_3$, we have

(2.13.19)
$$\rho U_{12}^2 = \lambda_1{}^2(t_1 - t_2)/(\lambda_1{}^2 - \lambda_2{}^2), \qquad \rho U_{13}^2 = \lambda_1{}^2(t_1 - t_3)/(\lambda_1{}^2 - \lambda_3{}^2)$$

Formulas (2.13.18) and (2.13.19) are attributed to Ericksen (Truesdell [1961, Section 7]). These formulas are accessible to a simple interpretation for the existence of principal waves. Equation (2.13.18) implies that $U_{11}^2 > 0$ if the corresponding principal stress is a monotone increasing function of the corresponding principal stretch. From (2.13.19) we deduce that real transverse waves exist if the greater principal stress is associated with the greater principal stretch. This last condition is in fact known as the *B–E* inequality.[2]

[1] For a discussion of the current status of finite amplitude wave propagation experiments we refer the reader to an excellent monograph by Bell [1968, Chapter II] and an earlier paper [1964]. *Note added in proof*: see also Bell [1973].

[2] That is, Baker–Ericksen inequality, Eringen [1962, Section 47]; see also Section 1.18.

Using the relations

$$t_1 = \frac{1}{\lambda_2 \lambda_3} \frac{\partial \Sigma}{\partial \lambda_1}, \qquad t_2 = \frac{1}{\lambda_1 \lambda_3} \frac{\partial \Sigma}{\partial \lambda_2}, \qquad t_3 = \frac{1}{\lambda_1 \lambda_2} \frac{\partial \Sigma}{\partial \lambda_3}$$

simpler forms of (2.13.18) and (2.13.19) are obtained (Green [1963]):

$$\rho_0 U_{11}^2 = \lambda_1{}^2 \frac{\partial^2 \Sigma}{\partial \lambda_1{}^2}, \qquad \rho_0 U_{12}^2 = \frac{\lambda_1{}^2}{\lambda_1{}^2 - \lambda_2{}^2} \left(\lambda_1 \frac{\partial \Sigma}{\partial \lambda_1} - \lambda_2 \frac{\partial \Sigma}{\partial \lambda_2} \right) \tag{2.13.20}$$

where we used $\rho_0/\rho = \lambda_1 \lambda_2 \lambda_3$. Similar expressions for the other wave speeds can easily be written.

If the material is initially subjected to a hydrostatic stress $\mathbf{t} = -p\mathbf{I}$, then all principal stretches become equal to λ_0, which is given by the relation

(2.13.21)

$$p = -(h_0 + \lambda_0{}^2 h_1 + \lambda_0{}^4 h_2) = -2\left(\frac{1}{\lambda_0} \frac{\partial \Sigma}{\partial I} + 2\lambda_0 \frac{\partial \Sigma}{\partial II} + \lambda_0{}^3 \frac{\partial \Sigma}{\partial III} \right) \Bigg|_{\lambda_\alpha = \lambda_0}$$

Moreover, all waves propagating in this medium are now principal waves since every direction is a principal direction in the body. Therefore, all waves are either longitudinal or transverse and their speeds are calculated from (2.13.8) and (2.13.9):

$$\begin{aligned} \rho U_{\mathrm{L}}{}^2 &= 2\lambda_0{}^2 \left[h_1 + 2\lambda_0{}^2 h_2 + \sum_{a=0}^{2} \lambda_0^{2a} \left(\frac{\partial h_a}{\partial I} + 2\lambda_0{}^2 \frac{\partial h_a}{\partial II} + \lambda_0{}^4 \frac{\partial h_a}{\partial III} \right) \right] \\ \rho U_{\mathrm{T}}{}^2 &= \lambda_0{}^2 (h_1 + 2\lambda_0{}^2 h_2) \end{aligned} \tag{2.13.22}$$

In this case, $I = 3\lambda_0{}^2$, $II = 3\lambda_0{}^4$, $III = \lambda_0{}^6$, and

$$\frac{\partial}{\partial I} + 2\lambda_0{}^2 \frac{\partial}{\partial II} + \lambda_0{}^4 \frac{\partial}{\partial III} = \frac{1}{3} \frac{d}{d\lambda_0{}^2}$$

Therefore, we get from (2.13.21)

$$\sum_{a=0}^{2} \lambda_0{}^{2a}\, dh_a/d\lambda_0{}^2 = -(dp/d\lambda_0{}^2) - (h_1 + 2\lambda_0{}^2 h_2)$$

Furthermore, $\rho = \rho_0 \lambda_0^{-3}$ and we have $d\rho/d\lambda_0{}^2 = -\frac{3}{2}\rho_0 \lambda_0^{-5}$. Thus using $(2.13.22)_1$, the longitudinal wave velocity may be written as

$$U_{\mathrm{L}}{}^2 = \tfrac{4}{3}(\lambda_0{}^2/\rho)(h_1 + 2\lambda_0{}^2 h_2) + (dp/d\rho)$$

The two wave speeds are now related to each other by (Truesdell [1961])

$$U_{\mathrm{L}}{}^2 = \tfrac{4}{3} U_{\mathrm{T}}{}^2 + (dp/d\rho) \tag{2.13.23}$$

In real compressible materials, one expects $dp/d\rho > 0$. Thus we conclude that in the case of hydrostatic stress

$$|U_{\mathrm{L}}| > 2|U_{\mathrm{T}}|/\sqrt{3}$$

that is, *longitudinal waves always travel faster than transverse waves.*

If we regard an elastic fluid as a special class of elastic material in which $h_0 = -p(\rho)$, $h_1 = h_2 = 0$, we have $U_{\mathrm{T}} = 0$, and U_{L} becomes the classical speed of sound:

$$U_{\mathrm{L}} = (dp/d\rho)^{1/2}$$

If the material is unstrained before the passage of the acceleration wave, formulas (2.13.8) and (2.13.9) should be evaluated at $\lambda_1 = \lambda_2 = \lambda_3 = 1$. Then these formulas reduce to

$$\rho U_{11}^2 = 4\left(\frac{\partial^2\Sigma}{\partial I^2} + 4\frac{\partial^2\Sigma}{\partial II^2} + \frac{\partial^2\Sigma}{\partial III^2} + 4\frac{\partial^2\Sigma}{\partial I\,\partial II} + 2\frac{\partial^2\Sigma}{\partial I\,\partial III} + 4\frac{\partial^2\Sigma}{\partial II\,\partial III}\right)\Bigg|_{I=3,\,II=3,\,III=1} \tag{2.13.24}$$

$$\rho U_{12}^2 = \rho U_{13}^2 = 2\left(\frac{\partial\Sigma}{\partial I} + \frac{\partial\Sigma}{\partial II}\right)\Bigg|_{I=3,\,II=3,\,III=1}$$

where we used the relation

$$\left(\frac{\partial\Sigma}{\partial I} + 2\frac{\partial\Sigma}{\partial II} + \frac{\partial\Sigma}{\partial III}\right)\Bigg|_{\overset{-1}{\mathbf{c}}=\mathbf{I}} = 0 \tag{2.13.25}$$

valid for the stress-free natural state. From the interrelations among the invariants of deformation and strain tensors, we write

$$\frac{\partial}{\partial I} + 2\frac{\partial}{\partial II} + \frac{\partial}{\partial III} = \frac{1}{2}\frac{\partial}{\partial I_E}, \qquad \frac{\partial}{\partial II} + \frac{\partial}{\partial III} = \frac{1}{4}\frac{\partial}{\partial II_E}$$

$$\frac{\partial}{\partial I} + \frac{\partial}{\partial II} = \frac{1}{2}\frac{\partial}{\partial I_E} - \frac{1}{4}\frac{\partial}{\partial II_E} \tag{2.13.26}$$

The use of this in (2.13.24) results in

$$\rho U_{11}^2 = \partial^2\Sigma/\partial {I_E}^2\Big|_{\mathbf{E}=\mathbf{0}}, \qquad \rho U_{12}^2 = \rho U_{13}^2 = -\tfrac{1}{2}\,\partial\Sigma/\partial II_E\Big|_{\mathbf{E}=\mathbf{0}} \tag{2.13.27}$$

Defining

$$\partial^2\Sigma/\partial {I_E}^2\Big|_{\mathbf{E}=\mathbf{0}} = \lambda + 2\mu, \qquad \partial\Sigma/\partial II_E\Big|_{\mathbf{E}=\mathbf{0}} = -2\mu \tag{2.13.28}$$

we get

$$U_{11}^2 = (\lambda + 2\mu)/\rho, \qquad U_{12}^2 = U_{13}^2 = \mu/\rho \tag{2.13.29}$$

If we assume that the strain energy admits a polynomial expansion in invariants, λ and μ may be identified as the conventional Lamé constants of the material.

We now consider elastic solids of degree two for which the third-order strain energy function is represented by expansion (1.16.14) and study the propagation of acceleration fronts. To this end we replace λ_α by $1 + E_\alpha$ in (2.13.8) and (2.13.9) and neglect the squares of the principal extension E_α:

$$\begin{aligned} \rho_0 U_{11}^2 \simeq 2(III)^{1/2}\Big[& h_1 + 2h_2 + \frac{1}{2}\frac{\partial}{\partial I_E}(h_0 + h_1 + h_2) + 4h_2 E_1 \\ & + \frac{1}{2}(E_2 + E_3)\frac{\partial}{\partial II_E}(h_0 + h_1 + h_2) + E_1 \frac{\partial}{\partial I_E}(h_1 + 2h_2) \\ & + E_1(2h_1 + 4h_2) + E_1 \frac{\partial}{\partial I_E}(h_0 + h_1 + h_2)\Big] \\ \rho_0 U_{12}^2 \simeq (III)^{1/2}[& h_1 + 2h_2 + 2(h_1 + 2h_2)E_1 + 2h_2(E_1 + E_2)] \end{aligned} \tag{2.13.30}$$

A similar expression may be derived for $\rho_0 U_{13}^2$. In deriving (2.13.30), we made use of (2.13.26). We now write

$$h_1 + 2h_2 \simeq \frac{1}{(III)^{1/2}}\left[\mu + \left(\lambda - \frac{m+n}{2}\right)I_E\right], \qquad h_2 \simeq \frac{1}{(III)^{1/2}}\left(\mu + \frac{n}{4}\right)$$

$$\frac{\partial}{\partial II_E}(h_0 + h_1 + h_2) \simeq \frac{1}{(III)^{1/2}}(m + n)$$

$$\frac{\partial}{\partial I_E}(h_1 + 2h_2) \simeq \frac{1}{(III)^{1/2}}\left(\lambda - \mu - \frac{m+n}{2}\right)$$

$$\frac{\partial}{\partial I_E}(h_0 + h_1 + h_2) \simeq \frac{1}{(III)^{1/2}}[\lambda + (6l + 2m - \lambda)I_E]$$

After some manipulations we arrive at the following expressions linear in principal extensions (cf. Truesdell [1964]):

$$\begin{aligned} \rho_0 U_{11}^2 &= \lambda + 2\mu + (\lambda + 6l + 2m)(E_1 + E_2 + E_3) + 2(2\lambda + 5\mu - m)E_1 \\ \rho_0 U_{12}^2 &= \mu + 2\mu E_1 + [\lambda - \tfrac{1}{2}(m + n)](E_1 + E_2 + E_3) + (2\mu + \tfrac{1}{2}n)(E_1 + E_2) \\ \rho_0 U_{13}^2 &= \mu + 2\mu E_1 + [\lambda - \tfrac{1}{2}(m + n)](E_1 + E_2 + E_3) + (2\mu + \tfrac{1}{2}n)(E_1 + E_3) \end{aligned} \tag{2.13.31}$$

We now consider an acceleration wave propagating in a prestressed linear elastic solid. From $(2.13.31)_{2,3}$ it follows that

$$U_{12} - U_{13} = (\mu/\rho_0)^{1/2}(\tilde{E}_2 - \tilde{E}_3) \tag{2.13.32}$$

where E_α are replaced by $\tilde{E}_\alpha$, which are principal extensions of the infinitesimal strain tensor. Since

$$t_2 - t_3 = 2\mu(\tilde{E}_2 - \tilde{E}_3)$$

in linear elastic solids, we have at last

$$U_{12} - U_{13} = (1/2\mu)(\mu/\rho_0)^{1/2}(t_2 - t_3) \tag{2.13.33}$$

which implies that the difference in the speeds of transversely polarized waves is proportional to the difference between principal stresses. The sensitivity of these wave speeds to the existing stress field provides a method in experimental stress analysis which is similar to photoelastic methods. We shall return to this point later in Chapter IV.

2.14 ACCELERATION WAVES IN INCOMPRESSIBLE SOLIDS[1]

The propagation of acceleration waves in incompressible elastic solids may be obtained by specializing the analyses of Sections 2.12 and 2.13. We first notice from (2.12.8) that the constraint of incompressibility removes the possibility of longitudinal waves, since now

$$\mathbf{a} \cdot \mathbf{n} = 0 \tag{2.14.1}$$

Therefore, all acceleration waves propagating in incompressible materials will necessarily be transverse.

The constitutive relations are of the form

$$T_{Kk} = -pX_{K,k} + (\partial\Sigma/\partial x_{k,K}) \tag{2.14.2}$$

where p is the unknown pressure function and the deformation gradients are subject to the constraint $\det(x_{k,K}) = 1$. We infer from (2.14.2) that the pressure is continuous across the singular surface, namely $[p] = 0$. Therefore, in view of (2.4.5),

$$[\nabla p] = A\mathbf{n} \tag{2.14.3}$$

where

$$A = [p_{,i}]n_i = [\partial p/\partial n]$$

[1] See Ericksen [1953].

The differential equations of motion read

$$-p_{,k} + T_{Kk,K} + \rho f_k = \rho \ddot{x}_k$$

If we remember that $(X_{K,k})_{,K} = 0$ [Eq. (1.4.6)] and follow the procedure explored in Section 2.12, we obtain on the surface of discontinuity

$$-An_k + (Q_{kl} - \rho U^2 \delta_{kl})a_l = 0 \tag{2.14.4}$$

where the amplitude vector **a** was defined in $(2.12.6)_3$ and the tensor Q_{kl} is given by

$$Q_{kl} = A_{KkLl} x_{m,K} x_{n,L} n_m n_n$$

Multiplying both sides of (2.14.4) by n_k and recalling (2.14.1), we determine A:

$$A = Q_{kl} n_k a_l \tag{2.14.5}$$

Thus (2.14.4) can be written as

$$\left(\overset{*}{Q}_{kl} - \rho U^2 \delta_{kl}\right)a_l = 0 \tag{2.14.6}$$

where

$$\overset{*}{Q}_{kl} = Q_{kl} - Q_{ml} n_k n_m \tag{2.14.7}$$

The local speed of propagation is found from

$$\det\left(\overset{*}{Q}_{kl} - \rho U^2 \delta_{kl}\right) = 0$$

or as roots of the following bicubic equation:

$$-(\rho U^2)^3 + I_{\overset{*}{Q}}(\rho U^2)^2 - II_{\overset{*}{Q}}\,\rho U^2 + III_{\overset{*}{Q}} = 0 \tag{2.14.8}$$

However, it is readily seen from (2.14.7) that

$$\overset{*}{Q}_{kl} n_k = 0 \tag{2.14.9}$$

which implies that the matrix $\overset{*}{\mathbf{Q}}$ is singular, that is,

$$III_{\overset{*}{Q}} = \det \overset{*}{\mathbf{Q}} = 0$$

This property of $\overset{*}{\mathbf{Q}}$ reduces (2.14.8) to a biquadratic equation after canceling the root $\rho U^2 = 0$ which has no physical significance:

$$(\rho U^2)^2 - I_{\overset{*}{Q}}\,\rho U^2 + II_{\overset{*}{Q}} = 0$$

The two roots of this equation give the local speeds of propagation:

(2.14 10)
$$\rho U^2 = \tfrac{1}{2}\left[I_{\overset{*}{Q}} \mp \left(I_{\overset{*}{Q}}^{\,2} - 4II_{\overset{*}{Q}}\right)^{1/2}\right]$$

Corresponding to these, directions of discontinuities are obtained from (2.14.6). Relation (2.14.9) excludes the possibility of finding a component of $\mathbf{a}$ in the direction of propagation $\mathbf{n}$ as it should be. If $\mathbf{a}$ is a proper vector of $\overset{*}{\mathbf{Q}}$, then the corresponding wave speed is obtained to be

$$\rho U^2 = \overset{*}{Q}_{kl}\, v_k\, v_l$$

where $\mathbf{v}$ is a unit vector in the direction of the transverse discontinuity.

For isotropic incompressible elastic solids the foregoing analysis may be pursued further. The elasticities corresponding to the solids of this type are read from (2.13.2) by omitting terms in h_0 and terms containing derivatives with respect to the third invariant III. With these provisions (2.13.4) reads

$$\begin{aligned}
\mathbf{Q}(\mathbf{n}) = {} & \left(h_1 \mathbf{n}\cdot \overset{-1}{\mathbf{c}}\mathbf{n} + h_2\, \mathbf{n}\cdot \overset{-2}{\mathbf{c}}\mathbf{n}\right)\mathbf{I} + h_2\left(\mathbf{n}\cdot \overset{-1}{\mathbf{c}}\mathbf{n}\right)\overset{-1}{\mathbf{c}} + h_2\, \overset{-1}{\mathbf{c}}\mathbf{n}\otimes \overset{-1}{\mathbf{c}}\mathbf{n} \\
& + 2\left(\frac{\partial h_1}{\partial I} + I\,\frac{\partial h_2}{\partial II}\right)\overset{-1}{\mathbf{c}}\mathbf{n}\otimes \overset{-1}{\mathbf{c}}\mathbf{n} - 2\,\frac{\partial h_1}{\partial II}\left(\overset{-1}{\mathbf{c}}\mathbf{n}\otimes \overset{-2}{\mathbf{c}}\mathbf{n} + \overset{-2}{\mathbf{c}}\mathbf{n}\otimes \overset{-1}{\mathbf{c}}\mathbf{n}\right) \\
& - 2\,\frac{\partial h_2}{\partial II}\,\overset{-2}{\mathbf{c}}\mathbf{n}\otimes \overset{-2}{\mathbf{c}}\mathbf{n} + h_1\, \overset{-1}{\mathbf{c}}\mathbf{n}\otimes \mathbf{n} + h_2\, \overset{-2}{\mathbf{c}}\mathbf{n}\otimes \mathbf{n}
\end{aligned}$$

where we employed the compatibility condition

$$\frac{\partial h_1}{\partial II} = -\left(\frac{\partial h_2}{\partial I} + I\,\frac{\partial h_2}{\partial II}\right)$$

From (2.14.7) it then follows that

(2.14.11)
$$\begin{aligned}
\overset{*}{\mathbf{Q}}(\mathbf{n}) = {} & \left(h_1 \mathbf{n}\cdot \overset{-1}{\mathbf{c}}\mathbf{n} + h_2\, \mathbf{n}\cdot \overset{-2}{\mathbf{c}}\mathbf{n}\right)\mathbf{I} + h_2\left(\mathbf{n}\cdot \overset{-1}{\mathbf{c}}\mathbf{n}\right)\overset{-1}{\mathbf{c}} \\
& + 2\left(\frac{h_2}{2} + \frac{\partial h_1}{\partial I} + I\,\frac{\partial h_1}{\partial II}\right)\overset{-1}{\mathbf{c}}\mathbf{n}\otimes \overset{-1}{\mathbf{c}}\mathbf{n} - 2\,\frac{\partial h_1}{\partial II}\left(\overset{-1}{\mathbf{c}}\mathbf{n}\otimes \overset{-2}{\mathbf{c}}\mathbf{n} + \overset{-2}{\mathbf{c}}\mathbf{n}\otimes \overset{-1}{\mathbf{c}}\mathbf{n}\right) \\
& - 2\,\frac{\partial h_2}{\partial II}\,\overset{-2}{\mathbf{c}}\mathbf{n}\otimes \overset{-2}{\mathbf{c}}\mathbf{n} - 2\left(h_1 \mathbf{n}\cdot \overset{-1}{\mathbf{c}}\mathbf{n} + h_2\, \mathbf{n}\cdot \overset{-2}{\mathbf{c}}\mathbf{n}\right)\mathbf{n}\otimes \mathbf{n} \\
& + h_1\, \overset{-1}{\mathbf{c}}\mathbf{n}\otimes \mathbf{n} + h_2\, \overset{-2}{\mathbf{c}}\mathbf{n}\otimes \mathbf{n} - 2\left[\left(h_2 + \frac{\partial h_1}{\partial I} + I\,\frac{\partial h_1}{\partial II}\right)\left(\mathbf{n}\cdot \overset{-1}{\mathbf{c}}\mathbf{n}\right)\right. \\
& \left. - \frac{\partial h_1}{\partial II}\,\mathbf{n}\cdot \overset{-2}{\mathbf{c}}\mathbf{n}\right]\mathbf{n}\otimes \overset{-1}{\mathbf{c}}\mathbf{n} + 2\left(\frac{\partial h_1}{\partial II}\,\mathbf{n}\cdot \overset{-1}{\mathbf{c}}\mathbf{n} + \frac{\partial h_2}{\partial II}\,\mathbf{n}\cdot \overset{-2}{\mathbf{c}}\mathbf{n}\right)\mathbf{n}\otimes \overset{-2}{\mathbf{c}}\mathbf{n}
\end{aligned}$$

To make the physical picture clear we select a local rectangular system attached to the wave surface in such a way that x_3-axis is directed along the normal **n**. In this local coordinate system the expression of $\overset{*}{\mathbf{Q}}$ becomes

(2.14.12)

$$\begin{aligned}
\overset{*}{Q}_{kl} = {} & \left(h_1 \overset{-1}{c}_{33} + h_2 \overset{-2}{c}_{33}\right)\delta_{kl} + h_2 \overset{-1}{c}_{33}\,\overset{-1}{c}_{kl} + 2\left(\frac{h_2}{2} + \frac{\partial h_1}{\partial I} + I\,\frac{\partial h_1}{\partial II}\right)\overset{-1}{c}_{k3}\,\overset{-1}{c}_{l3} \\
& - 2\,\frac{\partial h_1}{\partial II}\left(\overset{-1}{c}_{k3}\,\overset{-2}{c}_{l3} + \overset{-2}{c}_{k3}\,\overset{-1}{c}_{l3}\right) - 2\,\frac{\partial h_2}{\partial II}\,\overset{-2}{c}_{k3}\,\overset{-2}{c}_{l3} \\
& - 2\left(h_1 \overset{-1}{c}_{33} + h_2 \overset{-2}{c}_{33}\right)\delta_{k3}\,\delta_{l3} + h_1 \overset{-1}{c}_{k3}\,\delta_{l3} + h_2 \overset{-2}{c}_{k3}\;\delta_{l3} \\
& - 2\left[\left(h_2 + \frac{\partial h_1}{\partial I} + I\,\frac{\partial h_1}{\partial II}\right)\overset{-1}{c}_{33} - \frac{\partial h_1}{\partial II}\,\overset{-2}{c}_{33}\right]\delta_{k3}\,\overset{-1}{c}_{l3} \\
& + 2\left(\frac{\partial h_1}{\partial II}\,\overset{-1}{c}_{33} + \frac{\partial h_2}{\partial II}\,\overset{-2}{c}_{33}\right)\delta_{k3}\,\overset{-2}{c}_{l3}
\end{aligned}$$

In this coordinate system, clearly, $\overset{*}{Q}_{3l} = 0$. Also since $\mathbf{a} = \{a_1, a_2, 0\}$, (2.14.6) reduces to

$$\left(\overset{*}{Q}_{\alpha\beta} - \rho U^2\,\delta_{\alpha\beta}\right)a_\beta = 0$$

where Greek indices take on the values 1 and 2 only and $\overset{*}{Q}_{\alpha\beta}$ is given by

(2.14.13)

$$\begin{aligned}
\overset{*}{Q}_{\alpha\beta} = {} & \left(h_1 \overset{-1}{c}_{33} + h_2 \overset{-2}{c}_{33}\right)\delta_{\alpha\beta} + h_2 \overset{-1}{c}_{33}\,\overset{-1}{c}_{\alpha\beta} + 2\left(\frac{h_2}{2} + \frac{\partial h_1}{\partial I} + I\,\frac{\partial h_1}{\partial II}\right)\overset{-1}{c}_{\alpha 3}\,\overset{-1}{c}_{\beta 3} \\
& - 2\,\frac{\partial h_1}{\partial II}\left(\overset{-1}{c}_{\alpha 3}\,\overset{-2}{c}_{\beta 3} + \overset{-2}{c}_{\alpha 3}\,\overset{-1}{c}_{\beta 3}\right) - 2\,\frac{\partial h_2}{\partial II}\,\overset{-2}{c}_{\alpha 3}\,\overset{-2}{c}_{\beta 3}
\end{aligned}$$

We see that $\overset{*}{Q}_{\alpha\beta}$ is a symmetric 2×2 matrix. ρU^2 is therefore determined from the secular equation

$$\begin{vmatrix} \overset{*}{Q}_{11} - \rho U^2 & \overset{*}{Q}_{12} \\ \overset{*}{Q}_{12} & \overset{*}{Q}_{22} - \rho U^2 \end{vmatrix} = 0$$

whose roots are all real. The corresponding amplitudes may be found from

$$\left(\overset{*}{Q}_{11} - \rho U^2\right)a_1 + \overset{*}{Q}_{12}\,a_2 = 0$$

Just as in the case of compressible materials, explicit relations can be obtained for principal waves. If **n** is so chosen as to coincide with a principal direction of the deformation $\mathbf{n}_1$, then (2.14.11) is simplified to

$$(2.14.14) \qquad (1/\lambda_1{}^2)\overset{*}{\mathbf{Q}}(\mathbf{n}_1) = (h_1 + \lambda_1{}^2 h_2)\mathbf{I} - (h_1 + 2\lambda_1{}^2 h_2)\mathbf{n}_1 \otimes \mathbf{n}_1 + h_2 \overset{-1}{\mathbf{c}}$$

Since $\overset{*}{\mathbf{Q}}\mathbf{n}_1 = \mathbf{0}$, with $\mathbf{n}_1$ corresponding to a zero eigenvalue, it does not give rise to a propagating discontinuity. But one can verify that the other two principal directions $\mathbf{n}_2$ and $\mathbf{n}_3$ are proper vectors of $\overset{*}{\mathbf{Q}}$ with the associated speeds of propagation given by

(2.14.15)

$$\rho U_{12}^2 = \lambda_1{}^2[h_1 + (\lambda_1{}^2 + \lambda_2{}^2)h_2], \qquad \rho U_{13}^2 = \lambda_1{}^2[h_1 + (\lambda_1{}^2 + \lambda_3{}^2)h_2]$$

subject to the constraint $\lambda_1\lambda_2\lambda_3 = 1$. If we compare (2.14.15) with (2.13.9), we see that they have the same form.

The foregoing analysis shows that there may exist six transverse principal acceleration waves in an isotropic incompressible elastic solid—two waves associated with every principal direction.

2.15 GROWTH OF THE ACCELERATION WAVES

In previous sections we learned how to determine the speed of propagation and the amplitude of an acceleration wave in terms of the existing deformation field ahead of the wave front. The knowledge of the speed of propagation U_N as a function of the surface coordinates of the singular surface and time, together with the initial configuration of the wave surface at a given time, determines the configuration of the wave front at subsequent times through the differential relations (2.2.36) and (2.2.43) or (2.2.44), which may be written as

(2.15.1)

$$\delta X_K/\delta t = U_N N_K, \qquad \delta N_K/\delta t = -A^{\alpha\beta} X_{K\alpha} U_{N,\beta} = (N_K N_L - \delta_{KL})U_{N,L}$$

The integral of these equations subject to initial conditions gives the subsequent configurations of the initial wave surface. The problem that remains is how to determine the change in the amplitude of an acceleration wave with a given initial amplitude as it travels in the material[1]

[1] This problem has been studied by many investigators for various special material constitutions. Except for Varley's work [1965a], all previous problems treated are concerned with special situations. Here we follow a general formulation for elastic materials given by Şuhubi [1970] and study the growth of plane waves in homogeneous deformation (Green [1964]) and of waves of arbitrary shape propagating into an undisturbed and uniformly dilated medium (Chen [1968a,b]) as special cases. For further references see Şuhubi [1970].

(*Note added in proof:* The method of bicharacteristics was recently explored by Wright [1973]).

In order to derive the growth equation, we differentiate the equations of motion (2.12.9) with respect to time and evaluate the jumps of the resulting equations on the material wave front $\Sigma(t)$. Thus

$$A_{KkLlMm}[\dot{x}_{m,M}\, x_{l,KL}] + A_{KkLl}[\dot{x}_{l,KL}] = \rho_0[\dddot{x}_k] \tag{2.15.2}$$

where the coefficients of jumps were defined by (2.11.3). The jump in the third-order derivatives may be evaluated from (2.5.8) by recalling that $x_{k,K}$ and $\dot{x}_k$ are continuous across the wave surface $\Sigma(t)$. Hence writing $\phi = \dot{x}_k$ and $B = -U_N S_k$ in (2.5.8) and (2.4.14) we obtain

(2.15.3)

$$[\dddot{x}_k] = U_N{}^2 C_k + 2U_N{}^2\,(\delta S_k/\delta t) + 3U_N S_k\,(\delta U_N/\delta t)$$

$$[\dot{x}_{k,KL}] = C_k N_K N_L - A^{\alpha\beta}(N_K X_{L\beta} + N_L X_{K\beta})(U_N S_k)_{;\alpha} + B^{\alpha\beta} X_{K\alpha} X_{L\beta} U_N S_k$$

where

$$C_k = [\dot{x}_{k,KL}] N_K N_L$$

is an unknown vector, $B_{\alpha\beta}$ is defined by (2.6.15), and $B^{\alpha\beta} = A^{\alpha\gamma} A^{\beta\delta} B_{\gamma\delta}$. Now if we use the identity

$$[\dot{x}_{m,M}\, x_{l,KL}] = \overset{+}{\dot{x}}_{m,M}[x_{l,KL}] + \overset{+}{x}_{l,KL}[\dot{x}_{m,M}] - [\dot{x}_{m,M}][x_{l,KL}]$$

(2.12.6), and (2.15.3), Eqs. (2.15.2) can be transformed into the form:

$$\begin{aligned} 2\rho_0 U_N{}^2 \frac{\delta S_k}{\delta t} &+ 3\rho_0 U_N S_k \frac{\delta U_N}{\delta t} \\ &= (\mathcal{Q}_{kl} - \rho_0 U_N{}^2 \delta_{kl}) C_l + U_N \mathcal{Q}_{klm} S_l S_m \\ &\quad - A_{KkLl}[A^{\alpha\beta}(N_K X_{L\beta} + N_L X_{K\beta})(U_N S_l)_{;\alpha} + U_N B^{\alpha\beta} X_{K\alpha} X_{L\beta} S_l] \\ &\quad + A_{KkLlMm}(\overset{+}{\dot{x}}_{m,M} N_K N_L S_l - U_N \overset{+}{x}_{l,KL} N_M S_m) \end{aligned}$$

where $\mathcal{Q}_{kl}$ was defined in (2.12.12) and

$$\mathcal{Q}_{klm} = A_{KkLlMm} N_K N_L N_M \tag{2.15.4}$$

The unknown vector $\mathbf{C}$ may be eliminated by multiplying the above expression by S_k, recalling (2.12.11) and the symmetry of $\mathcal{Q}_{kl}$. Thus

$$\begin{aligned} \rho_0 U_N{}^2 \left(\frac{\delta S^2}{\delta t} + 3S^2 \frac{1}{U_N}\frac{\delta U_N}{\delta t}\right) &\\ = U_N \mathcal{Q}_{klm} S_k S_l S_m &- A_{KkLl} A^{\alpha\beta}(N_K X_{L\beta} + N_L X_{K\beta})(U_N S_l)_{;\alpha} S_k \\ &+ U_N A_{KkLl} B^{\alpha\beta} X_{K\alpha} X_{L\beta} S_k S_l \\ &+ A_{KkLlMm}(\overset{+}{\dot{x}}_{m,M} N_K N_L S_k S_l - U_N \overset{+}{x}_{l,KL} N_M S_k S_m) \end{aligned}$$

where $S^2 = S_k S_k$. Now define a unit vector $\mathbf{v}$ in the direction of the acceleration jump such that $\mathbf{v} \cdot \mathbf{n} \geq 0$. We may then write $\mathbf{S} = S\mathbf{v}$, where $|S| \equiv |\mathbf{S}|$ is the magnitude of $\mathbf{S}$. If we note that

$$\frac{\delta S^2}{\delta t} + \frac{3S^2}{U_N} \frac{\delta U_N}{\delta t} = S^2 \frac{\delta}{\delta t} \log(U_N{}^3 S^2)$$

and [cf. (2.2.37)]

$$\frac{\delta}{\delta t} = U_N \frac{d}{dN} = U_N \left(N_K \frac{\partial}{\partial X_K} + \frac{1}{U_N} \frac{\partial}{\partial t} \right)$$

where the total derivative in the normal direction to the surface $\Sigma(t)$ is denoted by d/dN, we finally arrive at the growth equation for the amplitude vector:

(2.15.5)

$$\begin{aligned} \rho_0 U_N{}^2 \frac{d}{dN} \log(U_N{}^3 S^2) &= \mathcal{Q}_{klm} v_k v_l v_m S - A_{KkLl} A^{\alpha\beta}(N_K X_{L\beta} + N_L X_{K\beta}) \\ &\quad \cdot [\log(U_N S)_{;\alpha} v_k v_l + v_k v_{l;\alpha}] + A_{KkLl} B^{\alpha\beta} X_{K\alpha} X_{L\beta} v_k v_l \\ &\quad + A_{KkLlMm} \left(\frac{1}{U_N} \dot{\overset{+}{x}}_{m,M} N_K N_L v_k v_l - \overset{+}{x}_{l,KL} N_M v_k v_m \right) \end{aligned}$$

The spatial form of the growth equation may then be obtained from (2.15.5) by using (2.6.15), (2.6.18), (2.12.7), the relation $\rho = j\rho_0$, and

$$dn/dN = U/U_N = C_n$$

where dn is the total differential along the normal direction $\mathbf{n}$ to the wave front $\sigma(t)$ traveling with speed U relative to the material, or

$$\frac{d}{dn} = n_k \frac{\partial}{\partial x_k} + \frac{1}{U} \frac{\partial}{\partial t}$$

Thus we get

(2.15.6)

$$\begin{aligned} \rho U^2 \frac{d}{dn} \log(C_n U^3 a^2) &= Q_{klm} v_k v_l v_m a - j^{-1} A_{KkLl}(x_{m,K} X_L{}^\alpha + x_{m,L} X_K{}^\alpha) \\ &\quad \cdot [(\log C_n Ua)_{;\alpha} v_k v_l + v_k v_{l;\alpha}] n_m + j^{-1} A_{KkLl} \\ &\quad \cdot [b_{\alpha\beta} - x_{p,MN} X_{M,m} X_{N,n} x_{m\alpha} x_{n\beta} n_p] X_K{}^\alpha X_L{}^\beta v_k v_l \\ &\quad + j^{-1} A_{KkLlMm} \left[\frac{1}{U} \dot{\overset{+}{x}}_{m,M} x_{p,K} x_{q,L} n_p n_q v_k v_l \right. \\ &\quad \left. - \overset{+}{x}_{l,KL} x_{p,M} n_p v_k v_m \right] \end{aligned}$$

where Q_{kl} is defined by (2.12.15),

$$Q_{klm} = j^{-1} A_{KkLlMm} x_{p,K} x_{q,L} x_{r,M} n_p n_q n_r \tag{2.15.7}$$

and a is given by the relation $\mathbf{a} = a\mathbf{v}$, $|a|$ being the magnitude of the amplitude. We also used the notations

$$X_K{}^\alpha = A^{\alpha\beta} X_{K\beta}, \qquad A^{\alpha\beta} = (\det a_{\lambda\mu}/\det A_{\nu\omega}) e^{\alpha\gamma} e^{\beta\delta} c_{mn} x_{m\gamma} x_{n\delta}$$

If the material ahead of the wave front is at rest, we deduce from (2.12.11) or (2.12.14) that U_N (or U) and directions $\mathbf{v}$ become functions of the position of the wave only, independent of time. If we also reasonably assume that S (or a) is a function of the position of wave only, then the derivatives d/dN (or d/dn) in (2.15.5) [or (2.15.6)] have a meaning of the ordinary derivative along $\mathbf{N}$ (or $\mathbf{n}$) and the term $\overset{+}{\dot{x}}_{m,M}$ disappears.

If the material is isotropic in the undistorted state, A_{KkLl} are given by (2.13.1) and A_{KkLlMm} follow from the same expression:

$$\begin{aligned} A_{KkLlMm} = 2j[&(X_{[K,|n|} X_{L],l} X_{M,m} - X_{[K,|m|} X_{L],l} X_{M,n} \\ &- X_{[K,|n|} X_{L],m} X_{M,l}) t_{nk} + 2X_{[K,|n|} X_{L],l} x_{p,M} F_{knmp} \\ &+ 2X_{[K,|n|} X_{M],m} x_{p,L} F_{knlp} + \delta_{LM} X_{K,n} F_{knlm} \\ &+ 2X_{K,n} x_{p,L} x_{r,M} F_{knlpmr}] \end{aligned} \tag{2.15.8}$$

where

(2.15.9)

$$\begin{aligned} F_{klmnpq} &= \partial F_{klmn}/\partial \overset{-1}{c}_{pq} = \partial t_{kl}/\partial \overset{-1}{c}_{mn}\, \partial \overset{-1}{c}_{pq} \\ &= (h_2/4)[(\delta_{kp}\delta_{qr} + \delta_{kq}\delta_{pr})(\delta_{rm}\delta_{ln} + \delta_{rn}\delta_{lm}) \\ &\quad + (\delta_{lq}\delta_{pr} + \delta_{lp}\delta_{qr})(\delta_{nr}\delta_{km} + \delta_{mr}\delta_{kn})] + \delta_{kl}\Phi^{(0)}_{mnpq} + \overset{-1}{c}_{kl}\Phi^{(1)}_{mnpq} \\ &\quad + \overset{-2}{c}_{kl}\Phi^{(2)}_{mnpq} + \tfrac{1}{2}(\delta_{kp}\delta_{lq} + \delta_{kq}\delta_{lp})\Psi^{(1)}_{mn} + \tfrac{1}{2}(\delta_{km}\delta_{ln} + \delta_{kn}\delta_{lm})\Psi^{(1)}_{pq} \\ &\quad + \tfrac{1}{2}\left(\delta_{lq}\overset{-1}{c}_{kp} + \delta_{lp}\overset{-1}{c}_{kq} + \delta_{kq}\overset{-1}{c}_{lp} + \delta_{kp}\overset{-1}{c}_{lq}\right)\Psi^{(2)}_{mn} \\ &\quad + \tfrac{1}{2}\left(\delta_{ln}\overset{-1}{c}_{km} + \delta_{lm}\overset{-1}{c}_{kn} + \delta_{kn}\overset{-1}{c}_{lm} + \delta_{km}\overset{-1}{c}_{ln}\right)\Psi^{(2)}_{pq} \end{aligned}$$

with the definitions

$$\Psi_{mn}^{(a)} = \frac{\partial h_a}{\partial I}\,\delta_{mn} + \frac{\partial h_a}{\partial II}\left(\delta_{mn} I - \overset{-1}{c}_{mn}\right) + III\,\frac{\partial h_a}{\partial III}\,c_{mn}$$

$$\Phi_{klmn}^{(a)} = \frac{\partial h_a}{\partial II}\left[\delta_{kl}\,\delta_{mn} - \frac{1}{2}(\delta_{km}\,\delta_{ln} + \delta_{kn}\,\delta_{lm})\right] + \frac{\partial h_a}{\partial III}\Big[I\,\delta_{kl}\,\delta_{mn} - \delta_{kl}\,\overset{-1}{c}_{mn}$$

$$- \delta_{mn}\,\overset{-1}{c}_{kl} - \frac{I}{2}(\delta_{km}\,\delta_{ln} + \delta_{kn}\,\delta_{lm}) + \frac{1}{2}\Big(\delta_{km}\,\overset{-1}{c}_{ln} + \delta_{kn}\,\overset{-1}{c}_{lm} + \delta_{ln}\,\overset{-1}{c}_{km}$$

$$+ \delta_{lm}\,\overset{-1}{c}_{kn}\Big)\Big] + \Big[\frac{\partial^2 h_a}{\partial I^2}\,\delta_{mn} + \frac{\partial^2 h_a}{\partial I\,\partial II}\left(I\,\delta_{mn} - \overset{-1}{c}_{mn}\right)$$

$$+ III\,\frac{\partial^2 h_a}{\partial I\,\partial III}\,c_{mn}\Big]\,\delta_{kl} + \Big[\frac{\partial^2 h_a}{\partial I\,\partial II}\,\delta_{mn} + \frac{\partial^2 h_a}{\partial II^2}\left(I\,\delta_{mn} - \overset{-1}{c}_{mn}\right)$$

$$+ III\,\frac{\partial^2 h_a}{\partial II\,\partial III}\,c_{mn}\Big]\left(I\,\delta_{kl} - \overset{-1}{c}_{kl}\right) + \Big[\frac{\partial^2 h_a}{\partial I\,\partial III}\,\delta_{mn}$$

$$+ \frac{\partial^2 h_a}{\partial II\,\partial III}\left(I\,\delta_{mn} - \overset{-1}{c}_{mn}\right) + III\,\frac{\partial^2 h_a}{\partial III^2}\,c_{mn}\Big]\,III c_{kl}$$

$(a = 0, 1, 2)$. The use of (2.13.1) and (2.15.8) in the growth equation (2.15.6), after some manipulations, yields the following result:

(2.15.10)

$$\rho U^2 \frac{d}{dn}\log(C_n U^3 a^2)$$

$$= 2\left(\overset{-1}{c}_{qr}\,n_q\,n_r\,F_{kplm} + 2\,\overset{-1}{c}_{qt}\,\overset{-1}{c}_{ru}\,F_{kplqmr}\,n_t\,n_u\right)n_p\,v_k\,v_l\,v_m\,a$$

$$- 2A^{\alpha\beta}\left(\delta_{mr}\,x_{p\beta} + \overset{-1}{c}_{pm}\,c_{rq}\,x_{q\beta}\right)F_{krlp}[(\log C_n U a)_{;\alpha}\,v_k\,v_l + v_k\,v_{l;\alpha}]n_m$$

$$+ 2A^{\alpha\gamma}A^{\beta\delta}c_{mn}\,x_{p\delta}\,x_{m\gamma}\,F_{knlp}(b_{\alpha\beta} + X_{Q,qr}\,x_{s,Q}\,x_{q\alpha}\,x_{r\beta}\,n_s)v_k\,v_l$$

$$+ 2\Big(2\,\overset{-1}{c}_{pr}\,\overset{+}{X}_{K,r[m}\,\delta_{n]q}\,F_{knlp} + \overset{-1}{c}_{qr}\,\overset{+}{X}_{K,nr}\,F_{knlm}$$

$$+ 2\,\overset{-1}{c}_{ps}\,\overset{-1}{c}_{qr}\,\overset{+}{X}_{K,ns}\,F_{knlpmr}\Big)x_{l,K}\,n_q\,v_k\,v_m$$

$$+ \frac{2}{U}\Big(2\,\overset{-1}{c}_{pq}\,n_{[n}\,\overset{+}{\dot{x}}_{|m|,m]}\,F_{knlp}\,n_q + \overset{-1}{c}_{pq}\,F_{knlm}\,n_n\,n_p\,\overset{+}{\dot{x}}_{m,q}$$

$$+ 2\,\overset{-1}{c}_{pq}\,\overset{-1}{c}_{rs}\,F_{knlpmr}\,n_n\,n_q\,\overset{+}{\dot{x}}_{m,s})v_k\,v_l$$

The existing motion ahead of the acceleration wave determines U and $\mathbf{v}$ (or U_N and $\mathbf{v}$) through (2.12.14) [or (2.12.11)] at every instant. Therefore, the growth equations (2.15.5), (2.15.6), or (2.15.10) give the variation of the amplitude of the wave as it travels in the material. In general, the growth equation is quite complicated for the deduction of tangible results except in some special cases which are now investigated in detail.

A. Plane Acceleration Waves[1]

As an illustrative example to the application of the growth equation (2.15.10), we consider the propagation of plane acceleration waves in a homogeneously deformed, isotropic elastic solid at rest. The *constant* principal stretches of the deformation are denoted by λ_1, λ_2, and λ_3, and coordinate axes are chosen so as to coincide with the *fixed* principal directions

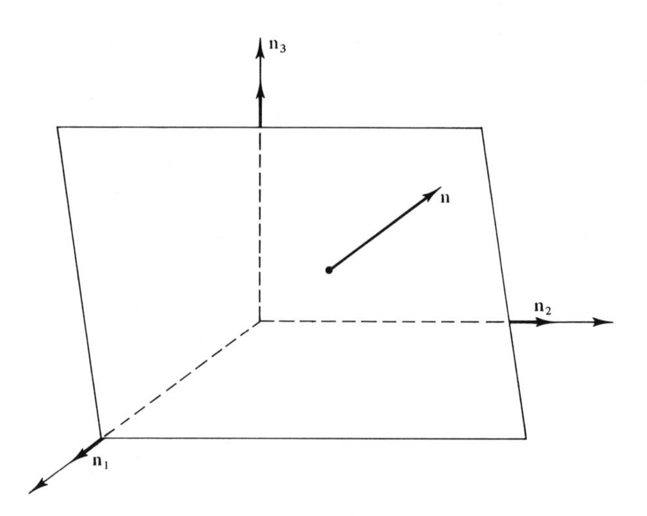

Fig. 2.15.1 Principal directions of the deformation.

of the deformation (Fig. 2.15.1). The *constant* speed of propagation is determined from (2.12.17) and the corresponding direction of amplitudes from (2.12.14). Here we consider only the uniform discontinuities; that is, we suppose that the magnitude of the discontinuity a is independent of the position in the wave plane. Because of the assumptions made (plane wave,

[1] Green [1964] investigated this problem somewhat differently from the present treatment.

homogeneous deformation, material initially at rest, discontinuities independent of the position in the wave front) only the first term in (2.15.10) survives, yielding the result

$$(\rho U^2/a^2)\, da^2/dn = 2Ka \tag{2.15.11}$$

where

$$K = \left(\overset{-1}{c}_{tu} F_{kplm} + 2\, \overset{-1}{c}_{qt}\, \overset{-1}{c}_{ru} F_{kplqmr}\right) n_t n_u n_p v_k v_l v_m \tag{2.15.12}$$

may be presumed to be a known constant calculated from the initial state of deformation, the material properties, the direction of propagation, and the direction of the acceleration discontinuity. The solution of the equation

$$da/dn = (K/\rho U^2) a^2 \tag{2.15.13}$$

derived from (2.15.11), subjected to the initial condition $a = a_0$ at the initial configuration $n = 0$ (or $t = 0$, since $n = Ut$), is

$$a(n) = a_0[1 - (Ka_0/\rho U^2) n]^{-1} \tag{2.15.14}$$

If we replace n by Ut in (2.15.14), we get the alternate form

$$a(t) = a_0[1 - (Ka_0/\rho U) t]^{-1}$$

The growth or decay of the acceleration discontinuity is now clearly administered by the sign of the constant K. If $K > 0$, the amplitude of compressive waves ($a_0 < 0$) will steadily decay to zero as $n \to \infty$ (or $t \to \infty$), whereas the amplitude of expansive waves ($a_0 > 0$) will grow steadily, and after a finite time lapse

$$t = \rho U/Ka_0, \qquad U > 0 \tag{2.15.15}$$

it will become infinite. If $K < 0$, the behaviors of expansive and compressive waves are reversed. If $K = 0$, then the amplitude of the wave remains unchanged as it travels in the material.

The evaluation of the constant K becomes simpler for principal waves. Assume that the plane acceleration wave propagates along the principal direction $\mathbf{n}_1$. Then $\mathbf{n} \equiv \mathbf{n}_1$ and $\mathbf{v}$ is either $\mathbf{n}_1$ (longitudinal wave) or $\mathbf{n}_2$ or $\mathbf{n}_3$ (transverse waves). Hence

$$K = \lambda_1{}^2 (F_{kplm} + 2\lambda_1{}^2 F_{kplqmr} n_q n_r) n_p v_k v_l v_m \tag{2.15.16}$$

For longitudinal waves $\mathbf{v} = \mathbf{n}_1$, and we have

$$K_{11} = \lambda_1{}^2\left(h_1 + 2\lambda_1{}^2 h_2 + \sum_{a=0}^{2} \lambda_1^{2a}\, Dh_a\right) + 2\lambda_1{}^4\left(2h_2 + 2\,Dh_1 + 4\lambda_1{}^2\,Dh_2 + \sum_{a=0}^{2} \lambda_1^{2a}\, D^2 h_a\right) \tag{2.15.17}$$

where the differential operator D is defined by

$$D = (\partial/\partial I) + (\lambda_2{}^2 + \lambda_3{}^2)(\partial/\partial II) + \lambda_2{}^2\lambda_3{}^2(\partial/\partial III)$$

Care should be taken in applying this operator. The coefficients should be considered as pure numbers although the invariants on which differentiations would be applied are expressed in terms of them. Comparing (2.15.17) with (2.13.8), we realize that the first parenthesis is just $\rho U_{11}^2/2$. Therefore, we can write

$$\frac{K_{11}}{\rho U_{11}^2} = \frac{1}{2} + \frac{\lambda_1{}^2(2h_2 + 2\,Dh_1 + 4\lambda_1{}^2\,Dh_2 + \sum_{a=0}^{2}\lambda_1^{2a}\,D^2h_a)}{h_1 + 2\lambda_1{}^2h_2 + \sum_{a=0}^{2}\lambda_1^{2a}\,Dh_a} \tag{2.15.18}$$

and the growth or decay of a longitudinal wave will be governed by the sign of the right-hand side of expression (2.15.18).

As in Section 2.13, K_{11} can be put in a more elegant form. One can show that

(2.15.19)

$$K_{11} = \frac{1}{2}\lambda_1{}^2\frac{\partial^2 t_1}{\partial\lambda_1{}^2} = \frac{\lambda_1}{2\rho}\frac{\partial}{\partial\lambda_1}(\rho^2U_{11}^2) = \frac{(1+e_1)^2}{2}\frac{\partial^2 t_1}{\partial e_1{}^2} = \frac{1+e_1}{2\rho}\frac{\partial}{\partial e_1}(\rho^2U_{11}^2)$$

where e_1 is the principal extension in the $\mathbf{n}_1$-direction.

To illustrate the physical meaning of the foregoing results, consider a one-dimensional case in which $\lambda_1 \neq 1$, $\lambda_2 = \lambda_3 = 1$, or $x_1 = x_1(X_1, t)$, and a longitudinal plane acceleration wave propagating in the direction of the principal stretch $\lambda_1 \equiv \partial x_1/\partial X_1$. From (2.12.6) it follows that

$$[\partial\dot{x}_1/\partial X_1] = [\dot{\lambda}_1] = [\dot{e}_1] = -U\lambda_1 a$$

which implies that the strain behind the acceleration wave tends to increase after the passage of the wave for expansion waves ($a > 0$) and tends to decrease for compression waves ($a < 0$).

If $K_{11} > 0$, we deduce from (2.15.19) that $\partial^2 t_1/\partial e_1{}^2 > 0$. Then materials satisfying this condition are called *hardening materials* (Fig. 2.15.2a). Equation (2.15.19) also implies that the speed of propagation increases with the increasing extension in hardening materials. Since the extension behind an expansion wave tends to increase after the passage of the wave ($de_1 > 0$), the speeds of propagation of the disturbances behind the wave exceed the speed of propagation of the wave front itself. Therefore, these disturbances created by the acceleration wave can catch up with the wave front and the magnitude of the amplitude of the wave begins to grow indefinitely due to this build-up process. In the case of a compressive wave, we have $de_1 < 0$ behind the acceleration wave and the speed of propagation decreases $[d(\rho^2U_{11}^2) < 0]$ for the disturbances behind the front, since $K_{11} > 0$.

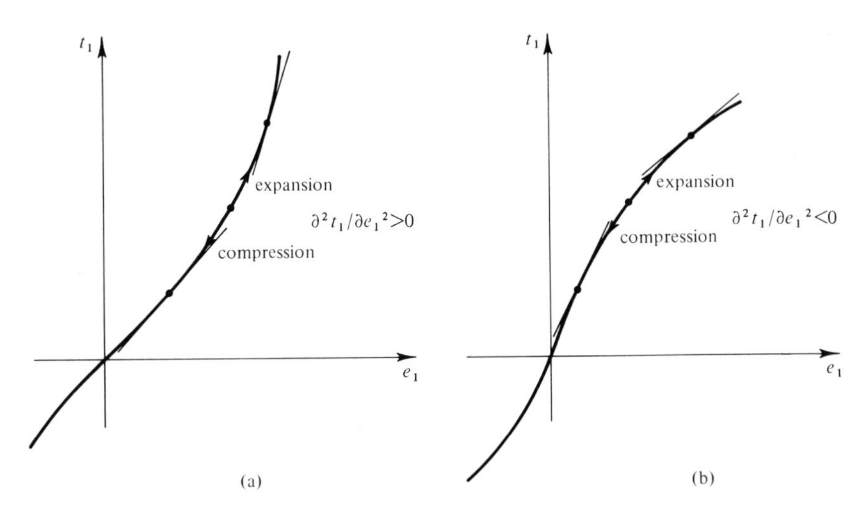

Fig. 2.15.2 Stress–strain curves for hardening (a) and softening (b) materials.

If $K_{11} < 0$, we have a *softening material* and one can easily see that the situation becomes reversed (Fig. 2.15.2b).

For transverse principal waves we have to take $\mathbf{v} = \mathbf{n}_2$ or $\mathbf{n}_3$. Then it is easily verified that

$$K_{12} = K_{13} = 0$$

which amounts to the fact that the transverse principal waves propagate without any change in their amplitudes.

B. Growth of Acceleration Waves of Arbitrary Form in Elastic Materials Subject to Uniform Dilatation[1]

In this case all principal stretches are equal, i.e., $\lambda_1 = \lambda_2 = \lambda_3 = \lambda$, and every direction is a principal direction of the state of deformation in the material. Hence all waves propagating in such a medium are necessarily principal waves and, consequently, are either longitudinal or transverse. It is straightforward to show that

$$B_{\alpha\beta} = (1/\lambda)b_{\alpha\beta}, \qquad A_{\alpha\beta} = (1/\lambda^2)a_{\alpha\beta}, \qquad A^{\alpha\beta} = \lambda^2 a^{\alpha\beta}, \qquad \overset{-1}{\mathbf{c}} = \lambda^2 \mathbf{I} \tag{2.15.20}$$

The longitudinal and transverse wave speeds are, respectively,

$$\rho U_L{}^2 = 2\lambda^2\left(h_1 + 2\lambda^2 h_2 + \sum_{a=0}^{2} \lambda^{2a}\, Dh_a\right), \qquad \rho U_T{}^2 = \lambda^2(h_1 + 2\lambda^2 h_2) \tag{2.15.21}$$

[1] See Şuhubi [1970].

where

$$D \equiv (\partial/\partial I) + 2\lambda^2 (\partial/\partial II) + \lambda^4 (\partial/\partial III)$$

The response functions h_0, h_1, and h_2 and their derivatives are to be evaluated for $\overset{-1}{\mathbf{c}} = \lambda^2 \mathbf{I}$ or $I = 3\lambda^2$, $II = 3\lambda^4$, $III = \lambda^6$. Substituting (2.15.20) into (2.15.10) we obtain the growth equation:

(2.15.22)

$$\begin{aligned}
2\rho U^2 \frac{da}{dn} = 2\lambda^2 \Bigg\{ h_1 + 2\lambda^2 h_2 + \sum_{a=0}^{2} \lambda^{2a} Dh_a + \lambda^2 \Bigg[3h_2 + 2\, Dh_1 \\
+ 4\lambda^2\, Dh_2 - \sum_{a=0}^{2} \lambda^{2a} \left(\frac{\partial h_a}{\partial II} + \lambda^2 \frac{\partial h_a}{\partial III} \right) \Bigg] \\
+ \lambda^2 \Bigg[h_2 + 2\, Dh_1 + 4\lambda^2\, Dh_2 + \sum_{a=0}^{2} \lambda^{2a} \Bigg(\frac{\partial h_a}{\partial II} + \lambda^2 \frac{\partial h_a}{\partial III} \\
+ 2\, D^2 h_a \Bigg) \Bigg] (v_k n_k)^2 \Bigg\} (v_l n_l) a^2 \\
- \lambda^2 \Bigg\{ \left(h_1 + 2\lambda^2 h_2 + 2 \sum_{a=0}^{2} \lambda^{2a}\, Dh_a \right) (x_{l\beta} v_k n_k + x_{k\beta} v_k n_l)(a v_l)_{;\alpha} a^{\alpha\beta} \\
- \Bigg[(h_1 + 2\lambda^2 h_2) b_\alpha^{\ \alpha} + \Bigg(h_1 + 2\lambda^2 h_2 \\
+ 2 \sum_{a=0}^{2} \lambda^{2a}\, Dh_a \Bigg) b^{\alpha\beta} x_{k\alpha} x_{l\beta} v_k v_l \Bigg] a \Bigg\}
\end{aligned}$$

where $b^{\alpha\beta} = a^{\alpha\gamma} a^{\beta\delta} b_{\gamma\delta}$ and response functions are to be evaluated at $\overset{-1}{\mathbf{c}} = \lambda^2 \mathbf{I}$.

For longitudinal waves $\mathbf{v} = \mathbf{n}$ and $U = U_{\mathrm{L}}$. Therefore, from (2.2.18) and (2.2.29) it follows that

(2.15.23)
$$v_k x_{k\alpha} = 0, \qquad (a v_k)_{;\alpha} x_{k\beta} = -b_{\alpha\beta}\, a$$

If we employ these identities in (2.15.22), we find the following growth equation (Chen [1968b]):

(2.15.24)
$$da/dn = \Omega(n) a + Q(\lambda) a^2$$

where $\Omega = \frac{1}{2} b_\alpha^{\ \alpha}$ is the mean curvature of the wave surface in the configuration corresponding to a distance n from the original configuration. The coefficient of the second term in (2.15.24) is

(2.15.25)

$$Q(\lambda) = \frac{1}{2}\left[1 + (4\lambda^4/\rho U_{\mathrm{L}}^2) \left(2h_2 + 2\, Dh_1 + 4\lambda^2\, Dh_2 + \sum_{a=0}^{2} \lambda^{2a}\, D^2 h_a \right) \right]$$

Since the wave speed U is constant, the wave surfaces $\sigma(t)$ form a family of parallel surfaces. If we denote the mean and Gaussian curvatures of the surface $\sigma(t_0)$ from which the travel distance $n \geq 0$ is measured by Ω_0 and K_0, for a family of parallel surfaces we may write[1]

(2.15.26)
$$\Omega(n) = (\Omega_0 - K_0 n)/(1 - 2\Omega_0 n + K_0 n^2), \qquad K(n) = K_0/(1 - 2\Omega_0 n + K_0 n^2)$$

For *transverse waves* we have $v_k n_k = 0$, and

$$(av_k)_{;\alpha} n_k = a_{;\alpha} v_k n_k + av_{k;\alpha} n_k = av_{k;\alpha} n_k$$

Differentiating $v_k n_k = 0$ with respect to the αth surface coordinate, we get $v_{k;\alpha} n_k = -v_k n_{k;\alpha}$. Hence

$$(av_k)_{;\alpha} n_k = -av_k n_{k;\alpha} = ab^{\beta}{}_{\alpha} v_k x_{k\beta}$$

The growth equation for transverse waves now follows from (2.15.22), with $U = U_T$:

$$da/dn = \Omega(n)a \tag{2.15.27}$$

An inspection of the differential equations (2.15.24) and (2.15.27) reveals that both waves may decay if and only if $\Omega(n) < 0$ for every n, i.e., if one of the following conditions is satisfied:

$$\text{(a)} \quad K_0 = 0, \qquad \Omega_0 < 0, \qquad \text{(b)} \quad K_0 > 0, \qquad \Omega_0 < 0$$

The first condition implies that at every instant the wave front is a *developable* surface, while the second one describes a *surface of positive curvature*. From (2.15.26) it then follows that if the wave front is either a developable surface or a surface of positive curvature at $n = 0$ (or $t = 0$) it remains so for all subsequent times. Henceforth we shall consider only wave surfaces satisfying either condition (a) or (b).

If we introduce (2.15.26) into (2.15.24) for *longitudinal waves*, we see that the resulting equation may be arranged into

$$(1 - 2\Omega_0 n + K_0 n^2)^{-1/2} (d/dn)(1 - 2\Omega_0 n + K_0 n^2)^{1/2} a = Q(\lambda)a^2$$

[1] In fact, (2.15.26) may be obtained by recalling the definitions of mean and Gaussian curvatures. If ρ_1 and ρ_2 are principal radii of curvature of the surface, then

$$\Omega = -\tfrac{1}{2}[(1/\rho_1) + (1/\rho_2)], \qquad K = 1/\rho_1\rho_2$$

For a member of a family of parallel surfaces one can write $\rho_1 = \rho_1{}^0 + n$, $\rho_2 = \rho_2{}^0 + n$, where $\rho_1{}^0$ and $\rho_2{}^0$ are principal radii of a typical surface from which the distance n is measured. It is now straightforward to derive the relations (2.15.26). (See also Thomas [1961a, Chapter 24].)

Since $(1 - 2\Omega_0 n + K_0 n^2)^{1/2}$ is a monotone increasing function for surfaces under consideration, we deduce that the amplitude of the expansive waves $(a > 0)$ will decay to zero if $Q(\lambda) < 0$, and the amplitude of the compressive waves $(a < 0)$ will decay to zero if $Q(\lambda) > 0$.

If these inequalities are not satisfied for a given material, the amplitude may still decay within a certain range of the distance n if

$$a(n) < -[1/Q(\lambda)](\Omega_0 - K_0 n)/(1 - 2\Omega_0 n + K_0 n^2) \tag{2.15.28}$$

when $a > 0$, $Q(\lambda) > 0$, or $a < 0$, $Q(\lambda) < 0$. However, in these cases the amplitude of the wave begins to increase after a critical distance.

It can be seen that transverse waves of the forms considered here will decay to zero in general.

Equation (2.15.24) together with (2.15.26) may be integrated to give the complete dependence of the amplitude of longitudinal waves on the travel distance. Thus

(2.15.29)

$$\frac{a(n)}{a_0} = (1 - 2\Omega_0 n + K_0 n^2)^{-1/2} \times \left[1 - \frac{Q(\lambda)a_0}{K_0^{1/2}} \log \frac{K_0^{1/2}(1 - 2\Omega_0 n + K_0 n^2)^{1/2} + K_0 n - \Omega_0}{K_0^{1/2} - \Omega_0}\right]^{-1}$$

for surfaces of positive curvature and

$$\frac{a(n)}{a_0} = (1 - 2\Omega_0 n)^{-1/2}\left\{1 + \frac{Q(\lambda)a_0}{\Omega_0}[(1 - 2\Omega_0 n)^{1/2} - 1]\right\}^{-1} \tag{2.15.30}$$

for developable surfaces, where a_0 is the amplitude of the wave at $n = 0$ (or $t = 0$). The first factors in (2.15.29) and (2.15.30) obviously represent the infinitesimal solutions.

If the surface is a plane ($\Omega_0 = 0$, $K_0 = 0$), then (2.15.30) reduces to

$$a/a_0 = [1 - Q(\lambda)a_0 n]^{-1}$$

If we identify in this case $Q(\lambda)$ as $K|_{\mathbf{c}=\lambda^2\mathbf{I}}/\rho U^2$, we see that this formula is identical to (2.15.14).

The insertion of (2.15.29) or (2.15.30) into (2.15.28) gives the range of n within which the amplitude of the longitudinal wave will decay if $a > 0$, $Q(\lambda) > 0$, or $a < 0$, $Q(\lambda) < 0$.

The growth equation for the transverse waves (2.15.27) can be integrated to give

$$a(n) = a_0(1 - 2\Omega_0 n + K_0 n^2)^{-1/2} \tag{2.15.31}$$

which shows that, similar to the infinitesimal shear waves (Thomas [1961b, Chapter III]), the decay law is independent of the state of uniform dilatation of the medium.

The general formulas will be applied now to wave surfaces of special form (Chen [1968b]). For instance, in *cylindrical waves* in which $\Omega_0 = -1/2r_0$, $K_0 = 0$, where r_0 is the initial radius of the wave,

(2.15.32) $$a(r)/a_0 = (r_0/r)^{1/2}\{1 - 2Q(\lambda)a_0 r_0[(r/r_0)^{1/2} - 1]\}^{-1}$$

where $n = r - r_0$, r being the actual radius of the wave surface.

For *spherical waves* $\Omega_0 = -1/r_0$, $K = 1/r_0{}^2$, and we get

(2.15.33) $$a(r)/a_0 = (r_0/r)[1 - Q(\lambda)a_0 r_0 \log(r/r_0)]^{-1}$$

The classical results are obtained for the transverse waves from (2.15.31):

(2.15.34) $$a(r)/a_0 = (r_0/r)^{1/2} \qquad \text{(cylindrical waves)}$$
$$a(r)/a_0 = r_0/r \qquad \text{(spherical waves)}$$

The growth formulation of acceleration waves propagating into an unstrained isotropic elastic medium may be carried out by simply observing that $\lambda = 1$ in the foregoing analysis corresponds to this case. Some numerical results due to Chen [1968a] concerning this case are now reproduced. The material is polyurethane foam rubber (47% by volume). Ko [1963] proposed the following expressions for response functions:

(2.15.35) $$h_0 = \mu[1 - (II/III^{3/2})], \qquad h_0 = \mu I/III^{3/2}, \qquad h_2 = -\mu/III^{3/2}$$

where μ is a constant. It can be verified that these functions satisfy the compatibility relations (2.13.5). The wave speeds are then found to be

(2.15.36) $$U_{\mathrm{L}} = (3\mu/\rho_0)^{1/2}, \qquad U_{\mathrm{T}} = (\mu/\rho_0)^{1/2}$$

and

(2.15.37) $$Q(1) = -0.75$$

Since $Q(1) < 0$, only expansion waves will decay to zero and accordingly we obtain

$$a/a_0 = (1 + 0.75a_0 n)^{-1}$$

for plane waves,

$$a/a_0 = (r_0/r)^{1/2}\{1 + 1.50a_0 r_0[(r/r_0)^{1/2} - 1]\}^{-1}$$

for cylindrical waves, and

$$a/a_0 = (r_0/r][1 + 0.75a_0 r_0 \log(r/r_0)]^{-1}$$

for spherical waves.

If the waves are compressive ($a_0 < 0$), then the plane, cylindrical, and spherical waves become infinite after a finite distance given, respectively, by

(2.15.38) $$n = 4/3|a_0|, \qquad r = r_0[1 + (2/3|a_0|r_0)]^2, \qquad r = r_0 \exp(\tfrac{4}{3}/|a_0|r_0)$$

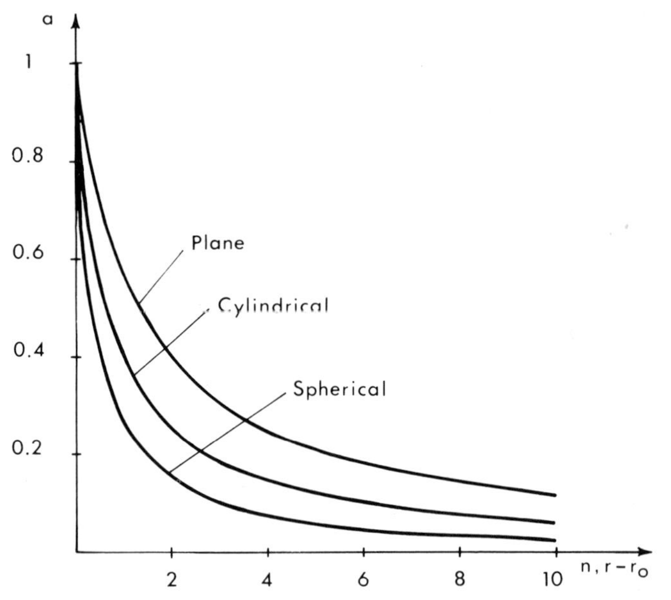

Fig. 2.15.3 Decay of expansion waves with $a_0 = 1$, $r_0 = 0.7$ (*after* Chen [1968a]).

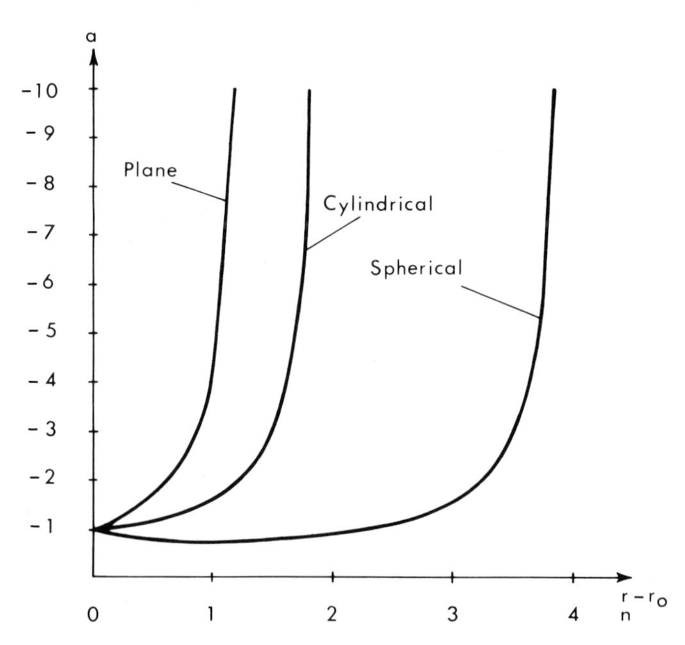

Fig. 2.15.4 Decay and growth of compressive waves with $a_0 = -1$, $r_0 = 0.7$ (*after* Chen [1968a]).

Figure 2.15.3 shows the decay of the amplitude of the expansion waves of plane, cylindrical, and spherical form plotted versus n. The initial amplitude is taken as $a_0 = 1$, and the initial radius for cylindrical and spherical waves as $r_0 = 0.7$. Figure 2.15.4, on the other hand, describes the behavior of compressive waves with initial amplitude $a_0 = -1$ and initial radius $r_0 = 0.7$. The amplitudes of plane and cylindrical waves increase steadily and become infinite at times (or distances) given by (2.15.38). However, it can be verified that (2.15.28) is satisfied for spherical waves with $a_0 = -1$, $r_0 = 0.7$, the inequality being $-1 < -1/1.05$. The amplitude decays initially and then increases indefinitely toward a distance given by $(2.15.38)_3$.

CHAPTER III

Finite Motions of Elastic Bodies

3.1 SCOPE OF THE CHAPTER

The study of the propagation of singular surfaces provides limited information about the dynamic response of materials. In general, problems concerning the oscillatory motions of bodies and their behavior under impact loads cannot be treated by the methods of singular surfaces. The finite motions of elastic bodies under initial and boundary conditions are governed by a set of nonlinear partial differential equations for which no rigorous and systematic theory has been developed. The few problems that have been treated in the literature are mostly concerned with incompressible bodies having special geometries and applied loads, e.g., one-dimensional problems, and/or with very special types of materials.

In this chapter we present a few selected problems to illustrate the inherent difficulties and to direct the attention of research workers to this important but barren area of mechanics. In Section 3.2 we present briefly a special class of motion in incompressible solids, called quasi-equilibrated motions. Within the context of this theory the radial oscillations of a cylindrical tube and a thick spherical shell are investigated in Sections 3.3 and 3.4, respectively. Section 3.5 is devoted to simple plane waves in compressible solids. Simple plane waves in incompressible solids are investigated in Section 3.6. The Riemann problem for an elastic half-space is the subject of Section 3.7. The general theory of simple waves is discussed in Section 3.8. In Section 3.9 we present an account of the reflection and transmission of one-dimensional waves at interfaces of a slab. Section 3.10 covers some special solutions appropriate to specific materials. A brief assessment of some wave solutions in elastic solids, existing in the literature, is given in Section 3.11.

3.2 QUASI-EQUILIBRATED MOTIONS OF INCOMPRESSIBLE BODIES

For some special class of motions of incompressible bodies it is possible to reduce the problem of motion to a corresponding problem of equilibrium. This possibility was explored by Truesdell [1962] in a paper written in Latin which was later reproduced in the monograph by Truesdell and Noll [1965].

Consider an incompressible elastic body subject to a given body force distribution $\mathbf{f}(\mathbf{x}, t)$. The motion $\mathbf{x} = \mathbf{x}(\mathbf{X}, t)$ is dynamically possible if one is able to determine a hydrostatic pressure function $p(\mathbf{x}, t)$ such that

$$-\operatorname{grad} p + \operatorname{div} \mathbf{T} + \rho \mathbf{f} = \rho \ddot{\mathbf{x}} \tag{3.2.1}$$

where the second-order tensor $\mathbf{T}$ is a function of deformation gradients $x_{k,K}(\mathbf{X}, t)$ and is related to the Cauchy stress tensor by

$$\mathbf{t} = -p\mathbf{I} + \mathbf{T} \tag{3.2.2}$$

If we can find a time-dependent scalar function $p_0(\mathbf{x}, t)$ so that

$$-\operatorname{grad} p_0 + \operatorname{div} \mathbf{T} + \rho \mathbf{f} = \mathbf{0} \tag{3.2.3}$$

is satisfied for all times t, then the motion for which (3.2.3) holds is called a *quasi-equilibrated motion*. Since $\mathbf{x}(\mathbf{X}, t)$ is regarded as a static deformation for each value of t, which plays now the role of a parameter, it is a solution of the equation of equilibrium (3.2.3), at time t with a given body force $\mathbf{f}$ and a pressure function p_0. The fictitious stress tensor corresponding to a quasi-equilibrated state may be expressed as

$$\mathbf{t}_0 = -p_0 \mathbf{I} + \mathbf{T} \tag{3.2.4}$$

where $\mathbf{T}$ is determined by the motion $\mathbf{x}(\mathbf{X}, t)$. However the quasi-equilibrated motion $\mathbf{x}(\mathbf{X}, t)$ is required to satisfy both (3.2.3) and (3.2.1) in order to represent the solution of a dynamical problem. This requires that the acceleration be derivable from a potential, namely,

$$\ddot{\mathbf{x}} = -\operatorname{grad} \zeta \tag{3.2.5}$$

where

$$\rho\zeta = p - p_0 \tag{3.2.6}$$

Therefore, the necessary condition for the existence of a quasi-equilibrated motion is

$$\operatorname{curl} \ddot{\mathbf{x}} = \mathbf{0} \tag{3.2.7}$$

If ζ is a single-valued function this condition is also sufficient. In a quasi-equilibrated motion the actual stress tensor is given by

$$\mathbf{t} = -\rho\zeta\mathbf{I} + \mathbf{t}_0 \tag{3.2.8}$$

where $\mathbf{t}_0$ is the stress tensor corresponding to an equilibrium state in the configuration at time t.

If it is desired that the quasi-equilibrated motion be dynamically possible for *all* elastic incompressible materials regardless of their constitution, the equilibrium deformation should be *controllable*, i.e., it should be sustained by the application *only* of surface tractions without requiring a special body force distribution to compensate the unbalanced terms in the equations of equilibrium. For isotropic incompressible materials subject to no body force Ericksen [1954] has found that there are only five allowable families of such deformations, namely: homogeneous deformations; bending, extension, and shear of a rectangular block; straightening, extension, and shear of a sector of a cylindrical tube; inflation, bending, torsion, extension, and shear of a cylindrical tube; and inflation and eversion of a spherical shell.[1] The acceleration potentials corresponding to each case may be determined by inspection and are listed in Truesdell and Noll [1965, Section 61].

In the two subsequent sections we consider two simple vibration problems falling into one of the five categories mentioned.

3.3 RADIAL OSCILLATIONS OF A CYLINDRICAL TUBE

Consider an infinitely long cylindrical tube made of a homogeneous, isotropic, and incompressible elastic material. Let the inner radius of the cylinder, in the undeformed state, be R_1 and the outer radius R_2. Here we

[1] Recently Kafadar [1972] treated the problem in which the invariants of the Cauchy deformation tensor $\mathbf{c}$ are constant. He proved that: (1) if $\mathbf{c}$ possesses a constant eigenvector, there are no new solutions; (2) if any two eigenvalues are equal, $\mathbf{c}$ possesses a constant eigenvector; and (3) if the eigenvalues of $\mathbf{c}$ are distinct, and if there are new solutions, there exists *no* holonomic coordinate system such that one of the eigenvectors of $\mathbf{c}$ is normal to a coordinate surface.

study radial oscillatory motion of this tube.[1] If we use the same frame of reference for the material and spatial descriptions, such a motion can be described, in cylindrical coordinates, by the equations of motion of the form

$$r = r(R, t), \qquad \theta = \Theta, \qquad z = Z \tag{3.3.1}$$

where (R, Θ, Z) are cylindrical coordinates of a typical point in the undeformed tube, and (r, θ, z) are cylindrical coordinates of the same point, at time t, in a fixed reference frame. In a rectangular coordinate system (3.3.1) reads

$$\begin{aligned} x_1 &= R^{-1}r(R, t)X_1 = r(R, t)\cos\theta \\ x_2 &= R^{-1}r(R, t)X_2 = r(R, t)\sin\theta \\ x_3 &= X_3 \end{aligned} \tag{3.3.2}$$

The matrix of deformation gradients is

$$[x_{k,K}] = \begin{bmatrix} r/R + [r' - (r/R)]\cos^2\theta & [r' - (r/R)]\sin\theta\cos\theta & 0 \\ [r' - (r/R)]\sin\theta\cos\theta & r/R + [r' - (r/R)]\sin^2\theta & 0 \\ 0 & 0 & 1 \end{bmatrix} \tag{3.3.3}$$

where primes denote differentiation with respect to R, i.e., $r' \equiv \partial r/\partial R$.

Since the material is incompressible, we have $\det(x_{k,K}) = 1$. This condition yields, from (3.3.3),

$$(r/R)\partial r/\partial R = 1 \tag{3.3.4}$$

which may be integrated to give

$$r^2 = R^2 + A(t) \tag{3.3.5}$$

where A is a function of time only. If we determine the function $A(t)$ then (3.3.5) gives the position of the particle occupying position R initially. If the inner radius of the tube is denoted by $r_1(t) = r(R_1, t)$, then

$$A(t) = r_1(t)^2 - R_1^{\,2}$$

and the motion is described by the function

$$r^2(R, t) = r_1^{\,2}(t) + R^2 - R_1^{\,2} \tag{3.3.6}$$

Therefore, the determination of the radial motion $r(R, t)$ of the tube is reduced to determination of the motion $r_1(t)$ of the inner wall.

[1] This problem was studied by Knowles [1960, 1962]. See also Huilgol [1967] for finite oscillations in curvilinearly aelotropic cylinders.

Substituting (3.3.3) in $(1.4.8)_1$ and $(1.4.9)_1$ we obtain the Cartesian components of the deformation tensors:

$$\mathbf{c}(\mathbf{x}, t) = \begin{bmatrix} \frac{r^2}{R^2} + \left[\left(\frac{R^2}{r^2}\right) - \left(\frac{r^2}{R^2}\right)\right] \sin^2\theta & -\left[\left(\frac{R^2}{r^2}\right) - \left(\frac{r^2}{R^2}\right)\right] \sin\theta\cos\theta & 0 \\ -\left[\left(\frac{R^2}{r^2}\right) - \left(\frac{r^2}{R^2}\right)\right] \sin\theta\cos\theta & \frac{r^2}{R^2} + \left[\left(\frac{R^2}{r^2}\right) - \left(\frac{r^2}{R^2}\right)\right] \cos^2\theta & 0 \\ 0 & 0 & 1 \end{bmatrix}$$

(3.3.7)

$$\overset{-1}{\mathbf{c}}(\mathbf{x}, t) = \begin{bmatrix} \frac{r^2}{R^2} + \left[\left(\frac{R^2}{r^2}\right) - \left(\frac{r^2}{R^2}\right)\right] \cos^2\theta & \left[\left(\frac{R^2}{r^2}\right) - \left(\frac{r^2}{R^2}\right)\right] \sin\theta\cos\theta & 0 \\ \left[\left(\frac{R^2}{r^2}\right) - \left(\frac{r^2}{R^2}\right)\right] \sin\theta\cos\theta & \frac{r^2}{R^2} + \left[\left(\frac{R^2}{r^2}\right) - \left(\frac{r^2}{R^2}\right)\right] \sin^2\theta & 0 \\ 0 & 0 & 1 \end{bmatrix}$$

The invariants of $\overset{-1}{\mathbf{c}}$ are

$$I = II = 1 + (r^2/R^2) + (R^2/r^2) \tag{3.3.8}$$

The physical components of $\mathbf{c}$ and $\overset{-1}{\mathbf{c}}$ in cylindrical coordinates are obtained by the usual transformation formulas

$$\begin{aligned} c_{rr} &= r^2/R^2, \quad c_{\theta\theta} = R^2/r^2, \quad c_{zz} = 1, \quad c_{r\theta} = c_{rz} = c_{\theta z} = 0 \\ \overset{-1}{c}_{rr} &= R^2/r^2, \quad \overset{-1}{c}_{\theta\theta} = r^2/R^2, \quad \overset{-1}{c}_{zz} = 1, \quad \overset{-1}{c}_{r\theta} = \overset{-1}{c}_{rz} = \overset{-1}{c}_{\theta z} = 0 \end{aligned} \tag{3.3.9}$$

The components of the stress tensor are determined by the use of the constitutive equations:

$$\begin{aligned} t_{rr} &= -p + 2\frac{\partial\Sigma}{\partial I}\frac{R^2}{r^2} - 2\frac{\partial\Sigma}{\partial II}\frac{r^2}{R^2}, \qquad t_{zz} = -p + 2\frac{\partial\Sigma}{\partial I} - 2\frac{\partial\Sigma}{\partial II} \\ t_{\theta\theta} &= -p + 2\frac{\partial\Sigma}{\partial I}\frac{r^2}{R^2} - 2\frac{\partial\Sigma}{\partial II}\frac{R^2}{r^2}, \qquad t_{r\theta} = t_{rz} = t_{\theta z} = 0 \end{aligned} \tag{3.3.10}$$

It is now straightforward to verify that the acceleration computed from (3.3.6) has a potential

$$-\zeta = (r_1\ddot{r}_1 + \dot{r}_1^{\,2})\log r + \tfrac{1}{2}r^{-2}r_1^{\,2}\dot{r}_1^{\,2} \tag{3.3.11}$$

such that

$$\ddot{r} = -\partial\zeta/\partial r \tag{3.3.12}$$

If we assume that p is a function of r and t only, then the equations of motion in the axial and circumferential directions are satisfied identically. The remaining equation in the radial direction is [see (1.20.31)]

$$(\partial t_{rr}/\partial r) + (t_{rr} - t_{\theta\theta})/r = \rho \ddot{r} = -\rho\, \partial\zeta/\partial r \tag{3.3.13}$$

Using (3.3.10) in (3.3.13) we solve for the normal stress t_{rr}:

$$t_{rr} = -\rho\zeta - 2\int^{r} \frac{1}{r}\left(\frac{R^2}{r^2} - \frac{r^2}{R^2}\right)\left(\frac{\partial\Sigma}{\partial I} + \frac{\partial\Sigma}{\partial II}\right) dr + C(t) \tag{3.3.14}$$

where $C(t)$ is an arbitrary function of time. The only compatible boundary tractions are the uniform tractions applied on the inner and outer walls of the cylinder. Therefore, boundary conditions that can be satisfied are:

$$t_{rr}|_{r=r_1} = -p_1(t), \qquad t_{rr}|_{r=r_2} = -p_2(t) \tag{3.3.15}$$

where $p_1(t)$ and $p_2(t)$ are prescribed functions of time. Evaluating (3.3.14) at $r = r_1$ and $r = r_2$ and subtracting the resulting expressions, we obtain

$$-[p_1(t) - p_2(t)] = -\rho(\zeta_1 - \zeta_2) - 2\int_{r_2}^{r_1} \frac{1}{r}\left(\frac{R^2}{r^2} - \frac{r^2}{R^2}\right)\left(\frac{\partial\Sigma}{\partial I} + \frac{\partial\Sigma}{\partial II}\right) dr \tag{3.3.16}$$

According to (3.3.11),

$$\begin{aligned} -\zeta_1 &= (r_1\ddot{r}_1 + \dot{r}_1{}^2)\log r_1 + (\dot{r}_1{}^2/2) \\ -\zeta_2 &= (r_1\ddot{r}_1 + \dot{r}_1{}^2)\log r_2 + (r_1{}^2\dot{r}_1{}^2/2r_2{}^2) \end{aligned}$$

If we set

$$x(t) = r_1(t)/R_1, \qquad u(R, t) = r^2/R^2, \qquad \gamma = (R_2{}^2/R_1{}^2) - 1 > 0 \tag{3.3.17}$$

then (3.3.6) gives

$$\begin{aligned} &r_2{}^2/R_2{}^2 = (\gamma + x^2)/(1 + \gamma), \qquad r_2{}^2/r_1{}^2 = 1 + (\gamma/x^2) \\ &u(R, t) = 1 + (R_1{}^2/R^2)[x^2(t) - 1] \end{aligned} \tag{3.3.18}$$

Introducing (3.3.17) and (3.3.18) into (3.3.16) we get

$$x \log\left(1 + \frac{\gamma}{x^2}\right)\ddot{x} + \left[\log\left(1 + \frac{\gamma}{x^2}\right) - \frac{\gamma}{\gamma + x^2}\right]\dot{x}^2 + f(x, \gamma) = 2\,\frac{p_1(t) - p_2(t)}{\rho R_1{}^2} \tag{3.3.19}$$

where

$$\begin{aligned} f(x, \gamma) &= \frac{4}{\rho R_1{}^2}\int_{r_2}^{r_1} \frac{1}{r}\left(\frac{R^2}{r^2} - \frac{r^2}{R^2}\right)\left(\frac{\partial\Sigma}{\partial I} + \frac{\partial\Sigma}{\partial II}\right) dr \\ &= \frac{2}{\rho R_1{}^2}\int_{(\gamma + x^2)/(1+\gamma)}^{x^2} (1 + u)\left(\frac{\partial\Sigma}{\partial I} + \frac{\partial\Sigma}{\partial II}\right)\frac{du}{u^2} \end{aligned} \tag{3.3.20}$$

and we used the relation

$$du = (2u/r)(1 - u)\, dr$$

In carrying out the integration $(3.3.20)_2$, we recall that $\partial\Sigma/\partial I$ and $\partial\Sigma/\partial II$ are functions of u only, since

$$I = II = 1 + u + (1/u)$$

and therefore

$$\frac{d\Sigma}{du} = \left(1 - \frac{1}{u^2}\right)\left(\frac{\partial\Sigma}{\partial I} + \frac{\partial\Sigma}{\partial II}\right)$$

We can thus write $f(x, \gamma)$ in another form which will be of use in future developments:

$$f(x, \gamma) = \frac{2}{\rho R_1{}^2} \int_{(\gamma + x^2)/(1+\gamma)}^{x^2} \frac{d\Sigma}{du} \frac{du}{u - 1} \tag{3.3.21}$$

Equation (3.3.19) is a second-order nonlinear ordinary differential equation to determine $x(t)$ subject to initial conditions

$$x(0) = x_0, \qquad \dot{x}(0) = v_0 \tag{3.3.22}$$

This implies that a motion is imparted on the inner wall of the tube by displacing it to a position $x_0 R_1$ with a velocity $v_0 R_1$. Once x is found, u is obtained directly from $(3.3.18)_3$, and the radial displacement u_r of a material point initially occupying position R is written as

$$u_r = r - R = R(u^{1/2} - 1)$$

Equation (3.3.19) is basically the same one obtained by Knowles [1960]. However, the derivation here is somewhat different.

It may be verified that (3.3.19) can be put into the form

$$\frac{d}{dx}\left[\frac{1}{2} x^2 \log\left(1 + \frac{\gamma}{x^2}\right) v^2\right] + xf(x, \gamma) = 2[p_1(t) - p_2(t)] \frac{x}{\rho R^2} \tag{3.3.23}$$

where $v \equiv \dot{x}$.

If the tube is very thin, an approximate form to (3.3.19) is obtained for the limiting case $\gamma \to 0$. In this case we write

$$R_1 \simeq R_2 = R, \qquad u \simeq x^2$$

and use the expansions

$$\log(1 + \gamma x^{-2}) \simeq \gamma/x^2, \qquad \gamma/(\gamma + x^2) \simeq \gamma/x^2$$

Likewise[1]

$$f(x, \gamma) \simeq (2\gamma/\rho R^2)\, d\Sigma/dx^2 = (\gamma/\rho R^2 x)\, d\Sigma/dx$$

Hence (3.3.19) is simplified to

$$\ddot{x} + (1/\rho R^2)(d\Sigma/dx) = 2[p_1(t) - p_2(t)]x/\gamma\rho R^2$$

where Σ is a known function of $1 + x^2 + x^{-2}$.

A. Free Oscillations[2]

In the case of free oscillations, the applied loads are assumed to vanish, i.e., $p_1(t) = p_2(t) = 0$, and a solution of the equation (3.3.19) with vanishing right-hand side is subject to the initial conditions (3.3.22). Define a function $F(x, \gamma)$ by

$$F(x, \gamma) = \int_1^x \xi f(\xi, \gamma)\, d\xi = \frac{2}{\rho R_1{}^2} \int_1^x \xi \int_{(\gamma+\xi^2)/(1+\gamma)}^{\xi^2} \frac{d\Sigma}{du} \frac{du}{u-1}\, d\xi \tag{3.3.24}$$

Evidently the function $F(x, \gamma)$ satisfies the relations

$$dF/dx = xf(x, \gamma), \qquad F(1, \gamma) = 0$$

We should remember that $x = 1$ corresponds to the undeformed state of the tube. By changing the order of integration in the shaded domain in Fig. 3.3.1,

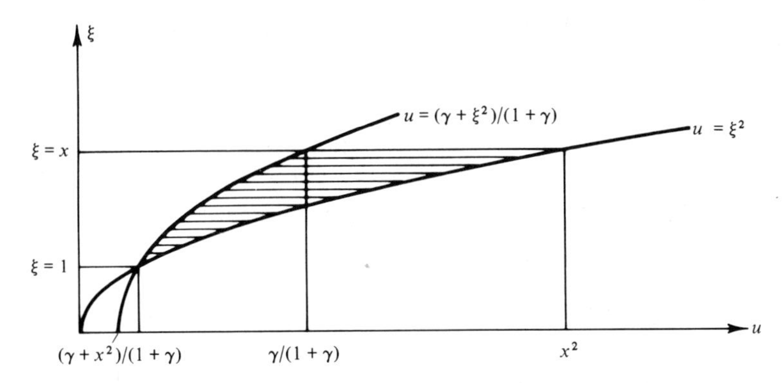

Fig. 3.3.1 Domain of integration in (u, ξ)-plane.

[1] For small γ, using (3.3.21) and the expansion

$$f(x, \gamma) = f(x, 0) + \gamma(\partial f/\partial\gamma)|_{\gamma=0} = \gamma(\partial f/\partial\gamma)|_{\gamma=0}$$

we have

$$(\partial f/\partial\gamma)|_{\gamma=0} = -(\rho R^2)^{-1} g(x^2)(1 - x^2)$$

where $g(u)$ is the integrand in (3.3.21).

[2] Knowles [1960].

we obtain after some manipulations

$$F(x, \gamma) = \frac{x^2 - 1}{\rho R_1{}^2} \int_{(\gamma + x^2)/(1+\gamma)}^{x^2} \frac{\Sigma(u)}{(u-1)^2}\, du \tag{3.3.25}$$

In deriving (3.3.25) we have further assumed that the strain energy vanishes in the undeformed state, i.e., $\Sigma(1) = 0$.

The integration of (3.3.23) with the vanishing right-hand side yields

$$\tfrac{1}{2}v^2 x^2 \log[1 + (\gamma/x^2)] + F(x, \gamma) = C \tag{3.3.26}$$

where C is an arbitrary constant which is to be determined from the initial conditions. Recalling that $v = v_0$ when $x = x_0$, we get from (3.3.26)

$$C = \tfrac{1}{2}v_0{}^2 x_0{}^2 \log[1 + (\gamma/x_0{}^2)] + F(x_0, \gamma) \tag{3.3.27}$$

thus replacing (3.3.26) by

$$v^2 x^2 \log[1 + (\gamma/x^2)] = v_0{}^2 x_0{}^2 \log[1 + (\gamma/x_0{}^2)] + F(x_0, \gamma) - F(x, \gamma)$$

In the problem under consideration, the principal stretch in the z-direction $\lambda_3 = 1$. Since the material is assumed to be incompressible, the other two principal stretches λ_1 and λ_2 in the radial and circumferential directions, respectively, are related by $\lambda_1 = 1/\lambda_2$. Therefore, all principal stretches are distinct and B–E inequalities (Section 1.18i) imply that

$$[(\partial\Sigma/\partial I) + (\partial\Sigma/\partial II)]_{I=II} = \tfrac{1}{2}(t_1 - t_2)/(\lambda_1{}^2 - \lambda_2{}^2) > 0 \tag{3.3.28}$$

It is worthwhile to note that inequality (3.3.28) is the condition for the existence of shear acceleration waves [cf., for instance, (2.13.9) and (2.13.19)].

With regard to (3.3.28) one can deduce that the sign of $d\Sigma/du$ is determined by u. Since $u > 0$ we see that $d\Sigma/du > 0$ if $u > 1$ and $d\Sigma/du < 0$ if $u < 1$. Consequently it follows from (3.3.21) that the function $f(x, \gamma)$ has the following properties:

$$f(x, \gamma) \begin{cases} <0 & \text{if } \ 0 < x < 1 \\ =0 & \text{if } \ x = 1 \\ >0 & \text{if } \ x > 1 \end{cases} \tag{3.3.29}$$

Inasmuch as the function $F(x, \gamma)$ defined by (3.3.24) satisfies $dF/dx = xf(x, \gamma)$, in light of (3.3.29) we infer that $F(x, \gamma)$ is a positive function. It is monotone decreasing when $0 < x < 1$, zero when $x = 1$, and monotone increasing when $x > 1$. On the other hand, inequality (3.3.28) clearly indicates that Σ becomes unbounded as u tends to infinity. Thus one can conclude that the function $F(x, \gamma)$ is also unbounded as $x \to \infty$.

Now we would like to find out the limit of the function $F(x, \gamma)$ as $x \to 0+$. For this purpose consider a change of variable $u \to 1/u$ in (3.3.25). Noting that $\Sigma(u) \to \Sigma(1/u)$, we can transform this equation into

$$F(x, \gamma) = \frac{1 - x^2}{\rho R_1^{\,2}} \int_{(1+\gamma)/(\gamma+x^2)}^{1/x^2} \frac{\Sigma(u)}{(u-1)^2}\, du$$

and evaluate the limit for $x \to 0+$:

$$F(0+, \gamma) = \frac{1}{\rho R_1^{\,2}} \int_{1+(1/\gamma)}^{\infty} \frac{\Sigma(u)}{(u-1)^2}\, du$$

This integral is also unbounded because $\Sigma(u)$ increases with u as $u \to \infty$ in accordance with inequality (3.3.28). Therefore, the function $F(x, \gamma)$ is infinite at both limits $x \to 0+$ and $x \to \infty$. Thus $F(x, \gamma)$ may be represented schematically as shown in Fig. 3.3.2.

We now return to Eq. (3.3.26) where the positive constant C is determined by (3.3.27). For real motions we must have $F(x, \gamma) < C$ except for those values of x for which $v = 0$. Thus

$$v = \mp \left[\frac{2C - 2F(x, \gamma)}{x^2 \log[1 + (\gamma/x^2)]} \right]^{1/2} \tag{3.3.30}$$

The motion $x(t)$ of the inner wall of the tube is periodic if and only if the so-called energy curves (3.3.30) in the (v, x) phase plane are closed. (See, for instance, Stoker [1950, Chapter II].) These curves, in turn, will be closed if and only if the equation

$$F(x, \gamma) = C = F(x_0, \gamma) + \tfrac{1}{2} v_0^{\,2} x_0^{\,2} \log[1 + (\gamma/x_0^{\,2})] \tag{3.3.31}$$

has exactly two roots in the interval $0 < x < \infty$. From the properties of the function $F(x, \gamma)$ discussed earlier we realize that this is indeed the case

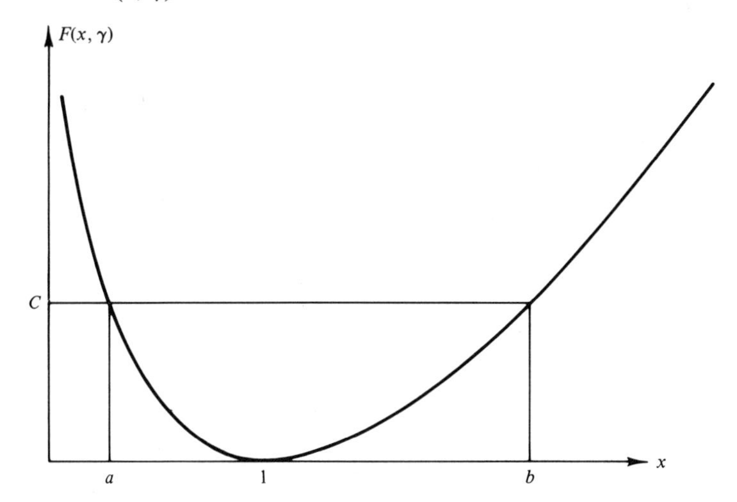

Fig. 3.3.2 Schematic form of $F(x, \gamma)$.

for every nonzero value of C (Fig. 3.3.2) and Eq. (3.3.31) has two simple roots a and b in the intervals $0 < x < 1$ and $1 < x < \infty$, respectively. Hence the energy curve corresponding to a given value of C will look like Fig. 3.3.3. The roots $a < 1$ and $b > 1$ are, respectively, the minimum and maximum amplitudes of the inner wall of the tube.

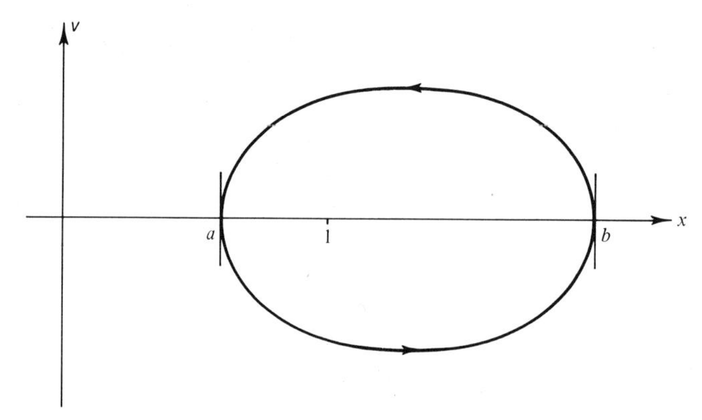

Fig. 3.3.3 Energy curve for $F(x, \gamma) = C$, as given in Fig. 3.3.2.

The period of motion is then

$$T = 2 \int_a^b v^{-1}\, dx = 2 \int_a^b \{x^2 \log[1 + (\gamma/x^2)]\}^{1/2}[2C - 2F(x, \gamma)]^{-1/2}\, dx \tag{3.3.32}$$

Since the first factor in (3.3.32) is bounded and the second factor vanishes at the simple roots $x = a$ and $x = b$, this integral converges yielding a finite value for the period of the oscillations corresponding to any given initial condition.

In the limiting case $\gamma \to 0$ corresponding to a very thin cylindrical shell, a simplification can be made in the analysis. One can show that for $\gamma \to 0$ Eq. (3.3.31) may be approximated by

$$\Sigma(x^2) = \Sigma(x_0{}^2) + (\rho v_0{}^2 R^2/2) \tag{3.3.33}$$

by noticing that

$$F(x, \gamma) \simeq (\gamma/\rho R^2)\Sigma(x^2), \qquad (v_0{}^2 x_0{}^2/2) \log[1 + (\gamma/x_0{}^2)] \simeq \gamma v_0{}^2/2$$

If $x = a_0 < 1$ is a root of (3.3.33) one may verify that $b_0 = 1/a_0 > 1$ is also a root of (3.3.33), since $\Sigma(x^2) = \Sigma(1/x^2)$. Therefore, a_0 and $b_0 = 1/a_0$ are the amplitudes in this limiting case, and $a_0 b_0 = 1$ or $r_{\min} \cdot r_{\max} = R^2$ in the free oscillations of thin shells. Likewise the period of the motion is approximated by

$$T_0 = 2 \int_{a_0}^{1/a_0} \{v_0{}^2 + (2/\rho R^2)[\Sigma(x_0{}^2) - \Sigma(x^2)]\}^{-1/2}\, dx \tag{3.3.34}$$

Another special case which leads to a manageable result corresponds to Mooney materials. The strain energy function for such materials has the form

$$\Sigma = \alpha(I - 3) + \beta(II - 3)$$

where α and β are positive constants. In this case $\Sigma(u)$ assumes the form

$$\Sigma(u) = (\alpha + \beta)(u + u^{-1} - 2)$$

and we are able to find explicit expressions for the functions $f(x, \gamma)$ and $F(x, \gamma)$:

$$f(x, \gamma) = K\left[\left(1 - \frac{1}{x^2}\right)\frac{\gamma}{\gamma + x^2} - \log\frac{1 + (\gamma/x^2)}{1 + \gamma}\right]$$
(3.3.35)
$$F(x, \gamma) = \frac{K(1 - x^2)}{2}\log\frac{1 + (\gamma/x^2)}{1 + \gamma}$$

where

$$K = 2(\alpha + \beta)/\rho R_1^2$$

In the limiting case of thin shells, i.e., $\gamma \to 0$, Eq. (3.3.33) assumes the form[1]

$$x^2 + (1/x^2) = (v_0^2/K) + x_0^2 + (1/x_0^2)$$

The root $a_0 < 1$ of this equation turns out to be

(3.3.36)
$$a_0 = \left[\frac{v_0^2}{2K} + \frac{x_0^2}{2} + \frac{1}{2x_0^2} - \frac{1}{2}\left(\frac{v_0^4}{K^2} + x_0^4 + \frac{1}{x_0^4} + \frac{2x_0^2v_0^2}{K} + \frac{2v_0^2}{Kx_0^2} - 2\right)^{1/2}\right]^{1/2}$$

and the other root is $b_0 = 1/a_0 > 1$.

If the initial conditions are so selected that $x_0 = 1$, $v_0 \neq 0$, that is, if the tube is set in motion by imparting to it an initial velocity in its undeformed state, then (3.3.36) becomes

$$a_0 = \left[1 + \frac{v_0^2}{2K} - \frac{1}{2}\left(\frac{v_0^4}{K^2} + 4\frac{v_0^2}{K}\right)^{1/2}\right]^{1/2}, \qquad b_0 = \frac{1}{a_0}$$

If the initial conditions are such that $x_0 \neq 1$, $v_0 = 0$, that is, if the tube is released from a deformed state without an initial velocity, we then find

$$a_0 = 1/x_0, \qquad b_0 = x_0$$

[1] This equation constitutes the basis for the use of Galerkin's method for an approximate analysis of a thick-walled tube made of Mooney materials. (See Nowinski and Wang [1966a].)

where a_0 stands for the minimum amplitude if initially $x_0 > 1$, that is, if the tube is in an expanded state initially. For $x_0 < 1$, the tube is released from a contracted state and b_0 becomes the minimum amplitude.

The period of oscillations is found from (3.3.34) and (3.3.36) as

$$T_0 = \frac{2}{K^{1/2}} \int_{a_0}^{1/a_0} \left(\frac{v_0^2}{K} + x_0^2 + \frac{1}{x_0^2} - x^2 - \frac{1}{x^2} \right)^{-1/2} dx$$

This integral can be easily evaluated to give

$$T_0 = \pi/K^{1/2} = \pi R[\rho/2(\alpha + \beta)]^{1/2} \tag{3.3.37}$$

(cf. Gradshteyn and Ryzhik [1965, p. 81, Eq. (2.261.3)]). Thus in the limiting case of a thin shell made of Mooney material the period proves to be independent of the amplitude or any other characteristic concerning the motion itself.

B. Forced Vibration of a Cylindrical Tube[1]

Another problem, which leads to an explicit solution and represents an illustration of the forced oscillation of a cylinder, arises from the application of an impulsive pressure difference to the walls of the tube, i.e.,

$$2[p_1(t) - p_2(t)]/\rho R_1^2 = pH(t) \tag{3.3.38}$$

where p is a given constant and $H(t)$ is the *Heaviside* unit step function. For the initial conditions we take $x(0) = 1$, $\dot{x}(0) = 0$, that is, the cylinder is initially in the undeformed state and is at rest. Equation (3.3.23) can be integrated to yield

$$\tfrac{1}{2}v^2 x^2 \log[1 + (\gamma/x^2)] + F(x, \gamma) = \tfrac{1}{2}p(x^2 - 1), \qquad t > 0 \tag{3.3.39}$$

and the velocity is thus obtained to be

$$v = \mp \left[\frac{p(x^2 - 1) - 2F(x, \gamma)}{x^2 \log[1 + (\gamma/x^2)]} \right]^{1/2}$$

From the properties of $F(x, \gamma)$ it is seen that the equation

$$F(x, \gamma) = \tfrac{1}{2}p(x^2 - 1) \tag{3.3.40}$$

has at most one root $x = a$ besides the trivial root $x = 1$.[2] If $p > 0$ (net outward pressure) then $a > 1$, and if $p < 0$ (net inward pressure) then $a < 1$

[1] See Knowles [1962].

[2] A detailed analysis of $F(x, \gamma) - \frac{1}{2}p(x^2 - 1)$ reveals that such a root exists provided that $\Sigma(u) \sim Ku^k$ as $u \to \infty$, where constants $K > 0$, $k > 1$. If $k = 1$, then (3.3.39) will have a root for all p satisfying $p < K \log(1 + \gamma)$.

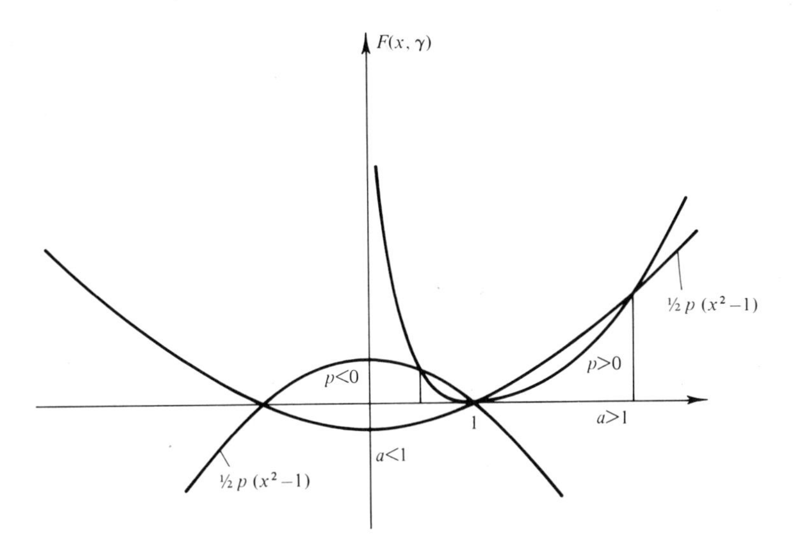

Fig. 3.3.4 Roots of $F(x, \gamma) = \frac{1}{2}p(x^2 - 1)$.

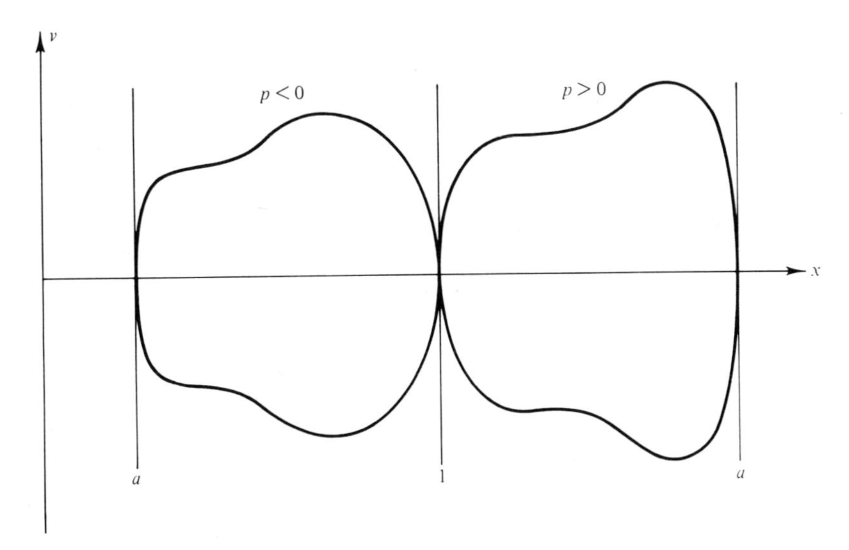

Fig. 3.3.5 Trajectories in the phase plane.

(cf. Fig. 3.3.4). Consequently the trajectories in the phase plane assume close forms which are depicted schematically in Fig. 3.3.5.

Since by definition $F'(x, \gamma) = xf(x, \gamma)$ and the expression

$$F(x, \gamma) - p(x^2 - 1)/2$$

vanishes both at $x = 1$ and $x = a$, there is a value $x = b$ between $x = 1$ and $x = a$ for which $F'(x, \gamma) - px$ vanishes. Therefore, we deduce that $x = b$ is a solution of the equation

$$f(x, \gamma) = p$$

It is now simple to verify that the root $x = b$ of the above equation is, in fact, the time-independent (or static) solution for an incompressible tube subjected to a pressure difference [cf. Eq. (3.3.19)]. Thus the inner surface of the cylinder oscillates about the equilibrium value $x = b$ between $x = 1$ and $x = a$.

The period of oscillations for the tube is then given by

(3.3.41)

$$T = 2\left|\int_1^a v^{-1}\,dx\right| = 2\left|\int_1^a \{x^2 \log[1 + (\gamma/x^2)]\}^{1/2}\,[p(x^2 - 1) - 2F(x, \gamma)]^{-1/2}\,dx\right|$$

We now specialize the foregoing analysis for the Mooney material. Using (3.3.35), Eq. (3.3.40) becomes

$$K \log\{(1 + \gamma)/[1 + (\gamma/x^2)]\} = p \tag{3.3.42}$$

It is easy to see that (3.3.42) has a unique positive root a for p satisfying

$$p < p_{cr} = K \log(1 + \gamma)$$

If $p \geq p_{cr}$ the periodic motion cannot exist. This condition may also be rewritten in the following form:

$$p_1 - p_2 < 2(\alpha + \beta)\log(R_2/R_1)$$

Therefore, it will always be satisfied for a net inward pressure, i.e., $p_1 - p_2 < 0$.

The root of (3.3.42) is

$$\begin{aligned} a &= \gamma^{1/2}[(1 + \gamma)e^{-p/K} - 1]^{-1/2} \\ &= ({R_2}^2 - {R_1}^2)^{1/2}\{{R_2}^2 \exp[(p_2 - p_1)/(\alpha + \beta)] - {R_1}^2\}^{-1/2} \end{aligned} \tag{3.3.43}$$

It follows from (3.3.43) that $a > 1$ if $p_1 - p_2 > 0$, and $a < 1$ if $p_1 - p_2 < 0$. Note that $a \to \infty$ as

$$p_1 - p_2 \to 2(\alpha + \beta)\log(R_2/R_1)$$

The general behavior of $p_1 - p_2$ versus a is shown in Fig. 3.3.6. If the pressure difference approaches the above value the tube expands indefinitely, never returning to the original state.

The period of oscillation is found from (3.3.41) as

$$T = \frac{1}{K^{1/2}}\int_1^{a^2}\left(\frac{\log[1 + (\gamma/u)]}{(u - 1)\log\{[1 + (\gamma/u)]/[1 + (\gamma/a^2)]\}}\right)^{1/2} du \tag{3.3.44}$$

where we made use of (3.3.35), (3.3.42), and the change of variable $x^2 \to u$.

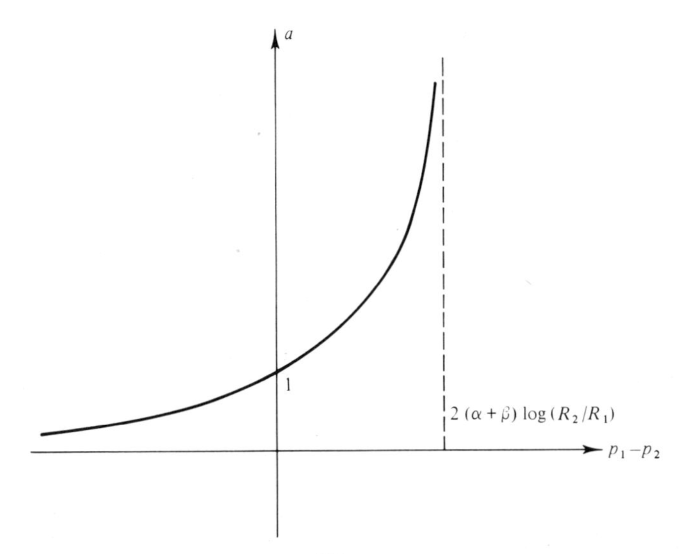

Fig. 3.3.6 Pressure difference $p_1 - p_2$ versus a.

By employing the inequalities

$$(x - y)/(1 + x) \leq \log[(1 + x)/(1 + y)] \leq (x - y)/(1 + y), \qquad \text{if} \quad x \geq y$$

$$x/(1 + x) \leq \log(1 + x) \leq x, \qquad \text{if} \quad x \geq 0$$

one can obtain from (3.3.44) an approximate expression for the period. For thin shells ($\gamma \to 0$)

(3.3.45)

$$T_0 \simeq \pi a/K^{1/2} = \pi R[\rho/2(\alpha + \beta)]^{1/2}(2\delta/R)^{1/2}\{\exp[(p_2 - p_1)/(\alpha + \beta)] - 1\}^{-1/2}$$

where $\delta = R_2 - R_1$ is the thickness of the tube. For small applied net pressure ($p \to 0$)

$$\text{(3.3.46)} \qquad T_1 \simeq (\pi/K^{1/2})[(1 + \gamma^{-1}) \log(1 + \gamma)]^{1/2}$$
$$= \pi R[\rho/2(\alpha + \beta)]^{1/2} [2R_2{}^2(R_2{}^2 - R_1{}^2)^{-1} \log(R_2/R_1)]^{1/2}$$

The last result is in complete agreement with the one corresponding to the solution in the classical linear theory of elasticity.

The combined radial and axial oscillations of a tube have been investigated approximately by Wesołowski [1964]. The radial oscillations, produced by a sudden setting of the cylinder into a uniform rotating motion about its axis, were investigated by Nowinski and Wang [1966b], for cylindrical tubes made of Mooney and Ishihara–Hashitsume–Tatibana (Ishihara *et al.* [1951]) materials [see (3.11.1)].

3.4 RADIAL OSCILLATIONS OF A SPHERICAL SHELL

In this section we discuss the radial oscillations of a spherical thick shell made of an incompressible isotropic elastic material. The inner and outer radii of the shell, in the undeformed state, are denoted by R_1 and R_2, respectively. The radial motion of the shell is represented by a set of relations

$$r = r(R, t), \qquad \theta = \Theta, \qquad \phi = \Phi$$

where R, Θ, and Φ are spherical coordinates of a typical material point, and r, θ, and ϕ are the spherical coordinates of the same point, at time t, referred to the same frame of reference. In Cartesian coordinates the equation of the radial motion is

$$x_k = R^{-1} r(R, t) X_k$$

The analysis from here on is completely analogous to the one presented for the cylindrical tube. Therefore, we omit the details on the discussion of strain field and refer the interested readers to Eringen [1962, Section 54].

The incompressibility condition is given by

$$(r^2/R^2)\, \partial r/\partial R = 1$$

Upon integration this gives

$$r^3 = R^3 + A(t)$$

where $A(t)$ is an arbitrary function of time. If the motion of the inner wall of the shell is denoted by $r_1(t) = r(R_1, t)$, then $A(t)$ can be written in terms of $r_1(t)$, and the radial equation of motion becomes

$$r^3(R, t) = R^3 + r_1{}^3(t) - R_1{}^3 \tag{3.4.1}$$

The acceleration of a material particle admits a potential

$$\zeta = r^{-1}(2r_1\dot{r}_1{}^2 + r_1{}^2\ddot{r}_1) - \tfrac{1}{2}r^{-4}r_1{}^4\dot{r}_1{}^2 \tag{3.4.2}$$

such that

$$\ddot{r} = -\partial\zeta/\partial r \tag{3.4.3}$$

The physical components of the stress tensor are

$$t_{rr} = -p + 2\frac{\partial\Sigma}{\partial I}\frac{R^4}{r^4} - 2\frac{\partial\Sigma}{\partial II}\frac{r^4}{R^4}$$

$$t_{\theta\theta} = t_{\phi\phi} = -p + 2\frac{\partial\Sigma}{\partial I}\frac{r^2}{R^2} - 2\frac{\partial\Sigma}{\partial II}\frac{R^2}{r^2}$$

$$t_{\theta r} = t_{\phi r} = t_{\theta\phi} = 0$$

The invariants of the deformation tensor $\overset{-1}{\mathbf{c}}$ are

$$I = (R^4/r^4) + 2(r^2/R^2), \qquad II = (r^4/R^4) + 2(R^2/r^2), \qquad III = 1$$

If the pressure p is selected to be a function of r and t only, it can be seen that the only equation of motion which is not satisfied identically is [see (1.20.39)]:

$$\frac{\partial t_{rr}}{\partial r} + \frac{2}{r}(t_{rr} - t_{\theta\theta}) = \rho \ddot{r} = -\rho \frac{\partial \zeta}{\partial r}$$

Integrating this equation with respect to r and employing the stress–strain relations we obtain

$$t_{rr} = -\rho\zeta + 4\int^{r} \left(\frac{r^2}{R^2} - \frac{R^4}{r^4}\right)\left(\frac{\partial \Sigma}{\partial I} + \frac{r^2}{R^2}\frac{\partial \Sigma}{\partial II}\right)\frac{dr}{r} + C(t) \tag{3.4.4}$$

where $C(t)$ is an arbitrary function of time only.

The compatible boundary conditions are evidently uniform tractions applied on the inner and outer surfaces of the shell. Thus writing

$$t_{rr}\Big|_{r=r_1} = -p_1(t), \qquad t_{rr}\Big|_{r=r_2} = -p_2(t) \tag{3.4.5}$$

we eliminate $C(t)$ to obtain

$$-[p_1(t) - p_2(t)] = -\rho(\zeta_1 - \zeta_2) + 4\int_{r_2}^{r_1} \left(\frac{r^2}{R^2} - \frac{R^4}{r^4}\right)\left(\frac{\partial \Sigma}{\partial I} + \frac{r^2}{R^2}\frac{\partial \Sigma}{\partial II}\right)\frac{dr}{r} \tag{3.4.6}$$

It follows from (3.4.2) that

$$\zeta_1 = r_1\ddot{r}_1 + \tfrac{3}{2}\dot{r}_1{}^2, \qquad \zeta_2 = r_2^{-1}(2r_1\dot{r}_1{}^2 + r_1{}^2\ddot{r}_1) - \tfrac{1}{2}r_2^{-4}r_1{}^4\dot{r}_1{}^2$$

Next we set

$$x(t) = r_1(t)/R_1, \qquad u(R, t) = r^3/R^3, \qquad \gamma = (R_2/R_1)^3 - 1 > 0 \tag{3.4.7}$$

Then

$$\begin{gathered} r_2{}^3/R_2{}^3 = (\gamma + x^3)/(1 + \gamma), \qquad r_2{}^3/r_1{}^3 = 1 + (\gamma/x^3) \\ u(R, t) = 1 + (R_1{}^3/R^3)(x^3 - 1) \end{gathered} \tag{3.4.8}$$

Substituting (3.4.7) and (3.4.8) into the acceleration potential and then into (3.4.6) we finally get

$$\begin{aligned} 2x\ddot{x}\{1 - [1 + (\gamma/x^3)]^{-1/3}\} + \dot{x}^2\{3 - 3[1 + (\gamma/x^3)]^{-1/3} \\ - (\gamma/x^3)[1 + (\gamma/x^3)]^{-4/3}\} + g(x, \gamma) = (2/\rho R_1{}^2)[p_1(t) - p_2(t)] \end{aligned} \tag{3.4.9}$$

with

$$(3.4.10)\qquad g(x,\gamma) = -\frac{8}{\rho R_1{}^2}\int_{r_2}^{r_1}\left(\frac{r^2}{R^2}-\frac{R^4}{r^4}\right)\left(\frac{\partial\Sigma}{\partial I}+\frac{r^2}{R^2}\frac{\partial\Sigma}{\partial II}\right)\frac{dr}{r}$$

$$= \frac{8}{3\rho R_1{}^2}\int_{(\gamma+x^3)/(1+\gamma)}^{x^3} u^{-7/3}(1+u)\left(\frac{\partial\Sigma}{\partial I}+u^{2/3}\frac{\partial\Sigma}{\partial II}\right)du$$

where we employed the relation

$$dr/r = du/3u(1-u)$$

It should be also remembered that $\Sigma(u)$ is a function of invariants I and II, hence of $2u^{2/3}+u^{-4/3}$, and $u^{4/3}+2u^{-2/3}$. Moreover, since

$$dI/du = \tfrac{4}{3}(u^{-1/3}-u^{-7/3}),\qquad dII/du = \tfrac{4}{3}(u^{1/3}-u^{-5/3})$$

and

$$\frac{d\Sigma}{du} = \tfrac{4}{3}u^{-1/3}(1-u^{-2})\left(\frac{\partial\Sigma}{\partial I}+u^{2/3}\frac{\partial\Sigma}{\partial II}\right)$$

the function $g(x,\gamma)$ can be cast into another form similar to (3.3.21):

$$(3.4.11)\qquad g(x,\gamma) = \frac{2}{\rho R_1{}^2}\int_{(\gamma+x^3)/(1+\gamma)}^{x^3}\frac{d\Sigma}{du}\frac{du}{u-1}$$

Equation (3.4.9) is a nonlinear second-order ordinary differential equation for the motion of the inner surface of the spherical shell and should be supplemented by the initial conditions in the form:

$$x(0) = x_0,\qquad \dot{x}(0) = v(0) = v_0$$

After the function $x(t)$ is determined, the motion of any point of the shell may be obtained from $(3.4.8)_3$. The radial displacement of any point is found from the relation

$$u_r = r - R = R(u^{1/3}-1)$$

Equation (3.4.9) can be written in the following form that leads to the direct evaluation of $\dot{x} = v(t)$:

$$(3.4.12)\qquad \frac{d}{dx}\left\{\left[1-\left(1+\frac{\gamma}{x^3}\right)^{-1/3}\right]x^3\dot{x}^2\right\} + x^2 g(x,\gamma) = \frac{2x^2[p_1(t)-p_2(t)]}{\rho R_1{}^2}$$

These equations are essentially similar to those obtained by Guo Zhong-Heng and Solecki [1963 a,b], in a different way.

For a thin spherical shell, a simplified equation has been obtained by Wang [1965]. When $\gamma \to 0$ we can write $R_1 \simeq R_2 = R$, $u \sim x^3$, and[1]

$$[1 + (\gamma/x^3)]^{-1/3} \simeq 1 - (\gamma/3x^3), \qquad [1 + (\gamma/x^3)]^{-4/3} \simeq 1 - (4\gamma/3x^3)$$
$$g(x, \gamma) \simeq (2/\rho R^2)(d\Sigma/dx^3) = (2\gamma/3\rho R^2 x^2)(d\Sigma/dx)$$

Hence (3.4.9) is approximated by the equation

$$\ddot{x} + \frac{1}{\rho R^2} \frac{d\Sigma}{dx} = \frac{3x^2[p_1(t) - p_2(t)]}{\gamma \rho R^2}$$

where Σ is a known function of $2x^2 + x^{-4}$ and $x^4 + 2x^{-2}$.

A. Free Oscillations

In the case of free oscillations, the right-hand side of (3.4.12) vanishes. We define a function $G(x, \gamma)$ by

$$G(x, \gamma) = \int_1^x \xi^2 g(\xi, \gamma)\, d\xi = \frac{2}{\rho R_1{}^2} \int_1^x \xi^2 \int_{(\gamma + \xi^3)/(1+\gamma)}^{\xi^3} \frac{d\Sigma}{du} \frac{du}{u - 1}\, d\xi$$

A change in the order of integration [as was done in Section 3.3 to arrive at (3.3.25)] results in

$$G(x, \gamma) = \frac{2(x^3 - 1)}{3\rho R_1{}^2} \int_{(\gamma + x^3)/(1+\gamma)}^{x^3} \frac{\Sigma(u)}{(u - 1)^2}\, du \tag{3.4.13}$$

Here now B–E inequalities (Section 1.18i) imply that

$$\frac{\partial \Sigma}{\partial I} + \frac{r^2}{R^2} \frac{\partial \Sigma}{\partial II} > 0 \tag{3.4.14}$$

Therefore, the sign of $d\Sigma/du$ would be solely determined by u. Following closely the same arguments as in Section 3.3 we can show that both functions $g(x, \gamma)$ and $G(x, \gamma)$ have exactly the same properties as the functions $f(x, \gamma)$ and $F(x, \gamma)$ of that section.

The first integral of (3.4.12) turns out to be

$$\{1 - [1 + (\gamma/x^3)]^{-1/3}\} x^3 v^2 + G(x, \gamma) = C$$

where C is an arbitrary constant to be determined from the initial conditions $v = v_0$ when $x = x_0$. Thus

$$C = \{1 - [1 + (\gamma/x_0{}^3)]^{-1/3}\} x_0{}^3 v_0{}^2 + G(x_0, \gamma) \tag{3.4.15}$$

Hence the velocity $v(t)$ of the inner surface of the shell becomes

$$v = \mp([C - G(x, \gamma)] x^{-3} \{1 - [1 + (\gamma/x^3)]^{-1/3}\}^{-1})^{1/2} \tag{3.4.16}$$

[1] Cf. footnote 1, p.153.

Following the arguments in Section 3.3 we can deduce that the energy curves in the (v, x) phase plane are closed, and the equation

$$G(x, \gamma) = C$$

with C given by (3.4.15) has exactly two roots $x = a < 1$ and $x = b > 1$, which are amplitudes of the oscillations of the inner surface. The period of the motion is calculated by

$$T = 2 \int_a^b v^{-1}\, dx \tag{3.4.17}$$

$$= 2 \int_a^b x^{3/2}\{1 - [1 + (\gamma/x^3)]^{-1/3}\}^{1/2}[C - G(x, \gamma)]^{-1/2}\, dx$$

An approximate expression can be obtained for thin shells by letting $\gamma \to 0$ in (3.4.13):

$$G(x, \gamma) \simeq (2\gamma/3\rho R^2)\Sigma(x^3)$$

The equation for the amplitudes of motion takes the following limiting form:

$$\Sigma(x^3) \simeq \Sigma(x_0^{\,3}) + (\rho R^2 v_0^{\,2}/2) \tag{3.4.18}$$

Since $\Sigma(u) = \Sigma(1/u)$, it is clear that if $a_0 < 1$ is a root of (3.4.18) then $b_0 = 1/a_0$ will be the second root. Therefore, the period of the free oscillations of a thin shell is

$$T_0 = 2 \int_{a_0}^{1/a_0} \{v_0^{\,2} + (2/\rho R^2)[\Sigma(x_0^{\,3}) - \Sigma(x^3)]\}^{1/2}\, dx \tag{3.4.19}$$

The motion of the shell due to an impulsive pressure may be examined in the same fashion as in the case of the cylinder problem.

B. Spherical Cavity in Infinite Medium

The radial motion of an infinite elastic medium with a spherical cavity can be dealt with as a special case of the radial motion of a spherical shell. The specialization[1] may be achieved by letting $R_2 \to \infty$, hence $\gamma \to \infty$.

The limiting form of (3.4.9) as $\gamma \to \infty$ and $p_2(t) \to 0$ is

$$2x\ddot{x} + 3\dot{x}^2 + \bar{g}(x) = 2p_1(t)/\rho R_1^{\,2} \tag{3.4.20}$$

where

$$\bar{g}(x) \equiv g(x, \infty) = \frac{2}{\rho R_1^{\,2}} \int_1^{x^3} (u - 1)^{-1} \frac{d\Sigma}{du}\, du$$

[1] Knowles and Jakub [1965] treated this problem independently of the shell problem.

In the same fashion we can write the limiting form of $G(x, \gamma)$ from (3.4.13):

$$\bar{G}(x) = G(x, \infty) = \frac{2(x^3 - 1)}{3\rho R_1^{\,2}} \int_1^{x^3} \frac{\Sigma(u)}{(u-1)^2}\, du \tag{3.4.21}$$

The circumferential stress $t_{\theta\theta}$ on the cavity wall $r = R_1 x$ is determined from

$$t_{\theta\theta} - t_{rr} = 2\left(\frac{r^2}{R^2} - \frac{R^4}{r^4}\right)\left(\frac{\partial \Sigma}{\partial I} + \frac{r^2}{R^2}\frac{\partial \Sigma}{\partial II}\right), \qquad t_{rr}\Big|_{r=R_1 x} = -p_1(t)$$

We find

$$t_{\theta\theta}\Big|_{r=R_1 x} = 2(x^2 - x^{-4})\left(\frac{\partial \Sigma}{\partial I} + x^2 \frac{\partial \Sigma}{\partial II}\right)\Big|_{r=R_1 x} - p_1(t) \tag{3.4.22}$$

or

$$\frac{t_{\theta\theta}}{\rho R_1^{\,2}}\Big|_{r=R_1 x} = \frac{1}{4}(x^3 - 1)x\,\frac{d\bar{g}(x)}{dx} - \frac{p_1(t)}{\rho R_1^{\,2}} \tag{3.4.23}$$

Equation (3.4.15) determining the integration constant C, in case of *free oscillations*, now becomes

$$C = x_0^{\,3} v_0^{\,2} + \frac{2(x_0^{\,3} - 1)}{3\rho R_1^{\,2}} \int_1^{x_0^{\,3}} \frac{\Sigma(u)}{(u-1)^2}\, du = x_0^{\,3} v_0^{\,2} + \bar{G}(x_0) \tag{3.4.24}$$

where x_0 and v_0 are the usual initial values. Therefore, the two amplitudes of the wall of the cavity are the roots $x = a$ and $x = b$ of the equation

$$(x^3 - 1)\int_1^{x^3} \frac{\Sigma(u)}{(u-1)^2}\, du = \tfrac{3}{2}\rho R_1^{\,2} v_0^{\,2} x_0^{\,3} + (x_0^{\,3} - 1)\int_1^{x_0^{\,3}} \frac{\Sigma(u)}{(u-1)^2}\, du \tag{3.4.25}$$

and the period follows from (3.4.17)

$$T = 2\int_a^b x\left[\frac{x}{C - \bar{G}(x)}\right]^{1/2} dx \tag{3.4.26}$$

As a special case, consider a medium made of Mooney material for which

$$\Sigma(u) = \alpha(2u^{2/3} + u^{-4/3} - 3) + \beta(u^{4/3} + 2u^{-2/3} - 3)$$

The functions $\bar{g}(x)$ and $\bar{G}(x)$ are determined to be

$$\begin{aligned} \bar{g}(x) &= \frac{2}{\rho R_1^{\,2}}\left[(5 - 4x^{-1} - x^{-4})\alpha + 2(2x - 1 - x^{-2})\beta\right] \\ \bar{G}(x) &= \frac{2(x^3 - 1)}{3\rho R_1^{\,2}}\,\frac{(5x^3 - x^2 - x - 3)\alpha + (3x^4 + x^3 + x^2 - 5x)\beta}{x^3 + x^2 + x} \end{aligned} \tag{3.4.27}$$

Hence it follows from (3.4.27) and (3.4.25) that amplitudes a and b are the roots of the equation

$$\begin{aligned}(3.4.28)\quad &\frac{x-1}{x}[(5x^3 - x^2 - x - 3)\alpha + (3x^4 + x^3 + x^2 - 5x)\beta]\\ &= \tfrac{3}{2}\rho R_1{}^2 x_0{}^3 v_0{}^2 + \frac{x_0 - 1}{x_0}\\ &\times [(5x_0{}^3 - x_0{}^2 - x_0 - 3)\alpha + (3x_0{}^4 + x_0{}^3 + x_0{}^2 - 5x_0)\beta]\end{aligned}$$

and the period may be obtained from (3.4.26). If we note that in lieu of (3.4.24) we can also write[1] $C = \bar{G}(b)$ then (3.4.26) takes the form

$$\begin{aligned}(3.4.29)\quad T = 2(\tfrac{3}{2}\rho R_1{}^2)^{1/2} \int_a^b x^2(b-x)^{-1/2}\{&[5x^3 + (5b-6)x^2 + b(5b-6)x\\ &- (3/b)]\alpha + [3x^4 + (3b-2)x^3 + b(3b-2)x^2\\ &+ (3b^3 - 2b^2 - 6)x]\beta\}^{-1/2}\, dx\end{aligned}$$

The circumferential stress on the wall of the cavity is therefore obtained to be

$$(3.4.30)\qquad t_{\theta\theta}|_x = 2(x^2 - x^{-4})(\alpha + \beta x^2)$$

The values of this stress component at the maximum and minimum displacements are obtained by replacing x by $b > 1$ and $a < 1$, respectively:

$$(3.4.31)\qquad \begin{aligned} t_{\theta\theta}|_a &= 2(a^2 - a^{-4})(\alpha + \beta a^2) < 0\\ t_{\theta\theta}|_b &= 2(b^2 - b^{-4})(\alpha + \beta b^2) > 0\end{aligned}$$

Another simple problem concerning the spherical cavity is the case of impulsive application of a uniform traction on the cavity wall which is at rest initially. In this case the pressure function is given by

$$2p_1(t)/\rho R_1{}^2 = pH(t)$$

where p is a constant and $H(t)$ is the Heaviside unit step function. Substitution of this function into (3.4.12) with $\gamma \to \infty$ yields

$$(3.4.32)\qquad (d/dx)(x^3\dot{x}^2) + x^2\bar{g}(x) = px^2, \qquad t \geq 0$$

On integrating (3.4.32) under the initial conditions

$$x(0) = 1, \qquad \dot{x}(0) = 0$$

[1] We have chosen the root $b > 1$ for reasons which will become evident in the subsequent analysis.

we get

$$x^3\dot{x}^2 + \bar{G}(x) = \tfrac{1}{3}p(x^3 - 1) \tag{3.4.33}$$

or

$$\dot{x} = v = \mp x^{-3/2}[\tfrac{1}{3}p(x^3 - 1) - \bar{G}(x)]^{1/2} \tag{3.4.34}$$

The complete analysis of this case is similar to the problem which we have discussed in the preceding section. Therefore we omit the details and give the results only.

For a periodic motion the equation

$$\bar{G}(x) = \tfrac{1}{3}p(x^3 - 1) \tag{3.4.35}$$

should have only one root a, in addition to the root $x = 1$. This is secured by appropriate conditions imposed on the strain energy function. Thus $x = a$ is the unique solution of the equation

$$p = \frac{2}{\rho {R_1}^2} \int_1^{x^3} \frac{\Sigma(u)}{(u-1)^2}\, du \tag{3.4.36}$$

if we employ (3.4.21) in (3.4.35). It is clear from the above equation that $a > 1$ if $p > 0$ and $a < 1$ if $p < 0$.

Since $\bar{G}'(x) = x^2\bar{g}(x)$ and $x = 1$ and $x = a$ both satisfy (3.4.35), there is a number b between 1 and a which satisfies the equation $x^2\bar{g}(x) = px^2$, or

$$\bar{g}(b) = p \tag{3.4.37}$$

If we introduce this into (3.4.32) we see that $x = b$ corresponds to the equilibrium configuration of the cavity wall under the action of the constant pressure $p\rho {R_1}^2/2$. Therefore, the cavity wall oscillates about an equilibrium configuration.

In order to obtain a periodic motion for every value of p, the integral in (3.4.36) should be unbounded for the limits $x \to 0$ and $x \to \infty$. If (3.4.36) is bounded in these limits, then the periodic motion becomes possible if and only if

$$-\frac{2}{\rho {R_1}^2} \int_0^1 \frac{\Sigma(u)}{(u-1)^2}\, du < p < \frac{2}{\rho {R_1}^2} \int_1^\infty \frac{\Sigma(u)}{(u-1)^2}\, du \tag{3.4.38}$$

The maximum circumferential stress on the cavity wall is calculated from (3.4.22) as

$$t_{\theta\theta}\Big|_a = 2(a^2 - a^{-4})\left(\frac{\partial \Sigma}{\partial I} + a^2 \frac{\partial \Sigma}{\partial II}\right) - \frac{1}{2} p\rho {R_1}^2$$

The maximum stress corresponding to the equilibrium configuration is simply

$$t_{\theta\theta}\Big|_b = 2(b^2 - b^{-4})\left(\frac{\partial\Sigma}{\partial I} + b^2\,\frac{\partial\Sigma}{\partial II}\right) - \frac{1}{2}\,p\rho R_1^{\,2}$$

Finally, the period of motion reads

$$T = 2\left|\int_1^a v^{-1}\,dx\right| = 2\left|\int_1^a x^{3/2}[\tfrac{1}{3}p(x^3-1) - \bar{G}(x)]^{-1/2}\,dx\right| \tag{3.4.39}$$

For a special medium made of Mooney material, utilizing (3.4.27) we find that a is the unique root of the equation

$$p = \frac{2}{\rho R_1^{\,2}}\,\frac{(5x^3 - x^2 - x - 3)\alpha + (3x^4 + x^3 + x^2 - 5x)\beta}{x^3 + x^2 + x}\Bigg|_{x=a} \tag{3.4.40}$$

Since the right-hand side of (3.4.40) is unbounded for $x \to 0$ and $x \to \infty$, the resulting motion is periodic for every value of p, positive or negative. The equilibrium solution is obtained from (3.4.37) and $(3.4.27)_1$:

$$p = \frac{2}{\rho R_1^{\,2}}\,[(5 - 4x^{-1} - x^{-4})\alpha + 2(2x - 1 - x^{-2})\beta]\,\big|_{x=b} \tag{3.4.41}$$

TABLE 3.4.1

Static and dynamic pressures corresponding to b and a, respectively

	Neo-Hookean material ($\beta/\alpha = 0$)				Mooney materials ($\beta/\alpha = 0.2$)			
a,b	Static $p\rho R_1^2/4\alpha$	Dynamic $p\rho R_1^2/4\alpha$	Static $t_{\theta\theta}/2\alpha$	Dynamic $t_{\theta\theta}/2\alpha$	Static $p\rho R_1^2/4\alpha$	Dynamic $p\rho R_1^2/4\alpha$	Static $t_{\theta\theta}/2\alpha$	Dynamic $t_{\theta\theta}/2\alpha$
1	0	0	0	0	0	0	0	0
1.1	0.340	0.189	0.187	0.337	0.415	0.229	0.234	0.425
1.2	0.592	0.343	0.366	0.614	0.733	0.421	0.500	0.813
1.3	0.786	0.481	0.554	0.858	0.988	0.596	0.805	1.196
1.4	0.941	0.603	0.758	1.097	1.199	0.754	1.167	1.612
1.5	1.068	0.710	0.985	1.342	1.379	0.897	1.597	2.079
1.6	1.174	0.807	1.234	1.601	1.536	1.029	2.104	2.611
1.7	1.264	0.893	1.507	1.877	1.674	1.149	2.697	3.222
1.8	1.341	0.971	1.803	2.173	1.800	1.262	3.383	3.921
1.9	1.409	1.042	2.124	2.491	1.914	1.366	4.171	4.718
2.0	1.469	1.107	2.469	2.830	2.019	1.464	5.069	5.623
2.1	1.522	1.166	2.837	3.192	2.117	1.557	6.086	6.646
2.2	1.570	1.221	3.228	3.576	2.208	1.644	7.233	7.797
2.3	1.613	1.272	3.642	3.983	2.295	1.727	8.519	9.087
2.4	1.652	1.318	4.078	4.412	2.377	1.805	9.954	10.525
2.5	1.687	1.362	4.537	4.863	2.455	1.881	11.550	12.124

After some manipulations the period T may be expressed as follows:

(3.4.42)
$$T = 2(\tfrac{1}{2}\rho R_1{}^2)^{1/2}\left|\int_1^a x^2[(x-1)(a-x)]^{-1/2}\right.$$
$$\times\left[\left(\frac{2a^2+2a+1}{a^3+a^2+a}x^2+\frac{2a^2+a+1}{a^3+a^2+a}x+\frac{1}{a}\right)\alpha\right.$$
$$\left.\left.+\left(x^3+\frac{a^2+a+2}{a^2+a+1}x^2+\frac{a^2+2a+2}{a^2+a+1}x\right)\beta\right]^{-1/2}dx\right|$$

The two asymptotic approximation to the period T can be deduced from (3.4.42) for large amplitudes (in the limit $a \to \infty$) and small motions (in the limit $a \to 1$):

(3.4.43) $$T \simeq 4(\rho R_1{}^2 a/2\beta)^{1/2}, \quad a \to \infty; \qquad T \simeq [\rho R_1{}^2/2(\alpha+\beta)]^{1/2}, \quad a \to 1$$

The last result is consistent with the one corresponding to linear incompressible solids. Knowles and Jakub [1965] have computed the static and dynamic pressures and also circumferential stresses corresponding to a given amplitude for $\beta/\alpha = 0$ (neo-Hookean materials) and $\beta/\alpha = 0.2$. Their results are reproduced in Table 3.4.1.

3.5 SIMPLE WAVES—SPECIAL THEORY

A special type of unsteady motions which bear a resemblance to plane wave solutions of the linear theory of elasticity can be constructed and used to obtain solutions of some interesting boundary and initial-value problems. The underlying method is based on the simple wave solutions of a system of hyperbolic partial differential equations. An account on the subject is presented in Appendix B for differential equations in two independent variables.

The simple wave theory has been widely utilized in constructing solutions of various problems in gas dynamics. In particular, one-dimensional nonsteady motion of inviscid gases has been studied extensively by means of the simple wave theory (cf. Courant and Friedricks [1948], Sedov [1970]). These studies, in fact, have motivated parallel investigations in nonlinear elastodynamics. In this section we present a discussion of one-dimensional motion, leaving the general theory to Section 3.8.

A motion will be called a *plane wave* if it is constant, for fixed times, on planes with fixed normal direction of propagation. Without loss in generality one may select a rectangular coordinate frame in such a way that the X_1-axis is directed along the direction of propagation. In a common rectangular frame of reference, for both material and spatial coordinates, a plane wave motion may be expressed in the form[1]

$$x_k(\mathbf{X}, t) = X_K\, \delta_{kK} + u_k(X_1, t) \tag{3.5.1}$$

where u_k are components of the displacement vector. For this particular motion, the deformation gradients, the Green deformation tensor, and its invariants are given by

(3.5.2)

$$\mathbf{F} = [x_{k,K}] = \begin{bmatrix} 1+p_1 & 0 & 0 \\ p_2 & 1 & 0 \\ p_3 & 0 & 1 \end{bmatrix} \qquad I = 2 + (1+p_1)^2 + p_2{}^2 + p_3{}^2$$

$$II = 1 + 2(1+p_1)^2 + p_2{}^2 + p_3{}^2$$

$$\mathbf{C} = \begin{bmatrix} (1+p_1)^2 + p_2{}^2 + p_3{}^2 & p_2 & p_3 \\ p_2 & 1 & 0 \\ p_3 & 0 & 1 \end{bmatrix} \qquad III = (1+p_1)^2$$

where

$$p_k(X_1, t) = \partial u_k / \partial X_1 \qquad \text{or} \qquad \mathbf{p}(X_1, t) = \partial \mathbf{u} / \partial X_1 \tag{3.5.3}$$

It follows from $(3.5.2)_1$ that the strain energy Σ is a function of $\mathbf{p}$. Therefore, the stresses $T_{1k}(X_1, t)$ appearing in the equations of motion

$$(\partial T_{1k} / \partial X_1) + \rho_0 f_k = \rho_0\, \partial^2 u_k / \partial t^2 \tag{3.5.4}$$

and acting on planes $X_1 =$ const are given by

$$T_{1k} = \partial \Sigma / \partial p_k \tag{3.5.5}$$

The corresponding Cauchy stress components are

$$t_{1k} = j^{-1}\, (\partial x_1 / \partial X_1) T_{1k} = T_{1k} \tag{3.5.6}$$

Substituting (3.5.5) into (3.5.4) and recalling (3.5.3), we obtain the equations of motion in terms of displacements, with vanishing body forces

$$A_{kl}\, \partial^2 u_l / \partial X^2 = \rho_0\, \partial^2 u_k / \partial t^2 \tag{3.5.7}$$

where the symmetric matrix A_{kl} is defined by

$$A_{kl}(\mathbf{p}) = A_{1k1l} = \partial^2 \Sigma / \partial p_k\, \partial p_l \tag{3.5.8}$$

[cf. (1.17.12)], and we replaced X_1 by X since there will be no ambiguity in this specification. Equations (3.5.7) constitute a *quasilinear* (linear in second-order derivatives) second-order system of partial differential equations with

[1] See Bland [1965b]. However, the presentation here is different to some extent.

three dependent variables u_1, u_2, and u_3, and two independent variables X and t. We are not able to describe a general treatment of these equations.

Nevertheless, the solution of a large class of problems of practical importance can be examined in a fairly general way by the method of *simple wave solutions* discussed in Appendix B. The applicability of the method is based on the hyperbolic nature of the system (3.5.7) which places restrictions on the coefficient functions A_{kl}. The method, as we shall see, enables us to use efficiently the classical method of characteristics, which is particularly suited to numerical computations. The simple wave solutions are invaluable in studying wave motions in infinite and semiinfinite media or in media with nonreflecting boundaries.

In order to apply directly the method explored in Appendix B we replace the second-order system (3.5.7) by an equivalent first-order larger system. Thus we set

$$\mathbf{p} = \partial \mathbf{u}/\partial X, \qquad \mathbf{v} = \partial \mathbf{u}/\partial t \tag{3.5.9}$$

so that the system now consists of six equations

$$\rho_0^{-1} A_{kl}(\mathbf{p}) \frac{\partial p_l}{\partial X} - \frac{\partial v_k}{\partial t} = 0, \qquad \frac{\partial p_k}{\partial t} - \frac{\partial v_l}{\partial X} = 0 \tag{3.5.10}$$

where the last equations are the consequence of definitions (3.5.9). The system (3.5.10) may be put in a more compact form by defining a column vector $\mathbf{U}$ by

$$\mathbf{U} = \begin{bmatrix} \mathbf{v} \\ \mathbf{p} \end{bmatrix} = \begin{bmatrix} v_1 \\ v_2 \\ v_3 \\ p_1 \\ p_2 \\ p_3 \end{bmatrix} \tag{3.5.11}$$

and a square, 6×6 matrix $\mathscr{A}$ by

$$\mathscr{A} = \left[\begin{array}{c|c} \mathbf{0} & -\rho_0^{-1}\mathbf{A} \\ \hline -\mathbf{I} & \mathbf{0} \end{array}\right] = \begin{bmatrix} 0 & 0 & 0 & -\rho_0^{-1}A_{11} & -\rho_0^{-1}A_{12} & -\rho_0^{-1}A_{13} \\ 0 & 0 & 0 & -\rho_0^{-1}A_{21} & -\rho_0^{-1}A_{22} & -\rho_0^{-1}A_{23} \\ 0 & 0 & 0 & -\rho_0^{-1}A_{31} & -\rho_0^{-1}A_{32} & -\rho_0^{-1}A_{33} \\ -1 & 0 & 0 & 0 & 0 & 0 \\ 0 & -1 & 0 & 0 & 0 & 0 \\ 0 & 0 & -1 & 0 & 0 & 0 \end{bmatrix} \tag{3.5.12}$$

The system (3.5.10) may now be rewritten as

$$\mathscr{A} \frac{\partial \mathbf{U}}{\partial X} + \frac{\partial \mathbf{U}}{\partial t} = \mathbf{0} \tag{3.5.13}$$

The characteristic determinant of the system is

$$Q = \det(\mathscr{A} - \lambda \mathbf{I}) = \det(\lambda \mathbf{I} - \mathscr{A}) = \begin{vmatrix} \lambda \mathbf{I} & \rho_0^{-1}\mathbf{A} \\ \mathbf{I} & \lambda \mathbf{I} \end{vmatrix}$$

(3.5.14)

$$= \begin{vmatrix} \lambda \mathbf{I} & \rho_0^{-1}\mathbf{A} - \lambda^2 \mathbf{I} \\ \mathbf{I} & \mathbf{0} \end{vmatrix} = \det[\rho_0^{-1}\mathbf{A}(\mathbf{p}) - \lambda^2 \mathbf{I}] = 0$$

where we used some well-known properties of partitioned matrices (cf. Bishop *et al.* [1965, Section 3.7]).

The six roots of this determinantal equation may be represented as

$$\lambda = \mp \lambda_i(\mathbf{p}), \qquad i = 1, 2, 3$$

Since for homogeneous materials Σ is assumed to be a function of displacement gradients only, the characteristic values λ are also functions of displacement gradients.

If all λ are *real* and *distinct* we call the system *totally hyperbolic*. In totally hyperbolic systems we can determine six real families of characteristics in the X,t-plane on integrating the following system of first-order, ordinary differential equations:

(3.5.15) $$dX/dt = \mp \lambda_i(\mathbf{p}), \qquad i = 1, 2, 3$$

Two families of characteristics C_i^+ and C_i^- corresponding to the same $|\lambda_i|$ are bisected by $X = \text{const}$ or $t = \text{const}$ lines in the X,t-plane (Fig. 3.5.1). It should be noted that (3.5.15) cannot determine characteristics without explicit knowledge of *the solution* $\mathbf{p} = \mathbf{p}(X, t)$. If the solution is known, then the right-hand side is a function of (X, t) and (3.5.15) becomes an ordinary

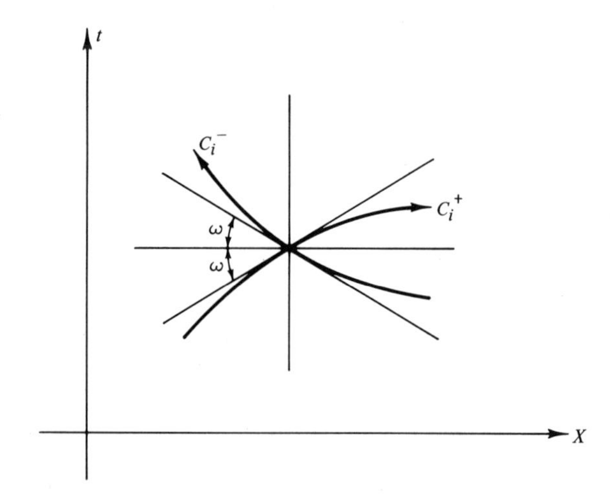

Fig. 3.5.1 Characteristics in the X,t-plane.

differential equation in X and t. A solution of (3.5.15) can be obtained as $\theta(X, t) = \text{constant}$, and this represents a plane wave front propagating in X-direction whose velocity (normal velocity) is either λ or $-\lambda$. Hence (3.5.15) expresses the fact that λ_i are the speed of propagation of plane waves in the material.

Since the nature of the system (3.5.13) evidently depends on the existence of *real* characteristic values λ_i, we must examine carefully the conditions leading to this conclusion. Because of the symmetry of A_{kl} all λ_i^2 are real. Whether they are also positive for all deformations relies upon the positive-definiteness of the same matrix. If we further assume that the strain energy function Σ is a convex function of its arguments, i.e.,

$$\Sigma(\bar{\mathbf{p}}) - \Sigma(\mathbf{p}) > (\bar{\mathbf{p}} - \mathbf{p}) \cdot \partial\Sigma/\partial\mathbf{p}$$

for all $\bar{\mathbf{p}} \neq \mathbf{p}$, then it can be shown that the matrix $A_{kl} = \partial^2\Sigma/\partial p_k \partial p_l$ is positive definite. The convexity of the strain energy function being a common assumption of the theory of elasticity (cf. Section 1.18iii), henceforth we suppose that all λ_i are real. The repeated roots (3.5.14) do not cause an undue complication in the formalism, so we may also assume that all λ_i are distinct in the present investigation.

We now look for *a simple wave solution* corresponding to a particular characteristic value $\lambda_{(k)}$ out of six characteristic roots. We have proved that (see Appendix B) the characteristics C_k corresponding to $\lambda_{(k)}$ are all straight lines and $\mathbf{U}$—consequently, $\mathbf{v}$ and $\mathbf{p}$—remains constant on these lines. The five independent *k-Riemann invariants* are obtained as the solution of

$$dv_1/r_1^{(k)} = dv_2/r_2^{(k)} = dv_3/dr_3^{(k)} = dp_1/r_4^{(k)} = dp_2/r_5^{(k)} = dp_3/r_6^{(k)} \tag{3.5.16}$$

where $\mathbf{r}^{(k)}(\mathbf{p})$ is the right eigenvector of $\mathscr{A}$ corresponding to $\lambda_{(k)}$, that is,

$$(\mathscr{A} - \lambda_{(k)}\mathbf{I})\mathbf{r}^{(k)} = \mathbf{0} \tag{3.5.17}$$

If we now write

$$\mathbf{r}^{(k)} = \begin{bmatrix} \boldsymbol{\mu}^{(k)} \\ \mathbf{v}^{(k)} \end{bmatrix} \qquad \boldsymbol{\mu}^{(k)} = \begin{bmatrix} r_1^{(k)} \\ r_2^{(k)} \\ r_3^{(k)} \end{bmatrix} \qquad \mathbf{v}^{(k)} = \begin{bmatrix} r_4^{(k)} \\ r_5^{(k)} \\ r_6^{(k)} \end{bmatrix}$$

and consider (3.5.12), we can put (3.5.17) into the form

$$-\lambda_{(k)}\,\boldsymbol{\mu}^{(k)} - \rho_0^{-1}\,\mathbf{A}\mathbf{v}^{(k)} = \mathbf{0}, \qquad -\boldsymbol{\mu}^{(k)} - \lambda_{(k)}\mathbf{v}^{(k)} = \mathbf{0}$$

Therefore, we have

$$\boldsymbol{\mu}^{(k)} = -\lambda_{(k)}\,\mathbf{v}^{(k)} \tag{3.5.18}$$

and $\mathbf{v}^{(k)}$ is obtained from

(3.5.19) $$(\rho_0^{-1}\mathbf{A} - \lambda_{(k)}^2\mathbf{I})\mathbf{v}^{(k)} = \mathbf{0}$$

In view of (3.5.18) we deduce from (3.5.16) that

(3.5.20) $$dv_i + \lambda_{(k)}\, dp_i = 0$$

for the k-simple wave. The relations

(3.5.21) $$dp_1/v_1^{(k)} = dp_2/v_2^{(k)} = dp_3/v_3^{(k)}$$

enable us to calculate the two k-Riemann invariants and to express p_1, p_2, and p_3 in terms of a single variable $\alpha_{(k)}$, as $p_i = p_i(\alpha_{(k)})$. $\alpha_{(k)}$ can, of course, be chosen as one of the displacement gradients p_i. The remaining three k-Riemann invariants are then found from (3.5.20) as

$$v_i + \int \lambda_{(k)}\, dp_i = \text{const}$$

or, using the first two k-Riemann invariants,

(3.5.22) $$v_i(\alpha_{(k)}) + \int \lambda_{(k)}(\alpha_{(k)})(dp_i/d\alpha_{(k)})\, d\alpha_{(k)} = \text{const}$$

We know that $\lambda_{(k)}$ are found in pairs of opposite signs. Since $\mathbf{v}^{(k)}$ is the same for $\mp\lambda_{(k)}$ the first two Riemann invariants associated with simple waves corresponding to either $\lambda_{(k)}$ or $-\lambda_{(k)}$ remain unchanged. The other three invariants are basically the same as (3.5.20) and (3.5.22) except for the $(-)$ sign in lieu of $(+)$ in that expression.

We conclude that there exist in general, six simple waves in an elastic material. We also know that simple waves should be adjacent to a *constant state* in which $\mathbf{p}$ and $\mathbf{v}$ take prescribed constant values $\mathbf{p}_0$ and $\mathbf{v}_0$. Such a situation may be realized, for instance, in the case of plane waves propagating into an elastic medium with homogeneous deformation and uniform velocity. The k-simple wave region is separated from the region of constant state by a straight boundary characteristic C_k. Since k-Riemann invariants have the same value in the regions of simple wave and constant state, the five constants which appear in the solutions of (3.5.21) and (3.5.22) can be determined in terms of given vectors $\mathbf{p}_0$ and $\mathbf{v}_0$ in the constant state. The values of $\alpha_{(k)}$ associated with the straight characteristics covering the k-simple wave region can only be specified by prescribed boundary or initial conditions.

After having obtained $\mathbf{v}$ and $\mathbf{p}$ we can calculate $\mathbf{u}$ upon integrating

$$\mathbf{u} = \int (\mathbf{p}\, dX + \mathbf{v}\, dt)$$

subject to some boundary and initial conditions.

The simple wave solution so obtained does not allow reflections since these violate the restrictions that simple waves can only be adjacent to constant states, and it is valid only in a domain in the X, t-plane in which the characteristic field of the k-simple wave is a family of nonintersecting straight lines. The intersecting k-characteristics evidently imply a discontinuity in $\mathbf{p}$ and $\mathbf{v}$ leading to a shock formation in the solid.

For isotropic solids further simplifications can be achieved leading to more tractable results. In such solids Σ is a function of the invariants I, II, and III. This implies that [see Eqs. (3.5.2)] $\Sigma = \Sigma(p, q^2)$, where

$$p \equiv p_1, \qquad q^2 \equiv p_2^{\,2} + p_3^{\,2}$$

The nonvanishing components of the stress tensor are [cf. (3.5.5) and (3.5.6)]

$$t_{11} = \partial\Sigma/\partial p, \qquad t_{12} = (p_2/p)\,\partial\Sigma/\partial q, \qquad t_{13} = (p_3/q)\,\partial\Sigma/\partial q$$

The components of A_{kl} are now given by

$$\begin{aligned}
A_{11} &= \frac{\partial^2\Sigma}{\partial p_1^{\,2}} = \frac{\partial^2\Sigma}{\partial p^2} = 2\,D\Sigma + 4(1+p_1)^2\,D^2\Sigma \\
A_{12} &= \frac{\partial^2\Sigma}{\partial p_1\,\partial p_2} = \frac{p_2}{q}\,\frac{\partial^2\Sigma}{\partial p\,\partial q} = 2p_2(1+p_1)\,D\,D_1\Sigma \\
A_{13} &= \frac{\partial^2\Sigma}{\partial p_1\,\partial p_3} = \frac{p_3}{q}\,\frac{\partial^2\Sigma}{\partial p\,\partial q} = 2p_3(1+p_1)\,D\,D_1\Sigma \\
A_{22} &= \frac{\partial^2\Sigma}{\partial p_2^{\,2}} = \frac{p_2^{\,2}}{q^2}\left(\frac{\partial^2\Sigma}{\partial q^2} - \frac{1}{q}\frac{\partial\Sigma}{\partial q}\right) + \frac{1}{q}\frac{\partial\Sigma}{\partial q} = 2\,D_1\Sigma + 4p_2^{\,2}\,D_1^{\,2}\Sigma \\
A_{23} &= \frac{\partial^2\Sigma}{\partial p_2\,\partial p_3} = \frac{p_2\,p_3}{q^2}\left(\frac{\partial^2\Sigma}{\partial q^2} - \frac{1}{q}\frac{\partial\Sigma}{\partial q}\right) = 4p_2\,p_3\,D_1^{\,2}\Sigma \\
A_{33} &= \frac{\partial^2\Sigma}{\partial p_3^{\,2}} = \frac{p_3^{\,2}}{q^2}\left(\frac{\partial^2\Sigma}{\partial q^2} - \frac{1}{q}\frac{\partial\Sigma}{\partial q}\right) + \frac{1}{q}\frac{\partial\Sigma}{\partial q} = 2\,D_1\Sigma + 4p_3^{\,2}\,D_1^{\,2}\Sigma
\end{aligned} \tag{3.5.23}$$

where the differential operators D and D_1 are defined by

$$D = \frac{\partial}{\partial I} + 2\,\frac{\partial}{\partial II} + \frac{\partial}{\partial III}, \qquad D_1 = \frac{\partial}{\partial I} + \frac{\partial}{\partial II}$$

A scrutiny of the coefficients A_{kl} in (3.5.23) reveals some facts concerning the nature of the waves. Since A_{12} and A_{13} vanish for $p_2 = p_3 = 0$, the first equation of (3.5.7) reduces to an equation in u_1 only and becomes uncoupled from the remaining equations determining solely u_2 and u_3 for a deformation field in which only $p_1 \neq 0$. This permits us to consider purely dilatational waves ($u_1 \neq 0$, $u_2 = u_3 = 0$) in elastic solids. On the other hand, the vanishing

p_1 does not lead to an uncoupling between u_1 and u_2 or u_3, in general, thereby indicating that the purely transverse waves generally cannot be generated in a nonlinear elastic solid. However, we shall see that such waves may exist in incompressible elastic solids. It is also common knowledge that they are present in linearly elastic solids.

With the help of (3.5.23), after some manipulations, (3.5.14) can be factored as

$$\left(\rho_0\lambda^2 - \frac{1}{q}\frac{\partial\Sigma}{\partial q}\right)\left[(\rho_0\lambda^2)^2 - \left(\frac{\partial^2\Sigma}{\partial p^2} + \frac{\partial^2\Sigma}{\partial q^2}\right)\rho_0\lambda^2 + \frac{\partial^2\Sigma}{\partial p^2}\frac{\partial^2\Sigma}{\partial q^2} - \left(\frac{\partial^2\Sigma}{\partial p\,\partial q}\right)^2\right] = 0$$

from which we deduce that the characteristic values are

$$\rho_0\lambda^2_{(1),(2)} = \frac{1}{2}\left\{\frac{\partial^2\Sigma}{\partial p^2} + \frac{\partial^2\Sigma}{\partial q^2} \mp \left[\left(\frac{\partial^2\Sigma}{\partial p^2} - \frac{\partial^2\Sigma}{\partial q^2}\right)^2 + 4\left(\frac{\partial^2\Sigma}{\partial p\,\partial q}\right)^2\right]^{1/2}\right\}$$

$$\rho_0\lambda^2_{(3)} = \frac{1}{q}\frac{\partial\Sigma}{\partial q} \tag{3.5.24}$$

Therefore, in an isotropic elastic solid, we can have six simple waves with characteristic speeds $\mp\lambda_{(1)}$, $\mp\lambda_{(2)}$, and $\mp\lambda_{(3)}$. We now study each case individually.

A. Case $\lambda^2 = \lambda^2_{(3)}$

If we substitute $\lambda = \lambda_{(3)}$ into (3.5.19), we get

$$\left(q\frac{\partial^2\Sigma}{\partial p^2} - \frac{\partial\Sigma}{\partial q}\right)v_1^{(3)} + \frac{\partial^2\Sigma}{\partial p\,\partial q}(p_2 v_2^{(3)} + p_3 v_3^{(3)}) = 0$$

$$q^2 p_2 \frac{\partial^2\Sigma}{\partial p\,\partial q}v_1^{(3)} + p_2\left(q\frac{\partial^2\Sigma}{\partial q^2} - \frac{\partial\Sigma}{\partial q}\right)(p_2 v_2^{(3)} + p_3 v_3^{(3)}) = 0$$

$$q^2 p_3 \frac{\partial^2\Sigma}{\partial p\,\partial q}v_1^{(3)} + p_3\left(q\frac{\partial^2\Sigma}{\partial q^2} - \frac{\partial\Sigma}{\partial q}\right)(p_2 v_2^{(3)} + p_3 v_3^{(3)}) = 0$$

the solution of which is found to be

$$v_1^{(3)} = 0, \qquad v_2^{(3)} = kp_3, \qquad v_3^{(3)} = -kp_2$$

where k is an arbitrary factor. Equation (3.5.21), in this case, reduces to

$$dp_1/0 = dp_2/p_3 = -dp_3/p_2 \tag{3.5.25}$$

The solutions to (3.5.25) gives two 3-Riemann invariants

$$p_1 = p = \text{const}, \qquad {p_2}^2 + {p_3}^2 = q^2 = \text{const} \tag{3.5.26}$$

Incidentally, they are the same for $\lambda = -\lambda_{(3)}$. Therefore, in the 3-simple wave region the displacement gradient component in the direction of propagation is constant, whereas the transverse component has a constant magnitude which can rotate in its plane (Fig. 3.5.2). The locus of the endpoint of the transverse component of the displacement gradient vector is a circle. Therefore, this simple wave is called a *circularly polarized wave*. Thus the displacement gradient vector has a constant magnitude and a constant component in the direction of propagation in circularly polarized simple waves. The component in the wave front is constant in magnitude but is in an arbitrary direction.

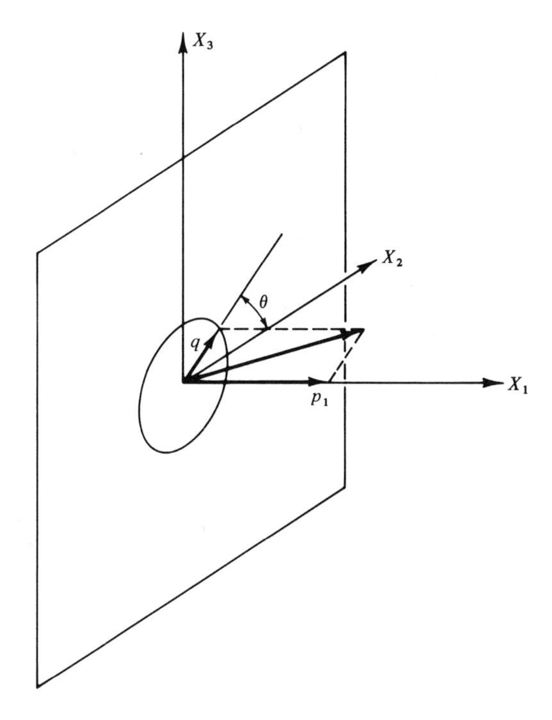

Fig. 3.5.2 Circularly polarized simple wave.

Since the wave speed $\lambda_{(3)}$ is a function of p and q, we conclude from (3.5.26) that it is constant throughout the simple wave region. From the equation of characteristic $dX/dt = \lambda_{(3)}$ we also deduce that all characteristics have the same slope. Hence all characteristics corresponding to a 3-simple wave are parallel straight lines (Fig. 3.5.3). Now this simple wave should be adjacent to a constant state in which the displacement gradient vector is a prescribed constant vector $\mathbf{p}_0(p_0, q_0)$. From (3.5.26) we therefore have

$$p_1 = p = p_0, \qquad p_2^2 + p_3^2 = q^2 = q_0^2$$

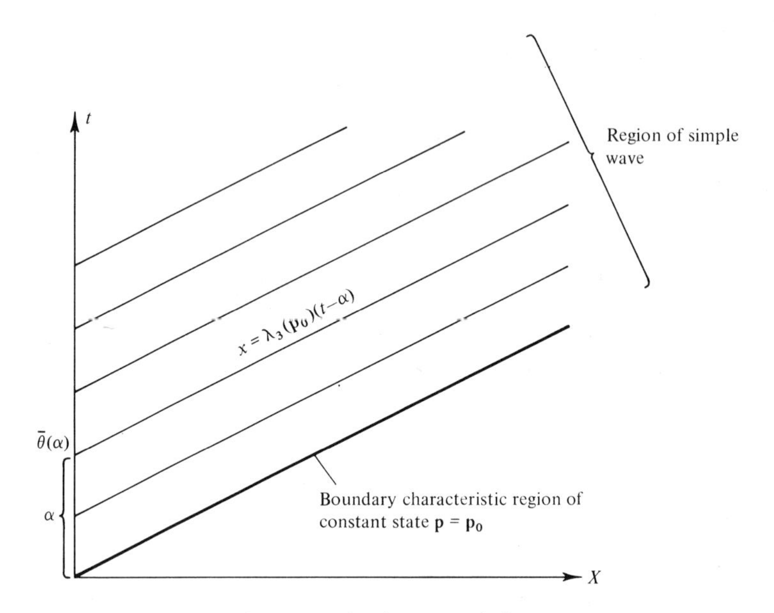

Fig. 3.5.3 Simple wave solution.

This situation is realized when the waves are propagating into a material homogeneously deformed by an amount $\mathbf{p}_0$.

Now define (Fig. 3.5.2)

$$p_2 = q \cos \theta, \qquad p_3 = q \sin \theta$$

Since all components of $\mathbf{p}$ should be constant on characteristic lines given by

$$X = \lambda_{(3)}(\mathbf{p}_0)(t - \alpha)$$

where α's are the intercepts of lines on the time axis, θ also takes constant values on individual characteristics. In order to determine these values associated with every characteristic we must specify a boundary condition for θ. We may, for instance, prescribe θ on the plane $X = 0$ such that

$$\theta(0, t) = \bar{\theta}(t)$$

where $\bar{\theta}(t)$ is a function of time only. The situation in this case is described in Fig. 3.5.4. Physically, this may be envisioned as a traction boundary condition with constant normal traction and rotating tangential traction of constant magnitude.

Since θ is constant on characteristics, the value of θ carried on individual straight lines will be $\bar{\theta}(\alpha)$, i.e.,

$$\theta = \bar{\theta}[t - X/\lambda_{(3)}(\mathbf{p}_0)], \qquad \text{for} \quad X \leq \lambda_{(3)}(\mathbf{p}_0)t$$

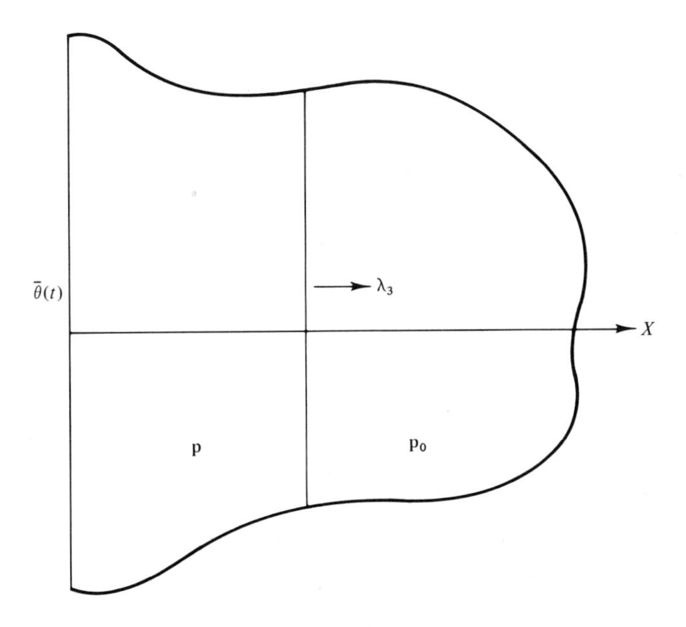

Fig. 3.5.4 Half-space.

If $X \geq \lambda_{(3)}(\mathbf{p}_0)t$ then $\theta = \theta_0 = \tan^{-1}(p_{30}/p_{20})$. The general solution, valid for $X \leq \lambda_{(3)}(\mathbf{p}_0)t$, thus becomes

$$p_1 = p_0, \qquad p_2 = q_0 \cos \bar{\theta}[t - X/\lambda_{(3)}(\mathbf{p}_0)], \qquad p_3 = q_0 \sin \bar{\theta}[t - X/\lambda_{(3)}(\mathbf{p}_0)]$$

$$\lambda_{(3)}(\mathbf{p}_0) = \left(\frac{1}{\rho_0 q_0} \frac{\partial \Sigma}{\partial q}\right)^{1/2}_{\mathbf{p}=\mathbf{p}_0} \tag{3.5.27}$$

From (3.5.22) we now get [for $X \leq \lambda_{(3)}(\mathbf{p}_0)t$]

$$\begin{aligned} v_1 &= v_{10} \\ v_2 &= v_{20} + \lambda_{(3)}(\mathbf{p}_0)q_0\{\cos \theta_0 - \cos \bar{\theta}[t - X/\lambda_{(3)}(\mathbf{p}_0)]\} \\ v_3 &= v_{30} + \lambda_{(3)}(\mathbf{p}_0)q_0\{\sin \theta_0 - \sin \bar{\theta}[t - X/\lambda_{(3)}(\mathbf{p}_0)]\} \end{aligned} \tag{3.5.28}$$

The displacement may now be obtained from (3.5.27) and (3.5.28).

B. Case $\lambda^2 = \lambda^2_{(1)}$ or $\lambda^2_{(2)}$

In order to discuss the 1- and 2-simple waves corresponding to the speeds $\lambda_{(1)}$ and $\lambda_{(2)}$, we first write $(3.5.24)_1$ in the form

$$\rho_0 \lambda^2 = \frac{1}{2}\left(\frac{\partial^2 \Sigma}{\partial p^2} + \frac{\partial^2 \Sigma}{\partial q^2} \mp \Lambda\right)$$

where

$$\Lambda^2 = \left(\frac{\partial^2\Sigma}{\partial p^2} - \frac{\partial^2\Sigma}{\partial q^2}\right)^2 + 4\left(\frac{\partial^2\Sigma}{\partial p\,\partial q}\right)^2$$

and then insert this into (3.5.19) to obtain

$$\frac{1}{2}\left(\frac{\partial^2\Sigma}{\partial p^2} - \frac{\partial^2\Sigma}{\partial q^2} \pm \Lambda\right)v_1 + \frac{p_2}{q}\frac{\partial^2\Sigma}{\partial p\,\partial q}v_2 + \frac{p_3}{q}\frac{\partial^2\Sigma}{\partial p\,\partial q}v_3 = 0$$

$$\frac{p_2}{q}\frac{\partial^2\Sigma}{\partial p\,\partial q}v_1 + \left[\frac{p_2{}^2}{q^2}\left(\frac{\partial^2\Sigma}{\partial q^2} - \frac{1}{q}\frac{\partial\Sigma}{\partial q}\right) + \frac{1}{q}\frac{\partial\Sigma}{\partial q} - \frac{1}{2}\left(\frac{\partial^2\Sigma}{\partial p^2} + \frac{\partial^2\Sigma}{\partial q^2} \mp \Lambda\right)\right]v_2$$
$$+ \frac{p_2 p_3}{q^2}\left(\frac{\partial^2\Sigma}{\partial q^2} - \frac{1}{q}\frac{\partial\Sigma}{\partial q}\right)v_3 = 0$$

$$\frac{p_3}{q}\frac{\partial^2\Sigma}{\partial p\,\partial q}v_1 + \frac{p_2 p_3}{q^2}\left(\frac{\partial^2\Sigma}{\partial q^2} - \frac{1}{q}\frac{\partial\Sigma}{\partial q}\right)v_2$$
$$+ \left[\frac{p_3{}^2}{q^2}\left(\frac{\partial^2\Sigma}{\partial q^2} - \frac{1}{q}\frac{\partial\Sigma}{\partial q}\right) + \frac{1}{q}\frac{\partial\Sigma}{\partial q} - \frac{1}{2}\left(\frac{\partial^2\Sigma}{\partial p^2} + \frac{\partial^2\Sigma}{\partial q^2} \mp \Lambda\right)\right]v_3 = 0$$

where the terms with the $(+)$ sign correspond to $\lambda_{(1)}$ and those with $(-)$ sign to $\lambda_{(2)}$ with respective eigenvectors $\boldsymbol{v}^{(1)}$ and $\boldsymbol{v}^{(2)}$. These equations yield, after some manipulations,

(3.5.29)

$$v_1^{(1,2)} = kq\left(\frac{\partial^2\Sigma}{\partial p\,\partial q}\right)^{-1}\left(\rho_0\lambda^2_{(1,2)} - \frac{\partial^2\Sigma}{\partial q^2}\right), \qquad v_2^{(1,2)} = kp_2, \qquad v_3^{(1,2)} = kp_3$$

where k is an arbitrary factor. Equation (3.5.21) now reduces to

$$q^{-1}\left(\rho_0\lambda^2 - \frac{\partial^2\Sigma}{\partial q^2}\right)^{-1}\frac{\partial^2\Sigma}{\partial p\,\partial q}\,dp = \frac{dp_2}{p_2} = \frac{dp_3}{p_3} \tag{3.5.30}$$

which constitutes two sets of equations for $\lambda^2 = \lambda^2_{(1)}$ and $\lambda^2 = \lambda^2_{(2)}$. The last two of (3.5.30) result in the Riemann invariant

$$p_2/p_3 = \text{const} \tag{3.5.31}$$

in the simple wave regions. If we define

$$p_2 = q\cos\theta, \qquad p_3 = q\sin\theta$$

as before, (3.5.31) implies that

$$\theta = \text{const} = \theta_0$$

that is, the component of the displacement gradient vector in a plane parallel to the wave front $X = \text{const}$ always has the same direction designated by θ_0

which is determined by the given constant state adjacent to the simple wave. In other words, the displacement gradient vector always lies in a plane containing the direction of propagation of the wave (Fig. 3.5.5). Such a wave is called a *plane polarized wave*.

It follows from (3.5.30) that

$$\frac{dp}{dq} = \left(\frac{\partial^2 \Sigma}{\partial p\, \partial q}\right)^{-1}\left(\rho_0 \lambda^2 - \frac{\partial^2 \Sigma}{\partial q^2}\right) = \frac{\partial^2 \Sigma}{\partial p\, \partial q}\left(\rho_0 \lambda^2 - \frac{\partial^2 \Sigma}{\partial p^2}\right)^{-1} \tag{3.5.32}$$

where we used the equation satisfied by $\rho_0 \lambda^2$ and the relation

$$dp_2/p_2 = dp_3/p_3 = dq/q$$

since θ is constant. For a given strain energy function $\Sigma(p, q^2)$, the right-hand side of (3.5.32) is a known function of p and q. Therefore, the solution to the first-order ordinary differential equation (3.5.32) is of the form

$$J(p, q) = \text{const}$$

where J is a Riemann invariant, and therefore the constant in this equation is evaluated from the constant state adjacent to the simple wave deformation.

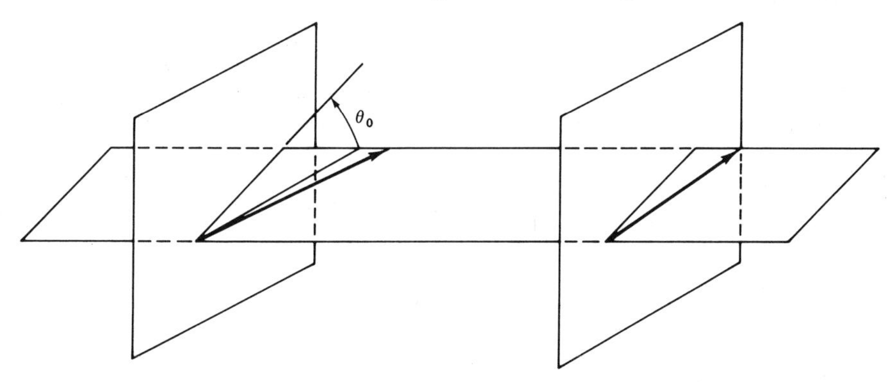

Fig. 3.5.5 Plane polarized simple wave.

Thus

$$J(p, q) = J(p_0, q_0] \equiv J_0 \tag{3.5.33}$$

where p_0 and q_0 are prescribed in the constant state ahead of the wave. Using (3.5.33) we can now express this simple wave deformation field as

$$p_1 = p = j(q), \qquad p_2 = q \cos \theta_0, \qquad p_3 = q \sin \theta_0 \tag{3.5.34}$$

This field may be created by applying time-dependent tractions on the boundary $X = 0$. However, the direction of the tangential traction always remains the same.

Since λ are functions of p and q we can express them in terms of a single variable q by using the relations (3.5.34). The velocity field is determined through the remaining Riemann invariants (3.5.22):

$$\begin{aligned} v_1 + \int_0^q \lambda(q)\,(dj/dq)\,dq &= \text{const} \\ v_2 + \cos\theta_0 \int_0^q \lambda(q)\,dq &= \text{const} \qquad (3.5.35) \\ v_3 + \sin\theta_0 \int_0^q \lambda(q)\,dq &= \text{const} \end{aligned}$$

The integration constants would be determined from the velocity of the constant state, i.e., $\mathbf{v} = \mathbf{v}_0$ when $q = q_0$. Evidently λ takes the values $\mp\lambda_{(1)}$ and $\mp\lambda_{(2)}$ in the foregoing expression thus generating four simple waves, two in each sense. However, it is clear that the Riemann invariant (3.5.33) is insensitive to the sign of λ, i.e., the sense of propagation of the simple wave, but the set (3.5.35) should be modified in compliance with the sign of the corresponding λ.

The variable q, and consequently all dependent variables, is constant on straight characteristics corresponding to a particular simple wave in accordance with the definition of simple wave solutions. The constants assigned to each characteristic covering the simple wave region can only be determined from a given boundary condition, say at $X = 0$. If the slope of the characteristics decreases as shown in Fig. 3.5.6 as the boundary condition evolves in time, then this solution presents no difficulty and the displacement gradients

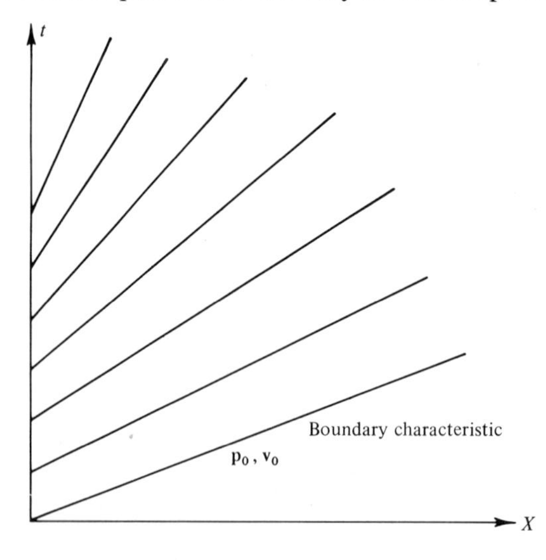

Fig. 3.5.6 Unloading wave.

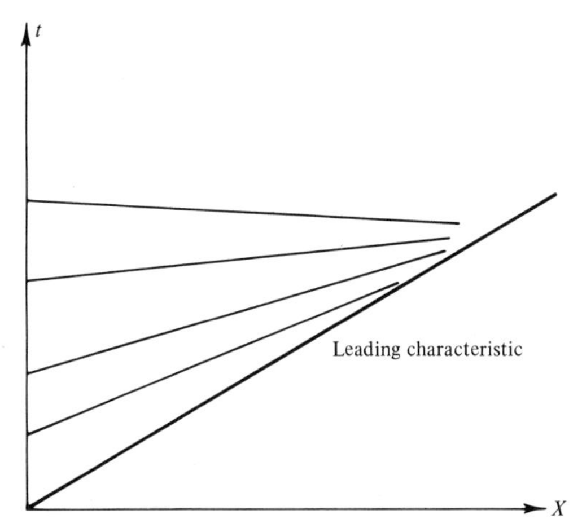

Fig. 3.5.7 Loading wave (shock formation).

and velocity components can be determined uniquely for the entire simple wave region. If the slope of the characteristics increases with the time evolution of the boundary conditions as shown in Fig. 3.5.7, then characteristics covering the simple wave region become coalescent in a certain subregion, leading to a shock formation. In this case, the simple wave solution ceases to be valid in such regions. Whether or not a shock will form in a material is decided by the dependence of λ on q. If we want to exclude the possibility of a shock formation we have to consider only cases in which $d\lambda < 0$. Therefore, if $dq > 0$ at the boundary (a loading wave) then $d\lambda/dq < 0$; if $dq < 0$ at the boundary (an unloading wave) then $d\lambda/dq > 0$. The reversed situation may lead to a shock formation. The condition $d\lambda < 0$ is also physically meaningful. It implies that the disturbances generated at a later time cannot catch up with disturbances emitted earlier. Thus, it is impossible for the disturbances to build up as they travel.

C. Special Cases

(i) *Pure Dilatational Wave*[1]. Consider the special case in which the deformation gradients are such that

$$p_1 = p, \qquad p_2 = p_3 = 0$$

Then according to the discussion presented on p. 177 we can freely choose $u_1 \neq 0$, $u_2 = u_3 = 0$ and the equations of motion reduce to the single equation

$$A_{11}\,\partial^2 u_1/\partial X^2 = \rho_0\,\partial^2 u_1/\partial t^2$$

[1] See Bland [1964b].

The characteristic value associated with this problem is

$$\rho_0 \lambda_1{}^2 = \partial^2\Sigma/\partial p^2$$

Clearly the waves corresponding to speeds λ_1 and $-\lambda_1$ represent *pure dilatational waves* in each sense provided the existing deformation ahead of the wave front (i.e., in the region of constant state) consists of a uniaxial strain in the X-direction and moves with a constant velocity along the X-axis. Then we have

$$v_1 + \int_0^p \lambda_1(\pi)\, d\pi = \text{const}, \qquad v_2 = v_3 = 0$$

or defining

$$P(p) = \int_0^p \left(\frac{1}{\rho_0}\frac{\partial^2\Sigma(\pi)}{\partial\pi^2}\right)^{1/2} d\pi, \qquad P_0 \equiv P(p_0)$$

we get

$$v_1 + P(p) = P_0 \tag{3.5.36}$$

where p_0 is the displacement gradient ahead of the wave front. We assume, here without much loss of generality, that $v_0 = 0$.

We now consider two kinds of boundary conditions.

(a) Normal stress prescribed on the boundary $X = 0$. This is equivalent to the prescription of the normal strain (Fig. 3.5.8): $p(0, t) = \bar{p}(t)$. Since $p(X, t)$ is constant on a straight characteristic determined by its intercept α on the time axis and its slope $\lambda_1(p)$,

$$X = \lambda_1(p)(t - \alpha)$$

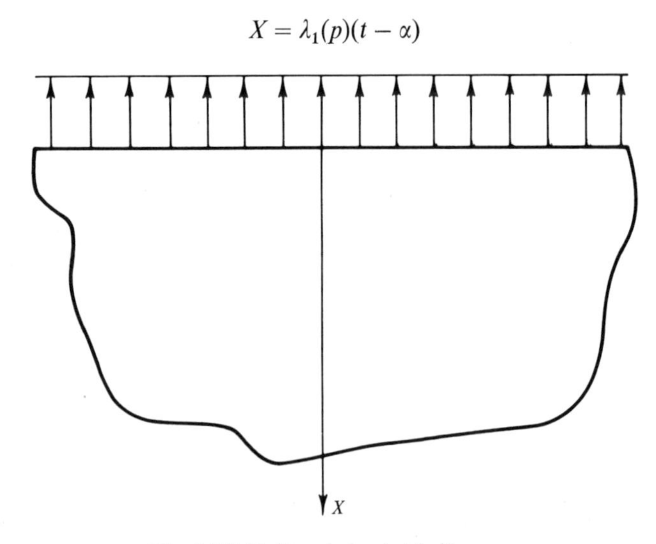

Fig. 3.5.8 Uniformly loaded half-space.

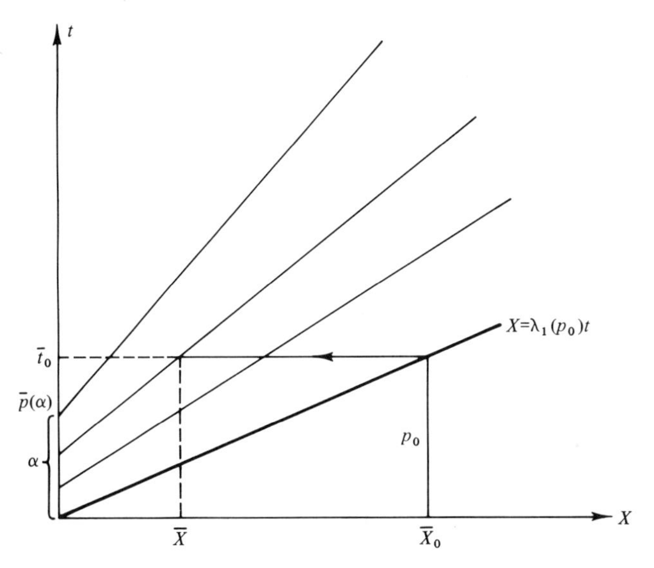

Fig. 3.5.9 Straight line characteristics.

then $p = \bar{p}(\alpha)$ is carried along this line (Fig. 3.5.9). Therefore, we can write

$$X = \lambda_1(\alpha)(t - \alpha) \tag{3.5.37}$$

where $\lambda_1(\alpha) = \lambda_1[\bar{p}(\alpha)]$ is a definite function of α for a given material. If the relation (3.5.37) is invertible for α, we get

$$\alpha = \alpha(X, t)$$

and the displacement gradients are simply obtained as

$$\begin{aligned} p &= p(X, t) = \bar{p}[\alpha(X, t)], && \text{for} \quad X < \lambda_1(p_0)t \\ p &= p_0, && \text{for} \quad X \geq \lambda_1(p_0)t \end{aligned} \tag{3.5.38}$$

Relation (3.5.36) now determines $v_1 = v_1(X, t)$.

The inversion of (3.5.37) is possible, at least in principle, if and only if the derivative of (3.5.37) with respect to α does not vanish everywhere in the X,t-plane with $X > 0$, that is, if there is no $t > \alpha$ satisfying

$$t - \alpha = \lambda_1/\lambda_1' > 0 \tag{3.5.39}$$

or no $X > 0$ satisfying

$$X = \lambda_1^2/\lambda_1' > 0 \tag{3.5.40}$$

where primes denote differentiation with respect to α. If there is a $t > \alpha$ satisfying (3.5.39) then the characteristics begin to coalesce into one another

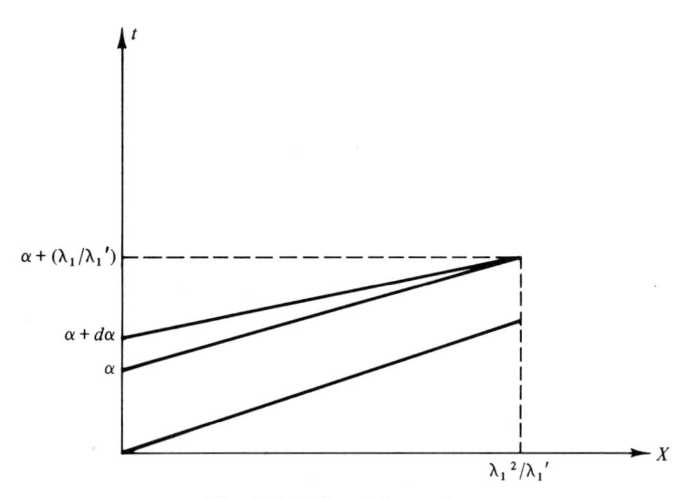

Fig. 3.5.10 Shock formation.

at that instant and in the place X given by (3.5.40), and a finite jump in p (i.e., a dilatational shock wave) begins to form (Fig. 3.5.10). The shock formation depends on the sign of λ_1'. Equations (3.5.39) and (3.5.40) require that $d\lambda_1/d\alpha > 0$ be necessary for the shock formation. If $d\lambda_1/d\alpha > 0$, then the signal corresponding to the characteristic $\alpha + d\alpha$ travels faster than the signal corresponding to the characteristic α. Therefore, a signal emitted at the boundary at a certain time can catch up with the signals emitted earlier. Hence disturbances tend to grow, giving rise to the formation of a shock wave. Since

$$\frac{d\lambda_1}{d\alpha} = \frac{1}{2\rho_0 \lambda_1} \frac{d^3\Sigma}{dp^3} \frac{dp}{d\alpha}$$

a loading wave ($dp/d\alpha > 0$) may lead to a shock if $d^3\Sigma/dp^3 > 0$, and an unloading wave ($dp/d\alpha < 0$) may produce a shock if $d^3\Sigma/dp^3 < 0$ (cf. the discussion on p. 185). These are in agreement with what we have found in Section 2.15.

(b) Motion imparted by a normal velocity at the boundary $X = 0$. The boundary condition, in this case, is

$$v_1(0, t) = v(t)$$

Since v_1 is constant on a characteristic α, we may write

$$v_1 = v(\alpha)$$

Equation (3.5.36) now leads to

$$\int_0^p \lambda_1(\pi)\, d\pi = P_0 - v(\alpha)$$

If this equation can be solved for p, we get a relation $p = p(\alpha)$. Substitution of this into $\lambda_1 = \lambda_1(p)$ gives

$$\lambda_1 = \lambda_1(\alpha)$$

Therefore, α may be solved in terms of X and t from relation (3.5.37). The shock formation can now be examined in exactly the same way.

The displacement u_1 at a time $\bar{t}$ and at point $\bar{X}$ may be obtained by integration on a path in the X, t-plane. Choosing this path as a line $t = \bar{t}$, as shown in Fig. 3.5.9, we get

$$u_1 = \int_{\bar{X}_0}^{\bar{X}} p(X, \bar{t})\, dX + u_0 \tag{3.5.41}$$

where $\bar{X}_0 = \lambda_1(p_0)\bar{t}$ and u_0 is the displacement in the constant state.

(ii) *Waves of Small Amplitude*[1]. Since the integration of (3.5.32) for plane polarized waves presupposes the knowledge of explicit dependence of Σ on p and q, it is impossible to extract detailed information about the nature of these waves without specifying the form of the strain energy function. For illustrative purposes we consider here *quadratic solids* whose energy has the form

$$2\rho_0^{-1}\Sigma = c_1{}^2p^2 + c_2{}^2q^2 + \tfrac{1}{3}Ap^3 + Bpq^2 \tag{3.5.42}$$

where [cf. (2.11.26)]

$$\begin{aligned} c_1{}^2 &= (\lambda + 2\mu)/\rho_0, & c_2{}^2 &= \mu/\rho_0 \\ A &= 3(\lambda + 2\mu + 2l)/\rho_0, & B &= (\lambda + 2\mu - \tfrac{1}{2}m)/\rho_0 \end{aligned} \tag{3.5.43}$$

The form (3.5.42) is actually an isentropic (or isothermal if Σ is taken as the free energy) approximation of the real strain energy function for moderately small displacement gradients. The reader will also identify c_1 and c_2, respectively, as the classical dilatational and transverse wave speeds of the linear elastic solids.

Consider first $\lambda_{(1)}^2$ in $(3.5.24)_1$ [with (+) sign]. For a linear approximation in displacement gradients, we find that

$$\lambda_{(1)}^2 = U_1{}^2 = c_1{}^2 + Ap \tag{3.5.44}$$

and (3.5.32) becomes

$$dp/dq = [c_1{}^2 - c_2{}^2 + (A - B)p]/Bq \tag{3.5.45}$$

If the constant state in front of the wave is given by p_0 and q_0 then the integration of (3.5.45) yields

$$q = q_0\left[\frac{(A - B)p + c_1{}^2 - c_2{}^2}{(A - B)p_0 + c_1{}^2 - c_2{}^2}\right]^{B/(A-B)} \tag{3.5.46}$$

[1] See Bland [1965b].

which satisfies the condition $p = p_0$ when $q = q_0$. If now we specify a boundary condition on p at $X = 0$, such as $p(0, t) = \bar{p}(t)$, the technique which we discussed earlier helps us to determine $p = p(X, t)$ and $q = q(X, t)$. If $q_0 = 0$ then we see, from (3.5.46), that $q \equiv 0$ everywhere in the X,t-plane. Thus the wave is purely dilatational in this case. However, if the medium is sheared initially by an amount q_0, a transverse wave deformation $q(X, t)$ will accompany the dilatational wave deformation $p(X, t)$ with its magnitude given in (3.5.46). Such a wave will be called the *generalized dilatational wave*. It also follows from (3.5.44) that shocks tend to form for expansion waves if $A > 0$, and for compression waves if $A < 0$.

Now consider $\lambda_{(2)}^2$ in $(3.5.24)_1$ [with $(-)$ sign]. In the same fashion we get

$$\lambda_{(2)}^2 = U_2{}^2 = c_2{}^2 + Bp \tag{3.5.47}$$

and (3.5.32) now becomes

$$dp/dq = 0 \qquad \text{or} \qquad p = \text{const}$$

This proves that there is no coupling between the longitudinal and transverse deformation fields for waves propagating with velocity (3.5.47) linear in displacement gradients. Therefore, we can have $p = 0$ if the material is not strained initially in the longitudinal direction, and in this limit the waves are purely transverse. However, if we keep higher-order terms in $\lambda_{(2)}^2$ a coupling is introduced between the longitudinal and transverse components of the displacement gradients. One can establish that in such a second-order theory

$$U_2{}^2 = c_2{}^2 + Bp - B^2q^2/(c_1{}^2 - c_2{}^2), \qquad dp/dq = -Bq/(c_1{}^2 - c_2{}^2) \tag{3.5.48}$$

from which we obtain

$$p - p_0 = -B(q^2 - q_0{}^2)/2(c_1{}^2 - c_2{}^2) \tag{3.5.49}$$

and the wave velocity

$$U_2{}^2 = c_2{}^2 + Bp_0 - [B^2(3q^2 - q_0{}^2)/2(c_1{}^2 - c_2{}^2)] \tag{3.5.50}$$

where p_0 and q_0 determine the constant state adjacent to the simple wave region as before.

If $p_0 = q_0 = 0$, i.e., if waves propagate into an undeformed medium, we deduce from (3.5.49) that p is not equal to zero. Therefore, a second-order longitudinal wave is associated with the transverse wave. This also shows that, except for the linear solids, there is no possibility of finding a purely transverse wave.

The wave motion that we have considered here will be referred to then as a *quasi-transverse wave*, since the longitudinal component is of second order compared to the transverse component. The reader should also be reminded that the above analysis is correct only for elastic solids whose strain energy function is *exactly* given by (3.5.42). However if we regard this function as a polynomial approximation of the real strain energy function for sufficiently small deformations, higher-order terms in this expansion should be taken into account in an accurate second-order evaluation of (3.5.49) and (3.5.50). Second-order analysis of this type is included in Bland [1965, Chapter 3].

3.6 PLANE WAVES IN INCOMPRESSIBLE ELASTIC SOLIDS

For incompressible solids we have $III = 1$, so that (3.5.2) gives

$$p_1 = 0$$

Without loss of generality, we may assume that the displacement u_1 in the $X_1 \equiv X$-direction vanishes, so that all one-dimensional waves in incompressible elastic materials are of the transverse type having only the following nonzero displacement gradients:

$$p_\alpha = \partial u_\alpha / \partial X, \qquad \alpha = 2, 3 \tag{3.6.1}$$

The strain energy function depends only on p_2 and p_3:

$$\Sigma = \Sigma(p_2, p_3)$$

subject to certain invariance requirements.

Since $\mathbf{T} = \mathbf{T}(X, t)$ the equations of motion in the reference state are

$$\partial T_{1k} / \partial X = \rho_0 \, \partial^2 u_k / \partial t^2 \tag{3.6.2}$$

where

$$T_{1k} = -pX_{,k} + (\partial \Sigma / \partial p_k) \tag{3.6.3}$$

Here $p(X, t)$ is the unknown hydrostatic pressure function and

$$[X_{K,k}] = \begin{bmatrix} 1 & 0 & 0 \\ -p_2 & 1 & 0 \\ -p_3 & 0 & 1 \end{bmatrix}$$

Since $u_1 = 0$, for $k = 1$ (3.6.2) gives

$$T_{11} = T_{11}(t)$$

If we assume that $T_{11} = 0$ at a boundary $X = \text{const}$, then T_{11} vanishes everywhere, and it follows from (3.6.3) that

$$p = \partial \Sigma / \partial p_1 \,|_{p_1 = 0}$$

Finally writing

$$T_{12} = \partial \Sigma / \partial p_2, \qquad T_{13} = \partial \Sigma / \partial p_3$$

the last two equations of motion in terms of displacement components read

$$A_{\alpha\beta}\, \partial^2 u_\beta/\partial X^2 = \rho_0\, \partial^2 u_\alpha/\partial t^2, \qquad \alpha = 2, 3$$

where

$$A_{\alpha\beta}(p_2, p_3) = \partial^2\Sigma/\partial p_\alpha\, \partial p_\beta \tag{3.6.4}$$

Consequently plane wave motion in incompressible materials is governed by a system of two quasi-linear second-order partial differential equations with two dependent and two independent variables. If we assume that the 2×2 matrix $\mathbf{A}$ is positive definite with distinct eigenvalues, this system is totally hyperbolic and its simple wave solutions may be obtained by the same methods employed in Section 3.5. The only difference is that we now have only four simple waves instead of six, and all of these waves are transverse.

Since the analysis is entirely parallel to the one in Section 3.5 we omit the details.

Simple wave speeds are obtained from

$$\det(\rho_0^{-1} A_{\alpha\beta} - \lambda^2\, \delta_{\alpha\beta}) = 0 \tag{3.6.5}$$

and the three generalized Riemann invariants associated with the γ-*simple wave* are found from

$$dp_2/v_2^{(\gamma)} = dp_3/v_3^{(\gamma)}, \qquad v_\alpha + \int \lambda_{(\gamma)}(\mathbf{p})\, dp_\alpha = \text{const} \tag{3.6.6}$$

For isotropic incompressible elastic solids, the strain energy is a function of the invariants I and II of the deformation tensor only:

$$I = II = 3 + p_2{}^2 + p_3{}^2 = 3 + q^2 \tag{3.6.7}$$

with $q^2 = p_2{}^2 + p_3{}^2$. Thus in this case, $\Sigma = \Sigma(q^2)$, and from (3.6.4) we get

$$\begin{aligned}
A_{22} &= \frac{\partial^2\Sigma}{\partial p_2{}^2} = \frac{p_2{}^2}{q^2}\left(\frac{d^2\Sigma}{dq^2} - \frac{1}{q}\frac{d\Sigma}{dq}\right) + \frac{1}{q}\frac{d\Sigma}{dq} = 4p_2{}^2\, D_1{}^2\Sigma + 2\, D_1\Sigma \\
A_{23} &= \frac{\partial^2\Sigma}{\partial p_2\, \partial p_3} = \frac{p_2\, p_3}{q^2}\left(\frac{d^2\Sigma}{dq^2} - \frac{1}{q}\frac{d\Sigma}{dq}\right) = 4p_2\, p_3\, D_1{}^2\Sigma \\
A_{33} &= \frac{\partial^2\Sigma}{\partial p_3{}^2} = \frac{p_3{}^2}{q^2}\left(\frac{d^2\Sigma}{dq^2} - \frac{1}{q}\frac{d\Sigma}{dq}\right) + \frac{1}{q}\frac{d\Sigma}{dq} = 4p_3{}^2\, D_1{}^2\Sigma + 2\, D_1\Sigma
\end{aligned} \tag{3.6.8}$$

where

$$D_1 \equiv (\partial/\partial I) + (\partial/\partial II)$$

as before. We deduce that

$$\begin{aligned}
\rho_0\, \lambda_{(2)}^2 &\equiv \rho_0\, U_2{}^2 = 2\left(\frac{\partial\Sigma}{\partial I} + \frac{\partial\Sigma}{\partial II}\right) \\
\rho_0 \lambda_{(3)}^2 &\equiv \rho_0\, U_3{}^2 = 2\left(\frac{\partial\Sigma}{\partial I} + \frac{\partial\Sigma}{\partial II}\right) + 4q^2\left(\frac{\partial^2\Sigma}{\partial I^2} + 2\frac{\partial^2\Sigma}{\partial I\, \partial II} + \frac{\partial^2\Sigma}{\partial II^2}\right)
\end{aligned} \tag{3.6.9}$$

It can readily be verified that simple waves corresponding to $\pm U_2$ are again *circularly polarized*, i.e., the Riemann invariants are

$$q^2 = \text{const}, \qquad v_\alpha \pm U_2(q)p_\alpha = \text{const}$$

The constants are determined from the constant state ahead of the wave. Thus the characteristics of the simple wave are parallel straight lines in the X,t-plane and θ, defined by

$$p_2 = q \cos\theta, \qquad p_3 = q \sin\theta$$

on these characteristics takes constant values, which are determined by specifying a boundary condition on $X = 0$.

Again one can show that the simple waves corresponding to the wave speeds $\pm U_3$ are *plane polarized*, i.e.,

$$p_2/p_3 = \text{const} \qquad \text{or} \qquad \theta = \text{const} = \theta_0$$

and the remaining Riemann invariants are given by

$$v_2 \pm \cos\theta_0 \int_0^q U_3(q)\, dq = \text{const}$$

$$v_3 \pm \sin\theta_0 \int_0^q U_3(q)\, dq = \text{const}$$

If, in a special case, $\theta_0 = 0$ and the material is at rest initially, we have

$$p_2 = q, \qquad v_2 + \int_0^q U_3(q)\, dq = \text{const}$$

Introducing

$$Q(q) = \int_0^q U_3(\kappa)\, d\kappa$$

and denoting the initial shear in the medium by q_0, we get the velocity

$$v_2 = Q_0 - Q(q), \qquad Q_0 \equiv Q(q_0)$$

If we specify a boundary condition at $X = 0$ (Fig. 3.6.1), the preceding formulation permits us to determine the motion of an initially sheared incompressible half-space that is set into motion by the application of a time-dependent, spatially uniform shear stress or tangential velocity to the boundary. The analysis is completely analogous to the one corresponding to pure dilatation waves discussed in Section 3.5, p. 186. Hence we do not repeat it here. This problem has been studied independently by Chu [1964], who also considered step function loading and unloading problems and the occurrence of transverse shocks.

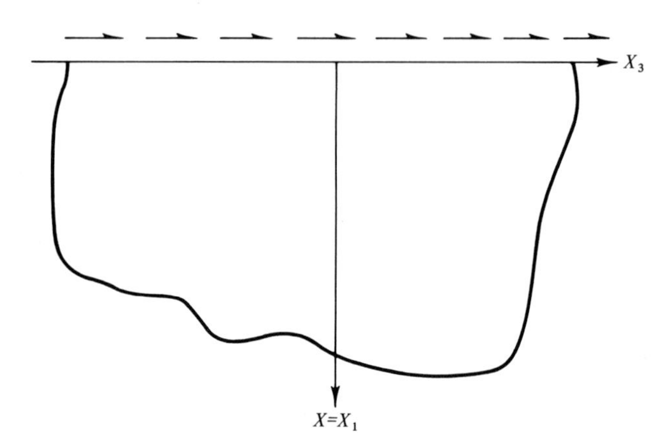

Fig. 3.6.1 Half-space subject to uniform shear.

3.7 RIEMANN PROBLEM FOR AN ISOTROPIC ELASTIC HALF-SPACE[1]

Consider an elastic half-space as shown in Fig. 3.7.1. The material is supposed to be at rest and homogeneously strained initially. We consider a motion of this medium in the form

$$x_1 = X_1 + u_1(X_1, t), \qquad x_2 = X_2 + u_2(X_1, t), \qquad x_3 = X_3 \tag{3.7.1}$$

Denoting displacement gradients by

$$p \equiv \partial u_1/\partial X_1, \qquad q \equiv \partial u_2/\partial X_1$$

initially we have

$$p = \text{const} = p_0, \qquad q = \text{const} = q_0$$

This deformation may be realized by the application of a constant normal and shear stress distribution at the boundary $X_1 = 0$. We consider now the following initial and boundary conditions:

$$\begin{aligned} &p(X_1, 0) = p_0, && q(X_1, 0) = q_0 \\ &p(0, t) = \begin{cases} p_0, & t \leq 0, \\ p_1, & t > 0, \end{cases} && q(0, t) = \begin{cases} q_0, & t \leq 0 \\ q_1, & t > 0 \end{cases} \end{aligned} \tag{3.7.2}$$

where p_1 and q_1 are given constants. The boundary conditions imply that the medium is set into motion by the sudden application of a *new* constant stress

[1] This problem has been thoroughly investigated by Davison [1966].

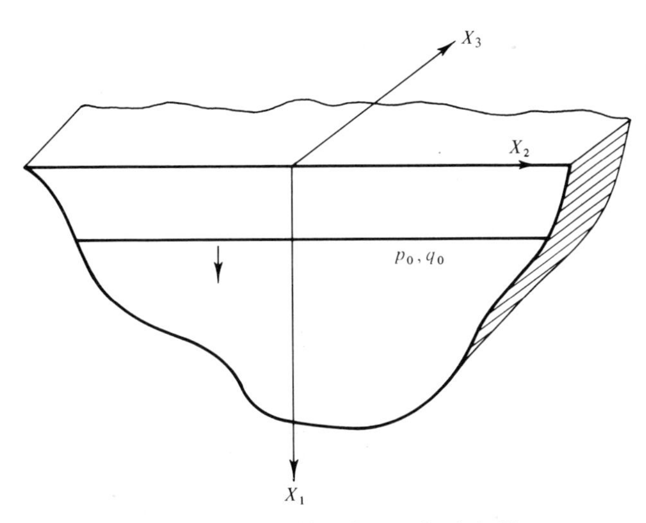

Fig. 3.7.1 Riemann problem for an elastic half-space.

distribution on the boundary $X_1 = 0$ (Fig. 3.7.2). If $(p_1, q_1) > (p_0, q_0)$, there would be a sudden loading; if $(p_1, q_1) < (p_0, q_0)$, a sudden unloading.

As indicated in Appendix B, this problem is known as the Riemann problem for governing differential equations, and it is shown there that it admits centered simple wave solutions, that is, p, q, and $\mathbf{v}$ are expressible as

$$p = p(X_1/t), \qquad q = q(X_1/t), \qquad \mathbf{v} = \mathbf{v}(X_1/t)$$

in some region of the X_1,t-plane. Since the field quantities are constant on the lines

$$X_1/t = \text{const} = U$$

emerging from the origin of the X_1, t-plane, the straight characteristics of the simple waves pass through the origin, hence the name, *centered simple waves.*

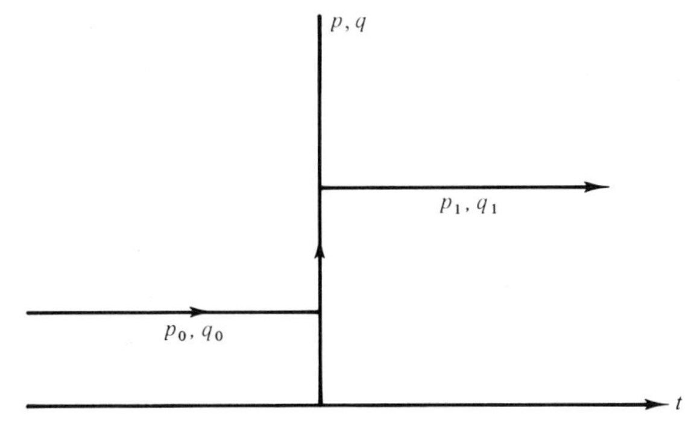

Fig. 3.7.2 The loading program.

There are four centered simple waves, the wave speeds of which are given by [cf. (3.5.24)]

(3.7.3)

$$U_1 = \frac{1}{(2\rho_0)^{1/2}} \left\{ \frac{\partial^2 \Sigma}{\partial p^2} + \frac{\partial^2 \Sigma}{\partial q^2} + \left[\left(\frac{\partial^2 \Sigma}{\partial p^2} - \frac{\partial^2 \Sigma}{\partial q^2} \right)^2 + 4 \left(\frac{\partial^2 \Sigma}{\partial p \, \partial q} \right)^2 \right]^{1/2} \right\}^{1/2}, \qquad -U_1$$

$$U_2 = \frac{1}{(2\rho_0)^{1/2}} \left\{ \frac{\partial^2 \Sigma}{\partial p^2} + \frac{\partial^2 \Sigma}{\partial q^2} - \left[\left(\frac{\partial^2 \Sigma}{\partial p^2} - \frac{\partial^2 \Sigma}{\partial q^2} \right)^2 + 4 \left(\frac{\partial^2 \Sigma}{\partial p \, \partial q} \right)^2 \right]^{1/2} \right\}^{1/2}, \qquad -U_2$$

where we wrote $\lambda_{(1)} = U_1$, $\lambda_{(2)} = U_2$. One can see that

$$U_1^2 = (\lambda + 2\mu)/\rho_0, \qquad U_2^2 = \mu/\rho_0$$

for infinitesimal deformations, where λ and μ are the classical Lamé constants. Thus U_1 is the longitudinal wave speed, whereas U_2 is the transverse wave speed of the linear elasticity. We retain, here too, the same names for U_1 and U_2 although this is somewhat misleading since nonlinearity causes, in general, couplings between these two kinds of waves thereby making it impossible to observe purely longitudinal or purely transverse waves. We further assume that the following inequality is satisfied for all admissible deformations, as it is in linearly elastic materials:

(3.7.4)
$$U_1^2 > U_2^2$$

This amounts to the assumption that at least one of the relations $\partial^2\Sigma/\partial p^2 \neq \partial^2\Sigma/\partial q^2$ or $\partial^2\Sigma/\partial p \, \partial q \neq 0$ is valid.

Since we are dealing with a region of infinite extent loaded at one end, we must consider only the outgoing waves generated at the boundary. Hence we are left with the two simple waves associated with positive U_1 and U_2. The waves propagating with velocities U_1 and U_2 will be called *longitudinal waves* and *transverse* or *shear waves*, respectively, with the reservations previously expressed on this terminology.

Since a simple wave can only connect two regions of constant state in the X_1, t-plane, we need to consider five regions, of which three regions of constant state connect two regions of simple waves. The first constant state (I) is determined by the prescription of p_0 and q_0, while the final constant state (V) is partly determined by given p_1 and q_1, since the velocity vector $\mathbf{v}_1$ in this final stage can only be known as the outcome of the solution. Two simple wave regions (II and IV) will be adjacent to the initial and final constant states, and these two regions of simple waves are connected, in general, with a constant state (III). The locations of regions II and IV are chosen in compliance with the assumed inequality (3.7.4). All succeeding regions are shown in Fig. 3.7.3. We now try to construct solutions that satisfy condition (B.3.4), in each region.

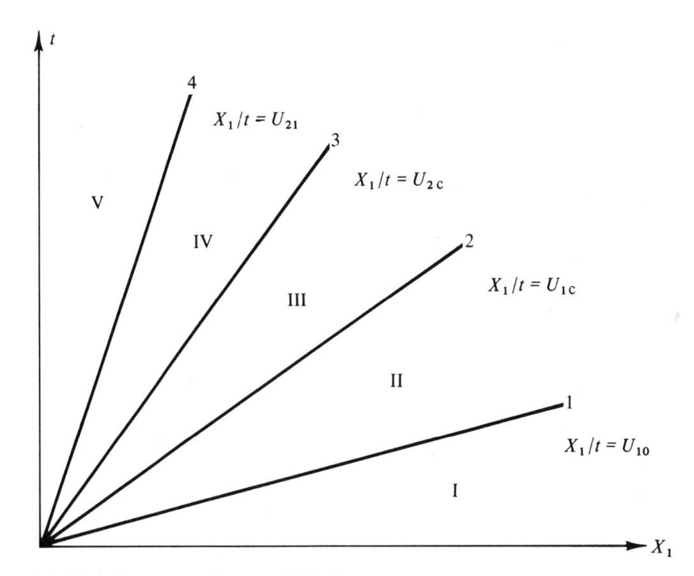

Fig. 3.7.3 Simple wave regions. I: initial constant state, $p = p_0, q = q_0$, $v = 0$; II: longitudinal wave; III: constant state, $p = p_c$, $q = q_c$, $v_1 = v_{1c}$, $v_2 = v_{2c}$; IV: shear wave; V: final constant state, $p = p_1$, $q = q_1$, $v_1 = v_{11}$, $v_2 = v_{21}$.

The slope of line 1, U_{10}, which separates the initial constant state from the longitudinal simple wave region and that of line 4, U_{21}, which separates the final constant state from the shear simple wave region, are determined at once by

$$U_{10} = U_1(p_0, q_0), \qquad U_{21} = U_2(p_1, q_1)$$

However, the values of p_c and q_c of the displacement gradients in the intermediate constant state, and consequently the slopes of lines 2 and 3, denoted, respectively, by

$$U_{1c} = U_1(p_c, q_c), \qquad U_{2c} = U_2(p_c, q_c)$$

need to be determined. We can accomplish this task if we recall the fact that regions of constant state can be connected by a simple wave if the generalized Riemann invariants of this simple wave have the same values at the constants states.

A. Longitudinal Wave Region

The Riemann invariants of the longitudinal wave are the solution of the equation [cf. (3.5.32–35)]

$$dp/dq = (\partial^2\Sigma/\partial p\,\partial q)^{-1}[\rho_0 U_1^2(p, q) - (\partial^2\Sigma/\partial q^2)] \tag{3.7.5}$$

and

(3.7.6) $$v_1 + \int U_1\, dp = \text{const}, \qquad v_2 + \int U_1\, dq = \text{const}$$

The solution of (3.7.5) is of the form

$$J_1(p, q) = \text{const} = J_1(p_0, q_0)$$

which we assume to be soluble for

(3.7.7) $$p = j_1(q)$$

Introducing this into $U_1 = U_1(p, q)$, solving the resulting equation for q, and again using (3.7.7) we get, in the longitudinal wave region, p and q as known functions of the independent variable U_1:

(3.7.8) $$p = P(U_1), \qquad q = Q(U_1), \qquad U_{1c} \le U_1 \le U_{10}$$

Equation (3.7.6) may now be written as

(3.7.9) $$v_1 + \int U_1 \frac{dP}{dU_1}\, dU_1 = \text{const}, \qquad v_2 + \int U_1 \frac{dQ}{dU_1}\, dU_1 = \text{const}$$

If we define

$$V_1(U_1) = \int_0^{U_1} U \frac{dP}{dU}\, dU, \qquad V_2(U_1) = \int_0^{U_1} U \frac{dQ}{dU}\, dU$$

then taking into account the values of the Riemann invariants in region I, we may write (3.7.9) as

(3.7.10)
$$v_1 + V_1(U_1) = V_1(U_{10}), \qquad v_2 + V_2(U_1) = V_2(U_{10}), \qquad U_{1c} \le U_1 \le U_{10}$$

Since the Riemann invariants of the longitudinal simple wave should be extended with the same value to region III, (3.7.8) and (3.7.10) are valid in region III, that is,

(3.7.11) $$\begin{aligned} p_c &= P(U_{1c}), & v_{1c} &= V_1(U_{10}) - V_1(U_{1c}) \\ q_c &= Q(U_{1c}), & v_{2c} &= V_2(U_{10}) - V_2(U_{1c}) \end{aligned}$$

These equations determine the constants p_c, q_c, v_{1c}, v_{2c} as given functions of a still unknown velocity U_{1c}.

B. Shear Wave Region

In the region of shear wave the velocity U_2 should be used in (3.7.5) and (3.7.6) instead of U_1. Considering the adjacent state V, the Riemann invariants become

$$J_2(p, q) = \text{const} = J_2(p_1, q_1)$$
$$v_1 + \int U_2\, dp = \text{const}, \qquad v_2 + \int U_2\, dq = \text{const} \tag{3.7.12}$$

Again $(3.7.12)_1$ is to be employed to express p as a function of q in region IV. If the resulting expression is introduced into the relation $U_2 = U_2(p, q)$ in this region, we can, at least in principle, relate p and q to U_2 as

$$p = \bar{P}(U_2), \qquad q = \bar{Q}(U_2), \qquad U_{21} \leq U_2 \leq U_{2c} \tag{3.7.13}$$

These relations enable one to write $(3.7.12)_{2,3}$ as

$$v_1 + \bar{V}_1(U_2) = \text{const}, \qquad v_2 + \bar{V}_2(U_2) = \text{const}$$

where

$$\bar{V}_1(U_2) = \int_0^{U_2} U \frac{d\bar{P}}{dU}\, dU, \qquad \bar{V}_2(U_2) = \int_0^{U_2} U \frac{d\bar{Q}}{dU}\, dU$$

The continuation of the above expressions into region V results in

$$v_1 + \bar{V}_1(U_2) = v_{11} + \bar{V}_1(U_{21})$$
$$v_2 + \bar{V}_2(U_2) = v_{21} + \bar{V}_2(U_{21}), \qquad U_{21} \leq U_2 \leq U_{2c} \tag{3.7.14}$$

Equations (3.7.13) and (3.7.14) give the displacement gradients and the velocity components in the region of shear wave, yet to be specified by U_{2c} which is still unknown in terms of the unknown velocity components v_{11} and v_{21} in the final constant state. The continuation of Riemann invariants into region III of the undetermined constant state removes the indeterminacies. Thus in region III we have

$$p_c = \bar{P}(U_{2c}), \qquad v_{1c} - v_{11} = \bar{V}_1(U_{21}) - \bar{V}_1(U_{2c})$$
$$q_c = \bar{Q}(U_{2c}), \qquad v_{2c} - v_{21} = \bar{V}_2(U_{21}) - \bar{V}_2(U_{2c}) \tag{3.7.15}$$

Equations (3.7.11) and (3.7.15) constitute eight equations to determine the eight unknowns U_{1c}, U_{2c}, p_c, q_c, v_{1c}, v_{2c}, v_{11}, and v_{21} in terms of the prescribed values p_0, q_0, p_1, q_1, and $U_{10}(p_0, q_0)$ and $U_{21}(p_1, q_1)$.

U_{1c} and U_{2c} may be evaluated first from the equations

$$P(U_{1c}) = \bar{P}(U_{2c}), \qquad Q(U_{1c}) = \bar{Q}(U_{2c}) \tag{3.7.16}$$

Then $(3.7.11)_{2,3}$ yield the components of the velocity in the intermediate constant state, and finally $(3.7.15)_{2,4}$ determine the velocity components in the final constant state as follows:

$$v_{11} = V_1(U_{10}) - \bar{V}_1(U_{21}) + \bar{V}_1(U_{2c}) - V_1(U_{1c})$$
$$v_{21} = V_2(U_{10}) - \bar{V}_2(U_{21}) + \bar{V}_2(U_{2c}) - V_2(U_{1c}) \tag{3.7.17}$$

Hence the simple wave solution to the boundary and initial value problem under consideration may be recapitulated as follows, in the various regions shown in Fig. 3.7.3:

(3.7.18)

$$\text{Region I:}\quad \left.\begin{array}{ll} p = p_0, & q = q_0 \\ v_1 = v_2 = 0 & \end{array}\right\} \qquad U_{10} \le X_1/t < \infty$$

$$\text{Region II:}\quad \left.\begin{array}{l} p = P(U_1), \qquad q = Q(U_1) \\ v_1 = V_1(U_{10}) - V_1(U_1) \\ v_2 = V_2(U_{10}) - V_2(U_1) \end{array}\right\} \qquad U_{1c} \le U_1 = X_1/t \le U_{10}$$

$$\text{Region III:}\quad \left.\begin{array}{l} p = p_c, \qquad q = q_c \\ v_{1c} = V_1(U_{10}) - V_1(U_{1c}) \\ v_{2c} = V_2(U_{10}) - V_2(U_{1c}) \end{array}\right\} \qquad U_{2c} \le X_1/t \le U_{1c}$$

$$\text{Region IV:}\quad \left.\begin{array}{l} p = \bar{P}(U_2), \qquad q = \bar{Q}(U_2) \\ v_1 = V_1(U_{10}) + \bar{V}_1(U_{2c}) \\ \qquad - V_1(U_{1c}) - \bar{V}_1(U_2) \\ v_2 = V_2(U_{10}) + \bar{V}_2(U_{2c}) \\ \qquad - V_2(U_{1c}) - \bar{V}_2(U_2) \end{array}\right\} \qquad U_{21} \le U_2 = X_1/t \le U_{2c}$$

$$\text{Region V:}\quad \left.\begin{array}{l} p = p_1, \qquad q = q_1 \\ v_1 = v_{11} = V_1(U_{10}) \\ \qquad - \bar{V}_1(U_{21}) + \bar{V}_1(U_{2c}) - V_1(U_{1c}) \\ v_2 = v_{21} = V_2(U_{10}) \\ \qquad - \bar{V}_2(U_{21}) + \bar{V}_2(U_{2c}) - V_2(U_{1c}) \end{array}\right\} \qquad 0 \le X_1/t \le U_{21}$$

In order to calculate p, q, v_1, and v_2 as functions of X_1 and t, the variables U_1 and U_2 which appear as the arguments of the functions P, Q, $\bar{P}$, $\bar{Q}$, V_1, V_2, $\bar{V}_1$, and $\bar{V}_2$ in regions II and IV should be replaced by X_1/t. Obviously the smooth solutions described in the foregoing analysis are valid if and only if the computed velocities satisfy the following inequalities:

$$\begin{array}{ll} 0 \le U_{21} \le U_{2c} \le U_{1c} \le U_{10} & \\ U_{1c} \le U_1(p, q) \le U_{10} & (p, q \text{ in region II}) \\ U_{21} \le U_2(p, q) \le U_{2c} & (p, q \text{ in region IV}) \end{array} \tag{3.7.19}$$

These conditions place restrictions on the material constitution unless we develop another formulation which will cover the cases that violate these inequalities.

As discussed previously, the shear waves disappear under the assumption $q_0 = q_1 = 0$, so that a pure longitudinal wave propagates in the medium. The regions in the X_1, t-plane are now as depicted in Fig. 3.7.4, with U_{10} and U_{11} given by

$$U_{10} = U_1(p_0, 0), \qquad U_{11} = U_1(p_1, 0)$$

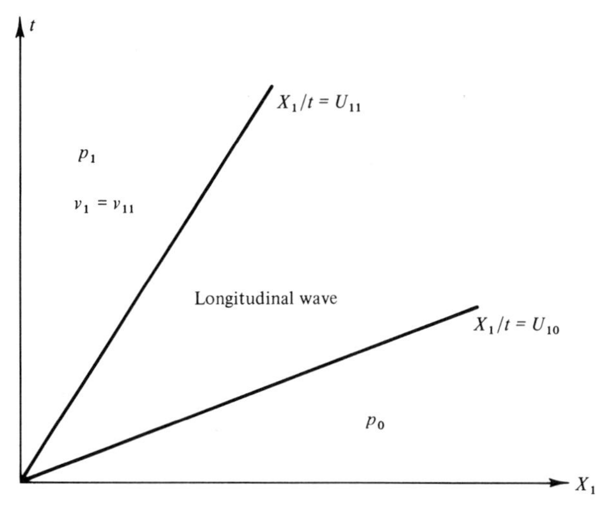

Fig. 3.7.4 Pure longitudinal wave.

The general solution reduces to

(3.7.20)

$$\left.\begin{array}{l} p = p_0 \\ v_1 = 0 \end{array}\right\} \qquad U_{10} \le X_1/t$$

$$\left.\begin{array}{l} p = P(U_1) = P(X_1/t) \\ v_1 = V_1(U_{10}) - V_1(U_1) = \int_{U_1}^{U_{10}} U(dP/dU)\,dU \\ \quad = \int_{X_1/t}^{U_{10}} U(dP/dU)\,dU \end{array}\right\} \qquad U_{11} \le X_1/t \le U_{10}$$

$$\left.\begin{array}{l} p = p_1 \\ v_{11} = V_1(U_{10}) - V_1(U_{11}) \end{array}\right\} \qquad 0 \le X_1/t \le U_{11}$$

The solution to the shear wave problem can be obtained by proper specialization of (3.7.18) by assuming $p_0 = p_1 = 0$. However, in this case a longitudinal wave region will also exist in general. But as in Section 3.5, it can be shown that for small amplitudes the longitudinal wave is of second order, as compared to the shear wave.

If the material is such that the *smoothness* conditions (3.7.19) are violated, then the regions of constant state cannot be connected by smooth waves. The solution corresponding to this case can only be obtained by considering the *weak* or *shock* solutions of the differential equations. In this case **p** and **v** suffer discontinuities. Again we call the faster wave the longitudinal and the slower wave the shear shock wave. The shocks connect three regions of constant state, as shown in Fig. 3.7.5.

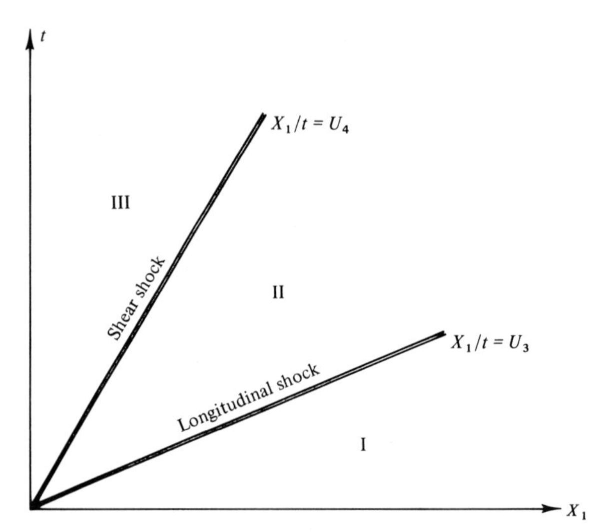

Fig. 3.7.5 Connecting shocks of three regions of constant state (weak solutions). I: $p_0 q_0$; II: p_c, q_c, v_{1c}, v_{2c}; III: p_1, q_1, v_{11}, v_{21}.

The plane shock conditions follow from Eqs. (2.7.5) and (2.9.1) as

(3.7.21)
$$[T_{11}] = -\rho_0 U[v_1] = \rho_0 U^2[p], \qquad [T_{12}] = -\rho_0 U[v_2] = \rho_0 U^2[q]$$

Hence Eqs. (3.7.21) evaluated across the longitudinal and shear shock waves give

$$U_3{}^2 = \frac{T_{11}^c - T_{11}^0}{\rho_0(p_c - p_0)}, \qquad U_4{}^2 = \frac{T_{12}^1 - T_{12}^c}{\rho_0(q_1 - q_c)}$$

(3.7.22)
$$(T_{11}^c - T_{11}^0)(q_c - q_0) - (T_{12}^c - T_{12}^0)(p_c - p_0) = 0$$
$$(T_{11}^1 - T_{11}^c)(q_1 - q_c) - (T_{12}^1 - T_{12}^c)(p_1 - p_c) = 0$$
$$v_{1c} = -U_3(p_c - p_0), \qquad v_{2c} = -U_3(q_c - q_0)$$
$$v_{11} = v_{1c} - U_4(p_1 - p_c), \qquad v_{21} = v_{2c} - U_4(q_1 - q_c)$$

The Piola–Kirchhoff stresses in these expressions are given by

$$T_{11}(p, q) = \partial\Sigma/\partial p, \qquad T_{12}(p, q) = \partial\Sigma/\partial q$$

If we can solve p_c and q_c from the second and third lines of (3.7.22), the other desired quantities are easily determined. We assume that such a solution exists. Of course, for the occurrence of the shock waves the stress potential Σ must satisfy certain requirements. For instance, we must have

$$(T_{11}^c - T_{11}^0)(p_c - p_0) > 0, \qquad (T_{12}^1 - T_{12}^c)(q_1 - q_c) > 0$$

for real shock velocities. Also we must recall our initial assumption $U_3 > U_4$. It is evident that some thermodynamical conditions must also be satisfied for shocks that are physically admissible. A purely mathematical motivation (Lax [1957]) suggests the following admissibility conditions:

$$U_1(p_c, q_c) > U_3 > U_1(p_0, q_0), \qquad U_2(p_1, q_1) > U_4 > U_2(p_c, q_c) \tag{3.7.23}$$

which are just the opposite of $(3.7.19)_{2,3}$. However, these conditions are quite doubtful, on physical grounds, especially for strong shocks,[1] But they turn out to be valid for the isentropic approximations considered in the present analysis (Bland [1964a,b]).

Combination of the smooth simple wave and shock solutions just discussed enables one to treat the Riemann problem under consideration for a large class of elastic materials. The conditions which specify the character of the wave solution are summarized in Table 3.7.1, which should be supplemented with our basic assumption $U_1 > U_2$ for all deformations.

TABLE 3.7.1

Inequalities		Longitudinal wave	Shear wave
$U_1(p_c, q_c) \leq U_1(p_0, q_0)$,	$U_2(p_1, q_1) \leq U_2(p_c, q_c)$	Smooth	Smooth
$U_1(p_c, q_c) \leq U_1(p_0, q_0)$,	$U_2(p_1, q_1) > U_2(p_c, q_c)$	Smooth	Shock
$U_1(p_c, q_c) > U_1(p_0, q_0)$,	$U_2(p_1, q_1) \leq U_2(p_c, q_c)$	Shock	Smooth
$U_1(p_c, q_c) > U_1(p_0, q_0)$,	$U_2(p_1, q_1) > U_2(p_c, q_c)$	Shock	Shock

As an illustrative example we consider a special material proposed by Ko [1963] [cf. Section 2.15, Eq. (2.15.35)]. The Cauchy stress tensor in this material is given by the relation

$$\mathbf{t} = \mu\left[\mathbf{I} - III^{-3/2}\left(III\mathbf{I} - I\overset{-1}{\mathbf{c}} + \overset{-2}{\mathbf{c}}\right)\right]$$

or noting that

$$III\mathbf{c} = \overset{-2}{\mathbf{c}} - I\overset{-1}{\mathbf{c}} + III\mathbf{I}$$

we have

$$\mathbf{t} = \mu(\mathbf{I} - III^{-1/2}\mathbf{c}) \tag{3.7.24}$$

where $\mu > 0$ is a material constant (the shear modulus for infinitesimal deformations). Since $T_{1k} = t_{1k}$ in the problem under discussion [cf. (3.5.6)], we get from (3.7.24)

$$T_{1k} = \mu(\delta_{1k} - III^{-1/2}c_{1k})$$

[1] This is evident from the shock theory in elastic compressible fluids.

From $(3.5.2)_1$ with $p_1 \equiv p$, $p_2 \equiv q$, $p_3 = 0$, it then follows that

$$\mathbf{c} = \begin{bmatrix} (1+q^2)/(1+p)^2 & -q/(1+p) & 0 \\ -q/(1+p) & 1 & 0 \\ 0 & 0 & 1 \end{bmatrix}, \qquad III^{1/2} = 1 + p$$

(3.7.25) $T_{11} = \mu[1 - (1+q^2)/(1+p)^3]$, $\quad T_{12} = \mu q/(1+p)^2$, $\quad T_{13} = 0$

and

$$\begin{aligned} \partial^2\Sigma/\partial p^2 &= \partial T_{11}/\partial p = 3\mu(1+q^2)/(1+p)^4 \\ \partial^2\Sigma/\partial q^2 &= \partial T_{12}/\partial q = \mu/(1+p)^2 \\ \partial^2\Sigma/\partial p\,\partial q &= \partial T_{11}/\partial q = \partial T_{12}/\partial p = -2\mu q/(1+p)^3 \end{aligned}$$

Hence the characteristic wave speeds are found from (3.7.3) as

$$\begin{aligned} U_1 &= \left(\frac{\mu}{2\rho_0}\right)^{1/2}\left\{\frac{3(1+q^2)}{(1+p)^4} + \frac{1}{(1+p)^2}\right. \\ &\quad \left. + \left[\left(3\frac{1+q^2}{(1+p)^4} - \frac{1}{(1+p)^2}\right)^2 + \frac{16q^2}{(1+p)^6}\right]^{1/2}\right\}^{1/2} \\ U_2 &= \left(\frac{\mu}{2\rho_0}\right)^{1/2}\left\{\frac{3(1+q^2)}{(1+p)^4} + \frac{1}{(1+p)^2}\right. \\ &\quad \left. - \left[\left(3\frac{1+q^2}{(1+p)^4} - \frac{1}{(1+p)^2}\right)^2 + \frac{16q^2}{(1+p)^6}\right]^{1/2}\right\}^{1/2} \end{aligned} \tag{3.7.26}$$

We immediately see that the inequality $q^2 < 3$ must be satisfied for real values of U_2. Therefore, we conclude that the differential field equations are not *totally* hyperbolic for this material unless $|q| < \sqrt{3}$.

C. Longitudinal Waves

Consider now the following initial and boundary value problem for the half-space made of Ko material:

$$p(X, 0) = 0, \qquad \mathbf{v}(X, 0) = \mathbf{0}$$

$$p(0, t) = \begin{cases} 0, & t \leq 0 \\ p_1, & t > 0 \end{cases}$$

$$q(X, t) \equiv 0$$

where $X \equiv X_1$. The smooth simple waves exist, according to Table 3.7.1, if

$$U_1(0, 0) > U_1(p_1, 0) \qquad \text{or} \qquad (3\mu/\rho_0)^{1/2} > (3\mu/\rho_0)^{1/2}(1 + p_1)^{-2}$$

Therefore, the simple wave solution corresponds to the case where $p_1 > 0$, $p_1 < -2$, and shocks correspond to $-2 < p_1 < 0$. In the former case, we find that

$$U_1 = (3\mu/\rho_0)^{1/2}(1 + p)^{-2}$$

or

$$p = P(U_1) = (3\mu/\rho_0)^{1/4} U_1^{-1/2} - 1$$

Hence

$$V_1(U_1) = \int_0^{U_1} U(dP/dU)\, dU = -(3\mu/\rho_0)^{1/4} U_1^{1/2}$$

$$U_{10} = (3\mu/\rho_0)^{1/2}, \qquad U_{11} = (3\mu/\rho_0)^{1/2}[1/(1 + p_1)^2]$$

The solution follows directly from (3.7.20):

$$p = \begin{cases} 0, & X/t \geq (3\mu/\rho_0)^{1/2} \\ (3\mu/\rho_0)^{1/4}(t/X)^{1/2} - 1, & (3\mu/\rho_0)^{1/2}(1 + p_1)^{-2} \leq X/t \leq (3\mu/\rho_0)^{1/2} \\ p_1, & 0 \leq X/t \leq (3\mu/\rho_0)^{1/2}(1 + p_1)^{-2} \end{cases}$$

(3.7.27)

$$v_1 = \begin{cases} 0, & X/t \geq (3\mu/\rho_0)^{1/2} \\ -(3\mu/\rho_0)^{1/4}[(3\mu/\rho_0)^{1/4} - (X/t)^{1/2}], & (3\mu/\rho_0)^{1/2}(1 + p_1)^{-2} \leq X/t \leq (3\mu/\rho_0)^{1/2} \\ -(3\mu/\rho_0)^{1/2}(1 - 1/|1 + p_1|), & 0 \leq X/t \leq (3\mu/\rho_0)^{1/2}(1 + p_1)^{-2} \end{cases}$$

If $-2 < p_1 < 0$, the shock solution will prevail. In this case, from (3.7.22) and (3.7.25) with $p_c = p_1$, $p_0 = 0$, we get

(3.7.28) $$U_3 = \{(\mu/\rho_0 p_1)[1 - (1 + p_1)^{-3}]\}^{1/2}, \qquad -1 < p_1 < 0$$

and

(3.7.29)

$$\left.\begin{aligned} p &= 0 \\ v_1 &= 0 \end{aligned}\right\}, \qquad X/t > U_3$$

$$\left.\begin{aligned} p &= p_1 \\ v_1 &= -U_3 p_1 \end{aligned}\right\}, \qquad X/t < U_3$$

D. Shear Waves

Here we consider the solution of the following boundary and initial value problem:

$$p(X, 0) = q(X, 0) = 0, \qquad \mathbf{v}(X, 0) = \mathbf{0}$$

$$p(0, t) = 0, \qquad t \in (-\infty, \infty)$$

$$q(0, t) = \begin{cases} 0, & t \leq 0 \\ q_1, & t > 0 \end{cases}$$

As we have frequently pointed out before, a longitudinal wave will always accompany a shear wave in a nonlinear solid. Therefore, in shear loading, all regions depicted in Fig. 3.7.3 will be present. However, some simplification may be achieved since the material here is not sheared in its initial configuration. Due to our basic assumption, shear disturbances will travel slower than longitudinal disturbances. We may, therefore, conjecture that $q_c = v_{2c} = 0$ in region III. The corresponding scheme is outlined in Fig. 3.7.6 representing

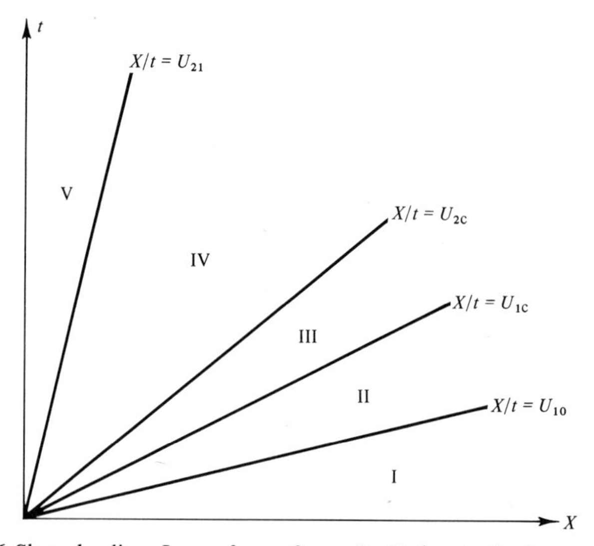

Fig. 3.7.6 Shear loading. I: $p = 0$, $q = 0$, $\mathbf{v} = \mathbf{0}$; II: longitudinal wave; III: $p = p_c$, $q = 0$, $v_1 = v_{1c}$, $v_2 = 0$; IV: shear wave; V: $p = 0$, $q = q_1$, $v_1 = v_{11}$, $v_2 = v_{21}$.

the smooth solutions only. Under these assumptions we find, from $(3.7.22)_1$ and (3.7.24), that

$$U_4^2 = \mu/\rho_0 \tag{3.7.30}$$

According to $(3.7.23)_2$, shear disturbances can propagate as shocks if

$$U_2(0, q_1) > U_4$$

or recalling (3.7.30) and $(3.7.26)_2$, if

$$\tfrac{1}{2}[4 + 3q_1^2 - (4 + 28q_1^2 + 9q_1^4)^{1/2}] > 1$$

This inequality cannot be satisfied for any choice of q_1. Hence shear waves are always smooth in this material. Unfortunately the shock conditions for accompanying longitudinal waves cannot be inferred at the outset without having detailed information about the shear wave. To this end we use an equation of the type (3.7.5) with U_2 to get

(3.7.31)
$$\frac{dp}{dq^2} = \frac{2(1+p)}{3(1+q^2)-(1+p)^2+\{[3(1+q^2)-(1+p)^2]^2+16q^2(1+p)^2\}^{1/2}}$$

Upon integration we have
$$p = \bar{P}(q^2)$$
subject to the condition
$$\bar{P}(q_1{}^2) = 0$$
where $q^2 < 3$.

It is not possible to give a closed form solution to differential equation (3.7.31). However, we can draw a few qualitative conclusions. First, it is possible to prove that
$$0 > p_c = \bar{P}(0) > -1$$

This implies that the longitudinal wave associated with the motion under consideration is a shock, as can be deduced if we recall the results obtained in the case of a pure longitudinal wave. Therefore, the X,t-diagram appropriate to this case should be as shown in Fig. 3.7.7. The slopes of the lines separating different regions are:

(3.7.32)
$$U_3 = (\mu/\rho_0)^{1/2}\{p_c^{-1}[1-(1-p_c)^{-3}]\}^{1/2}$$
$$U_{2c} = (\mu/\rho_0)^{1/2}(1+p_c)^{-1}$$
$$U_{21} = (\mu/2\rho_0)^{1/2}\{4+3q_1{}^2-[(2+3q_1{}^2)^2+16q_1{}^2]^{1/2}\}^{1/2}$$

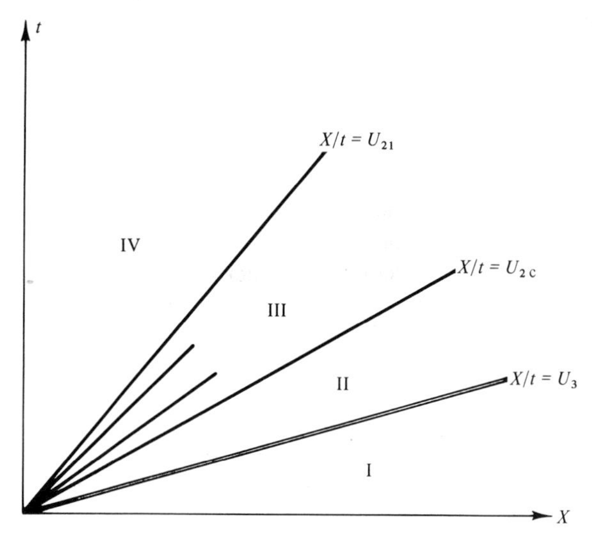

Fig. 3.7.7 Longitudinal shock in shear loading. I: $p=0$, $q=0$; II: longitudinal shock, $p=p_c$, $q=0$; III: shear wave; IV: $p=0$, $q=q_1$.

3.8 SIMPLE WAVES—GENERAL THEORY

We devote this section to a brief discussion of a general theory of simple wave motion. Although the general theory is presently far from producing tangible results, except for one-dimensional motions, it provides a mathematical basis on which a rational formulation of simple wave motions can be constructed. It is, of course, possible to advance the general theory of simple waves in parallel to the special theory discussed in Sections 3.5–3.7, but here we follow a somewhat indirect approach due to Varley [1965b].

The equations of motion are, as usual,

$$A_{KkLl}\, \partial^2 x_l/\partial X_K\, \partial X_L = \rho_0\, \partial^2 x_k/\partial t^2 \tag{3.8.1}$$

where A_{KkLl} are given functions of deformation gradients

$$A_{KkLl}(x_{m,M}) = \partial^2 \Sigma(x_{m,M})/\partial x_{k,K}\, \partial x_{l,L}$$

The second-order quasilinear partial differential equations (3.8.1) can be replaced by a quasilinear system of first order:

$$A_{KkLl}(\mathbf{F})\, \partial F_{lL}/\partial X_K - \rho_0\, \partial v_k/\partial t = 0, \qquad (\partial F_{kK}/\partial t) - (\partial v_k/\partial X_K) = 0$$

where

$$F_{kK} = x_{k,K} = \partial x_k/\partial X_K, \qquad v_k = \partial x_k/\partial t$$

Equations (3.8.1) determine three unknowns x_k which depend on four variables X_K and t. A well-posed Cauchy initial value problem is as follows: Find the integral of Eqs. (3.8.1) with given initial data on a hypersurface S in the four-dimensional space (X_K, t). The equation of the surface S may be written as $\Phi(\mathbf{X}, t) = 0$, which corresponds to a moving surface in the actual space of material points. Initial data on this surface consist of the values of functions sought and *one* exterior derivative of these functions, such as $x_{k,K}\Phi_{,K} + \dot{x}_k\dot{\Phi}$ on S. If S is not a *characteristic surface* of the system, all second- and higher-order partial derivatives of x_k are *uniquely* determined from the analytical initial data and the differential equations. Hence the solution to (3.8.1) is found in the full neighborhood of the surface S. On the other hand, if S is a characteristic surface, all second-order derivatives cannot be determined uniquely in a neighborhood of S. By generalizing the technique presented in Appendix B for a system of two independent variables, we can obtain the characteristic surfaces. To this end we write (3.8.1) in the following form:

$$\left(A_{KkLl}\frac{\partial^2}{\partial X_K\, \partial X_L} - \rho_0\, \delta_{kl}\frac{\partial^2}{\partial t^2}\right)x_l = 0 \tag{3.8.2}$$

The characteristic surface satisfies the equation

$$\det(A_{KkLl}\Phi_{,K}\Phi_{,L} - \rho_0\dot{\Phi}^2\, \delta_{kl}) = 0 \tag{3.8.3}$$

which is a first-order partial differential equation. Its solution for a given $\mathbf{x} = \mathbf{x}(\mathbf{X}, t)$ is a one-parameter family of surfaces in the space $(\mathbf{X}, t)$. It is clear that

$$N_K = \frac{\Phi_{,K}}{|\operatorname{grad} \Phi|}, \qquad U_N = -\frac{\partial \Phi/\partial t}{|\operatorname{grad} \Phi|}$$

are, respectively, the unit normal vector to the surface and its normal velocity, if we regard $\Phi(\mathbf{X}, t) = \text{const}$ as a moving surface in three-dimensional space $\mathbf{X}$. In four-dimensional space, $(N_K, -U_N)$ form the components of a normal vector (not unit) to a *fixed* surface $\Phi(\mathbf{X}, t) = \text{const}$. With this notation, the equation for characteristic surface takes the form

$$\det(A_{KkLl} N_K N_L - \rho_0 U_N^2 \delta_{kl}) = 0 \tag{3.8.4}$$

if we assume that $|\operatorname{grad} \Phi| \neq 0$. This equation for U_N is exactly the same as the one obtained in Section 2.12 for a propagating acceleration wave [cf. (2.12.11)]. Therefore the characteristic surface is essentially an acceleration front. This result should, of course, be expected. Since all the second-order derivatives of x_k are uniquely determined on every surface except on characteristics, a discontinuity at these derivatives can only occur along characteristic surfaces. Hence characteristic surfaces are actually wave fronts. The existence of real families of characteristics relies upon the positive roots of (3.8.4). If there are three distinct positive roots for U_N^2, system (3.8.1) is called *totally hyperbolic*. If there is at least one positive root for U_N^2, the system is hyperbolic. The nature of the system is closely related to the properties of the tensor A_{KkLl} (see the discussion in Section 2.12). In the following analysis we suppose that the system is at least hyperbolic for all deformations under consideration. In spatial coordinates the spatial form of the characteristic surfaces is given by $\phi(\mathbf{x}, t) = \Phi(\mathbf{X}(\mathbf{x}, t), t) = \text{const}$, and the characteristic condition (3.8.3) takes the form [cf. (2.12.14,15)]

$$\det(S_{kl}(\mathbf{F}, \mathbf{n}) - \rho_0 U^2 \delta_{kl}) = 0 \tag{3.8.5}$$

where $\mathbf{n}$ is the unit normal to the characteristic surface $\phi = \text{const}$ and U is the local speed of propagation given by

$$U = w - \mathbf{v} \cdot \mathbf{n} \tag{3.8.6}$$

Here w is the normal velocity of the surface, $v_n = \mathbf{v} \cdot \mathbf{n}$ the velocity of the material points in the normal direction to the surface upon the arrival of the characteristic surface, and

$$S_{kl} = A_{KkLl} x_{m,K} x_{n,L} n_m n_n \tag{3.8.7}$$

[cf. (2.12.15)]. Using (2.6.19), we may also transform (3.8.4) into

$$\det(A_{KkLl} M_K M_L - \rho_0 U^2 \delta_{kl}) = 0 \tag{3.8.8}$$

where

$$M_K = n_k x_{k,K} \tag{3.8.9}$$

For any scalar k we conclude that (3.8.5) may be written as

$$\det[S_{kl}(\mathbf{F}, k\mathbf{n}) - \rho_0(kU)^2 \delta_{kl}] = 0$$

in view of (3.8.7). Therefore, U and consequently the normal velocity w of the surface are homogeneous functions of $\mathbf{n}$ of degree one. Thus according to a well-known theorem of Euler on homogeneous functions

$$n_k \, \partial w/\partial n_k = w \tag{3.8.10}$$

Hence we deduce that any point $\mathbf{x}(t)$ moving with velocity

$$dx_k/dt = \partial w/\partial n_k = c_k \tag{3.8.11}$$

always stays on the front. At this point the unit normal satisfies the equation

$$dn_k/dt = (n_k n_l - \delta_{kl}) \, \partial w/\partial x_l \tag{3.8.12}$$

[cf. (2.2.44)]. (A derivation of Eq. (3.8.12) is given by Varley and Cumberbatch [1965].)

The solution of the system of ordinary nonlinear equations (3.8.11) and (3.8.12) gives a two-parameter curve in $\mathbf{x},t$-space and a two-parameter unit vector field which is normal to the moving characteristic surface. If at a certain time $t = t_0$ the front coincides with a surface σ_0, with a parametric representation

$$\mathbf{x}_0 = \mathbf{x}(t_0, a, b) \tag{3.8.13}$$

and unit normals

$$\mathbf{n}_0 = \mathbf{n}(t_0, a, b) \tag{3.8.14}$$

the solution

$$\mathbf{x} = \mathbf{x}(t, a, b), \qquad \mathbf{n} = \mathbf{n}(t, a, b) \tag{3.8.15}$$

of the system (3.8.11) and (3.8.12), subject to initial conditions, gives the parametric representation of the front at time t. The family of curves obtained in $\mathbf{x}$-space by taking $a = \text{const}$, $b = \text{const}$ is called *bicharacteristics* or *rays* of the original system of differential equations at the front, since these curves are transverse to the hypersurfaces $\phi(\mathbf{x}, t) = \text{const}$ in the four-dimensional space $(\mathbf{x}, t)$. For a detailed discussion the reader is referred to Courant and Hilbert [1962, Chapter VI].

If the system (3.8.1) is totally hyperbolic we obtain six families of bicharacteristics, corresponding to six propagating characteristic surfaces, three in each sense, associated with every real value of U, and consequently of w.

Motivated by the terminology common to linear wave theory, we name

$U(\mathbf{n}; \mathbf{x}, t)$ the *phase velocity* of the front, while $\mathbf{c}(\mathbf{n}; \mathbf{x}, t)$ is called the *group velocity* or *ray velocity*. It can be shown that (see Varley and Cumberbatch [1965]) the discontinuities in the derivatives of the material velocity $\mathbf{v}$ and the deformation gradients $\mathbf{F}$ at the front propagate with group velocity $\mathbf{c} = \partial w/\partial \mathbf{n}$ along rays.

The foregoing arguments applied to (3.8.8) with a vector $\mathbf{M}$ (not a unit vector) prove that the images of the bicharacteristics in the material description are given by

(3.8.16) $$dX_K/dt = \partial U/\partial M_K$$

and the group velocity by

(3.8.17) $$\mathbf{C} = \partial U/\partial \mathbf{M}$$

We now look for solutions of (3.8.1) such that $\mathbf{v}$ and $\mathbf{F}$ depend only on one variable, say α, so that they become constant on the same family of surfaces in some region of $\mathbf{X}$, t-space defined by $G(\mathbf{X}, t) = g(\mathbf{x}, t) = \alpha$. Such solutions, if they exist, will be called *simple wave solutions*. The region in which they prevail will be called a *simple wave region* in $\mathbf{X}$,t-space. The deformation associated with this kind of a solution will be called a *simple wave*.

Consider an arbitrary curve C in $\mathbf{X}$, t-space. Along this curve we may write

(3.8.18) $$dx_k = F_{kK}\, dX_K + v_k\, dt$$

We now introduce the Legendre transformation $\hat{\mathbf{x}}$ of $\mathbf{x}$ defined by

(3.8.19) $$\hat{x}_k = F_{kK} X_K + v_k t - x_k$$

so that along curve C relation (3.8.19) implies

$$d\hat{x}_k = X_K\, dF_{kK} + t\, dv_k$$

For a simple wave, $F_{kK} = x_{k,K}$ and v_k are functions of the variable α only, by definition. Thus we can write

(3.8.20) $$d\hat{x}_k = (X_K F'_{kK} + tv'_k)\, d\alpha$$

which means that $\hat{x}_k$ is a function of α only, and hence $d\hat{x}_k = \hat{x}'_k\, d\alpha$. In these expressions primes denote differentiation with respect to α. Equations (3.8.20) written in the form

(3.8.21) $$\hat{x}'_k(\alpha) = X_K F'_{kK}(\alpha) + tv'_k(\alpha)$$

provide three *implicit* relations for a single function $\alpha = G(\mathbf{X}, t)$. Therefore, in order to obtain a single function α of $\mathbf{X}$ and t we must assume that the coefficients of X_K and t in (3.8.21) are proportional to each other in all three equations. This may be expressed in the form

(3.8.22) $$U\hat{x}_k{}' + Lv_k{}' = 0, \qquad UF'_{kK} + M_K v_k{}' = 0$$

which are ordinary differential equations with α the independent variable. $U(\alpha)$, $L(\alpha)$, and $M_K(\alpha)$ in these expressions are five arbitrary (at least once differentiable) functions of α. Introducing (3.8.22) into (3.8.21) we find the following relation:

$$L(\alpha) = M_K(\alpha)X_K - U(\alpha)t \tag{3.8.23}$$

Thus if we specify the five functions L, U, and $\mathbf{M}$, the inversion of the implicit relation (3.8.23) determines a single function $\alpha = G(\mathbf{X}, t)$. It is clear that these functions cannot be selected arbitrarily.

Consider the equations

$$A_{KkLl}\,\partial F_{lL}/\partial X_K = \rho_0\,\partial v_k/\partial t$$

For a simple wave we may write

$$\partial F_{lL}/\partial X_K = F'_{lL}\,\partial G/\partial X_K, \qquad \partial v_k/\partial t = v_k'\,\partial G/\partial t$$

Hence the relations

$$A_{KkLl}F'_{lL}\,\partial G/\partial X_K = \rho_0 v_k'\,\partial G/\partial t \tag{3.8.24}$$

should be satisfied. Differentiating (3.8.23) we get

$$(L' + U't - M_K'X_K)\,d\alpha = M_K\,dX_K - U\,dt$$

which upon comparison with

$$d\alpha = (\partial G/\partial X_K)\,dX_K + (\partial G/\partial t)\,dt$$

gives

$$\partial G/\partial X_K = M_K/\Gamma, \qquad \partial G/\partial t = -U/\Gamma \tag{3.8.25}$$

where

$$\Gamma = L' + U't - M_K'X_K \tag{3.8.26}$$

Using (3.8.25) and $(3.8.22)_2$ in (3.8.24) we conclude that

$$(A_{KkLl}M_K M_L - \rho_0 U^2 \delta_{kl})v_k' = 0 \tag{3.8.27}$$

Therefore for the existence of a nontrivial solution v_k', $\mathbf{M}$ and U must satisfy condition (3.8.27) and

$$\det(A_{KkLl}M_K M_L - \rho_0 U^2 \delta_{kl}) = 0 \tag{3.8.28}$$

which is none other than the characteristic condition. We thus conclude that surfaces $\alpha =$ const are characteristic surfaces with a local speed of propagation U and normal direction $\mathbf{M}(|\mathbf{M}| \neq 1$, in general). Now any functions $\mathbf{F}(\alpha)$ and $\mathbf{v}(\alpha)$ which satisfy the ordinary differential equations (3.8.27) and $(3.8.22)_2$ for some $U(\alpha)$ and $\mathbf{M}(\alpha)$, satisfying (3.8.28), represent a simple wave solution of the equations of motion. The variable α is implicitly given by (3.8.23) for some $L(\alpha)$. Thus $\hat{\mathbf{x}}(\alpha)$ is determined from $(3.8.22)_1$, and $\mathbf{x} = \mathbf{x}(\alpha) = \mathbf{x}(\mathbf{X}, t)$ from (3.8.19). This concludes the scheme of the solution. However, it is apparent that the twelve equations (3.8.27) and $(3.8.22)_2$ are not sufficient to determine the sixteen unknowns $\mathbf{v}$, $\mathbf{F}$, $\mathbf{M}$, and U. Therefore, other relations are necessary in order to determine uniquely a simple wave deformation in a material. This freedom can only be removed by physical consideration in a given problem. Finally, it should be remarked that in simple wave deformation the surfaces $\alpha = \text{const}$ constitute a family of planes in the space $(\mathbf{X}, t)$ [cf. Eq. (3.8.23)].

So far we have not specified the nature of the independent variable α, which can be prescribed arbitrarily. We may, for example, select α as any one of the components of the deformation gradient tensor or the velocity vector. Another choice is to regard α as the value of time t at which the surface $\alpha = \text{const}$ traverses a particle $\mathbf{X} = \mathbf{X}_0$. In a general theory we leave the physical significance of α undefined so that it can be selected in the most convenient way to suit the problems under consideration.

The simple wave deformation can likewise be expressed in the spatial form. Defining a unit vector $\mathbf{n}$, as in (3.8.9), by

$$n_k = X_{K,k} M_K$$

we can transform (3.8.27) into

$$[S_{kl}(\mathbf{F}, \mathbf{n}) - \rho_0 U^2 \delta_{kl}]v_l' = 0$$

and the characteristic condition [cf. (3.8.5)] into

$$\det(S_{kl} - \rho_0 U^2 \delta_{kl}) = 0$$

Equation $(3.8.22)_2$ implies that

$$UF'_{kK} + F_{lK} n_l v_k' = 0 \tag{3.8.29}$$

The characteristic surfaces $\alpha = \text{const}$ in the spatial description may be obtained from (3.8.23). If we employ (3.8.19) to eliminate $F_{kK} X_K$, we get

$$l(\alpha) = n_k(\alpha) x_k - w(\alpha) t \tag{3.8.30}$$

where

$$l(\alpha) = L(\alpha) - n_k(\alpha)\hat{x}_k(\alpha), \qquad w(\alpha) = U(\alpha) + n_k(\alpha) v_k(\alpha) \tag{3.8.31}$$

We infer from Eq. (3.8.30) that the propagating surface $\alpha = \text{const}$ on which the particle velocity, deformation gradient, and hence stresses are constant is a plane $\Pi(\alpha, t)$ moving in the direction $\mathbf{n}(\alpha)$ with an absolute speed $w(\alpha)$ and a local speed $U(\alpha)$. Henceforth this surface will be referred to as the *wavelet* $\Pi(\alpha, t)$. The wavelets corresponding to different values of α are not, in general, parallel to each other and propagate with different speeds. Therefore, the resemblance to the general plane wave motion in linear elastodynamics is rather remote.

A simple wave region should be so chosen as not to include points of intersection of wavelets. At such a point, which belongs to, say, two wavelets with different α's, the deformation gradients and velocities will acquire two distinct sets of values. Therefore, a discontinuity should be considered on these quantities, which results in a shock formation.

We have at most twelve equations to determine the fifteen unknowns $\mathbf{v}$, $\mathbf{F}$, $\mathbf{n}$, and U (since $|\mathbf{n}| = 1$) at every point of the material. Thus some additional information is necessary to determine these quantities uniquely. Such information is usually extracted from the physical character of the problem under consideration, which provides hints about simple wave patterns.

Since w and $\mathbf{n}$ are constant on a particular wavelet, the corresponding bicharacteristics or rays are straight lines in the space $(\mathbf{x})$. In the space $(\mathbf{x}, t)$ wavelets constitute a one-parameter family of planes each of which carries a specific piece of information supplied, for instance, by the prescribed time-dependent deformations at the boundaries. However, simple wave deformations cannot allow reflections of wavelets from the boundaries. Thus the solutions obtained are valid until the occurrence of the first reflection.

The reflection phenomenon may, of course, be accounted for by an individual analysis. The reflected front, which is not a simple deformation itself, will propagate into the incoming wavelets, and the deformation ahead of it will be a simple wave deformation. Therefore, its speed W, evaluated according to (3.8.5), is a function of α and the unit normal $\mathbf{m}$ to the front. Thus the front can be determined by (3.8.11) and (3.8.12). These equations may now be written as

$$\frac{dx_k}{d\alpha} = \frac{\partial W}{\partial m_k}\frac{dt}{d\alpha}, \qquad \frac{dm_k}{d\alpha} = (m_k m_l - \delta_{kl})\frac{\partial W}{\partial \alpha}\frac{\partial \alpha}{\partial x_l}\frac{dt}{d\alpha} \tag{3.8.32}$$

if we represent the bicharacteristics by equations $\mathbf{x} = \mathbf{x}(\alpha, a, b)$, $t = t(\alpha, a, b)$. From (3.8.30) it follows that

$$l' = n_k' x_k + n_k x_k' - w't - wt'$$

or using $(3.8.32)_1$ we get

$$t' = \frac{dt}{d\alpha} = \frac{l' - n_k' x_k + w't}{n_l(\partial W/\partial m_l) - w}$$

In a similar way, from the partial differentiation of (3.8.30) with respect to x_k, we obtain

$$\partial\alpha/\partial x_k = n_k/(l' - n_l' x_l + w't)$$

Hence equations of the reflected front in terms of simple wave deformation become

$$\frac{dx_k}{d\alpha} = \frac{\partial W}{\partial m_k} \frac{l' - n_l' x_l + w't}{n_m(\partial W/\partial m_m) - w}, \qquad \frac{dm_k}{d\alpha} = \frac{n_k - (n_l m_l)m_k}{w - n_m(\partial W/\partial m_m)} \frac{\partial W}{\partial \alpha} \tag{3.8.33}$$

In general, once a simple wave deformation pattern is constructed, any material surface can be taken as the boundary of the material. Then the boundary conditions which lead to a simple wave deformation can be determined. Unfortunately, this inverse method usually results in quite complicated and unnatural boundary conditions.

Another particular situation arises when Γ in (3.8.25) vanishes. This means that a simple wave deformation with bounded particle accelerations and second deformation gradients (or stress gradients) in a certain time develops into a deformation with infinite accelerations and stress gradients in a later time. Using the relations (3.8.9), (3.8.31), (3.8.21), and (3.8.19) in (3.8.26), we see that Γ may be represented as follows:

$$\Gamma = l' + w't - n_k' x_k$$

Therefore, the foregoing argument implies the existence of a surface which satisfies

$$l(\alpha) = n_k(\alpha)x_k - w(\alpha)t, \qquad l'(\alpha) = n_k'(\alpha)x_k - w'(\alpha)t \tag{3.8.34}$$

These equations clearly define a propagating surface that is the envelope of a one-parameter family of plane wavelets. Since the wavelets are characteristic surfaces, the envelope is also a characteristic. This surface on which the particle accelerations and stress gradients become indefinite is called the *limit surface* and is a developable surface. The physically unrealistic occurrence of limit surfaces is generally prevented by the formation of a shock wave across which the velocity and displacement gradients, and consequently stresses, suffer finite jumps.

3.9 REFLECTION AND TRANSMISSION AT AN INTERFACE

In this section we discuss the reflection and transmission of large amplitude longitudinal waves in one dimension.[1] The motion is assumed to be subsonic so that no shock is accompanied. A wave arriving at an interface is partially transmitted and partially reflected. If two interfaces exist, then multiple reflections occur, resulting in decay of the amplitude of the pulse due to radiated energy across the interface to the surrounding medium. Quantitative

[1] This section is based largely on the work of Çekirge and Varley [1973].

results are obtained for a special class of nonlinear elastic materials. There is speculation that these results have applications in seismic waves at an interface in saturated soil, shock tubes, and other nonlinear elastic media.

Referred to the natural (undeformed reference) state, the Lagrangian equation of one-dimensional motion is

$$\partial T/\partial X = \rho_0 \, \partial v/\partial t \tag{3.9.1}$$

where $T(X, t)$ and $v(X, t)$ are, respectively, the normal traction per unit undeformed area and the velocity at X at time t, and ρ_0 is the density. The motion carries X to x at time t, i.e., $x = x(X, t)$. The velocity v and the strain e are given by

$$v = \partial x/\partial t, \qquad e = (\partial x/\partial X) - 1 \tag{3.9.2}$$

The density ρ in the deformed state is calculated by

$$\rho = \rho_0/(1 + e) \tag{3.9.3}$$

For an isotropic elastic solid, the one-dimensional stress constitutive equation is

$$T = T(e) \tag{3.9.4}$$

Substituting (3.9.4) into (3.9.1) we obtain

$$A^2(e) \, \partial e/\partial X = \partial v/\partial t \tag{3.9.5}$$

where

$$A(e) = [(1/\rho_0)\partial T/\partial e]^{1/2} \tag{3.9.6}$$

Differentiating $(3.9.2)_2$ with respect to t, we obtain the compatibility equation

$$\partial v/\partial X = \partial e/\partial t \tag{3.9.7}$$

Equations (3.9.5) and (3.9.7) are two first-order partial differential equations for the determination of $e(X, t)$ and $v(X, t)$. These equations may be transformed into a form first proposed by Riemann, by introducing

$$c = \int_0^e A(s) \, ds \tag{3.9.8}$$

Thus considering $T = T(c)$, then (3.9.5) and (3.9.7) may be written as

$$\partial v/\partial t = A(c) \, \partial c/\partial X, \qquad \partial c/\partial t = A(c) \, \partial v/\partial X \tag{3.9.9}$$

where the material function

$$A(c) \equiv (1/\rho_0) \, dT/dc > 0 \tag{3.9.10}$$

If we add and subtract the two equations (3.9.9), we obtain

$$[\partial(c \pm v)/\partial t] \mp [A(c)\, \partial(c \pm v)/\partial X] = 0$$

which shows that

$$f = \tfrac{1}{2}(c - v), \qquad g = \tfrac{1}{2}(c + v) \tag{3.9.11}$$

are invariant, respectively, at *characteristic wavelets* $\alpha(X, t) = \text{const}$ and $\beta(X, t) = \text{const}$ determined by

$$dX/dt\,|_\alpha = A(c), \qquad dX/dt\,|_\beta = -A(c) \tag{3.9.12}$$

At characteristic wavelets we have

$$\begin{aligned} &dx/dt = (\partial x/\partial t) + (\partial x/\partial X)(dX/dt) = v + a, && \text{at} \quad \alpha(X, t) = \text{const} \\ &dx/dt = v - a, && \text{at} \quad \beta(X, t) = \text{const} \\ &a(c) \equiv (1 + e)A \end{aligned} \tag{3.9.13}$$

where $a(c)$ is the local sound speed. We fix α and β characteristics by taking

$$\begin{aligned} \alpha &= t, \qquad \text{at the material boundary} \quad X = 0 \\ \beta &= t, \qquad \text{at the material boundary} \quad X = D \end{aligned}$$

so that

$$f = F(t), \quad \text{at} \quad X = 0; \qquad g = G(t), \quad \text{at} \quad X = D \tag{3.9.14}$$

According to (3.9.11) we then have at (X, t), with $0 \le X \le D$,

$$f = F(\alpha), \qquad g = G(\beta) \tag{3.9.15}$$

or

$$c = G(\beta) + F(\alpha), \qquad v = G(\beta) - F(\alpha) \tag{3.9.16}$$

Thus when $\alpha(X, t)$ and $\beta(X, t)$ are determined by solving (3.9.12), Eqs. (3.9.16) give the general solution of the one-dimensional wave.

A. The Linear Theory

For a medium obeying Hooke's law, we have

$$T = T_0 + E_0 e \tag{3.9.17}$$

Hence

$$A = (E_0/\rho_0)^{1/2} \equiv A_0 = \text{const} \tag{3.9.18}$$

$$\alpha = t - (X/A_0), \qquad \beta = t + (X - D)/A_0 \tag{3.9.19}$$

Thus Eqs. (3.9.16) indicate that, for these materials, the wave motion is a superposition of two noninteracting components: A *nondistorting, non-attenuated wave* (α-wave) moving to the right ($X > 0$) which carries the *signal* $F(\alpha)$, and a similar wave (β-wave) moving to the left (from $X = D$) which carries the *signal* $G(\beta)$. Since

(3.9.20)

$$t = \alpha = \beta + D/A_0, \quad \text{at} \quad X = 0; \qquad t = \beta = \alpha + D/A_0, \quad \text{at} \quad X = D$$

the determination[1] $F(\alpha)$ and $G(\beta)$ from prescribed conditions at $X = 0$ and $X = D$ usually reduces to solving a set of difference equations.

B. Nonlinear Theory, Decay of Free Vibrations

Suppose that in an elastic slab of length D, bounded by two elastic half planes, $X < 0$ and $X > D$, a longitudinal pulse is incident to the plane $X = 0$ from the right, carrying a signal $G(\beta)$ (Fig. 3.9.1). The width of the pulse is

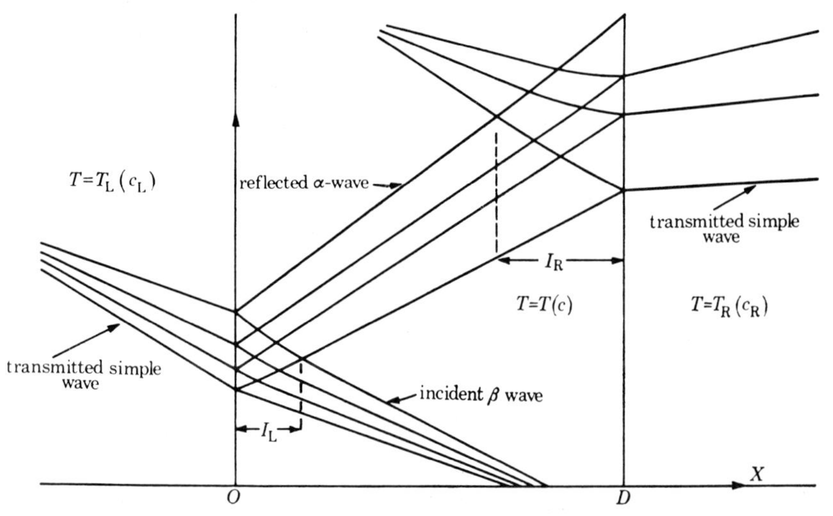

Fig. 3.9.1 The wave system that is set up when an incident pulse is partly reflected and partly transmitted at an interface between two elastic materials (*after* Çekirge and Varley [1973]).

considered smaller than D. The material is at rest ahead of the pulse before it reaches $X = 0$. Since the wave is moving into a region where $f = 0$, according to (3.9.11) and (3.9.12) the pulse is a simple wave, so that

$$(3.9.21) \qquad f = 0, \qquad v = c = g, \qquad \partial g/\partial t = G'(\beta)\, \partial\beta/\partial t, \qquad \text{at} \quad X$$

[1] We note that in Eulerian coordinates x the problem is much more complex, except for the linear theory.

Thus the stationary values of g and, in fact, of all state variables occur simultaneously in the incident pulse at the arrival of the wavelet at which the signal function G has stationary values. These stationary values can be considered measures of the amplitude of state variables in the pulse.

When the wave front reaches the interface $X = 0$, it produces a reflected α-wave which moves towards $X = D$ and a transmitted wave which moves into the surrounding medium $X < 0$. Since the reflected wave interacts with the incident wave, after it reaches $X = 0$ it is no longer a simple wave. Thus the stationary values of v at any X do not occur at the same time as those of T, c, and e. However, at a free interface $X = 0$ these stationary values do occur again, simultaneously. Since the stationary values can be determined in terms of those of g in the incident wave, the amplitudes of the reflected and transmitted pulses can be found at $X = 0$. To see this we observe that the wave transmitted into $X < 0$ is a simple wave. If a subscript L is used to denote the values of variables in the medium $X < 0$, then the equation of state reads

$$T_L = T_L(c_L), \qquad c_L = v_L = g_L \tag{3.9.22}$$

Since $X = 0$ is a material plane, we have

$$v = v_L, \qquad T = T_L \tag{3.9.23}$$

which together with (3.9.15) and (3.9.16) imply that

$$T(c) = T_L(c_L), \qquad \text{at} \quad X = 0 \tag{3.9.24}$$

where

$$c = g + f, \qquad c_L = g - f \tag{3.9.25}$$

These are the implicit equations for the *reflection function*

$$f = L(g), \qquad \text{at} \quad X = 0 \tag{3.9.26}$$

The *local reflection coefficient* is now given by

$$l(g) = L'(g) = [i(g) - 1]/[i(g) + 1] \tag{3.9.27}$$

where $i(g)$ is the *local impedance* of the interface, defined by

$$i(g) = \frac{dT_L/dc_L}{dT/dc} = \frac{\rho_L a_L}{\rho a} = \frac{\rho_{L0} A_L}{\rho_0 A_0} \tag{3.9.28}$$

If the interface is free, then irrespective of the amplitude of the incident pulse,

$$c = 0, \qquad L = -g, \qquad \text{at} \quad X = 0 \tag{3.9.29}$$

by (3.9.16). This corresponds to $i = 0$ in (3.9.27). If the interface is rigid, then

$$v = 0, \qquad L = g, \qquad \text{at} \quad X = 0 \tag{3.9.30}$$

which corresponds to $i = \infty$.

Once the reflection function $L(g)$ is determined, we have

$$f = L(g), \qquad c = g + L(g), \qquad v = g - L(g) = g_{\mathrm{L}}, \qquad \text{at} \quad X = 0 \tag{3.9.31}$$

$$\left(\frac{\partial f}{\partial t}, \frac{\partial c}{\partial t}, \frac{\partial v}{\partial t}\right) = \left(\frac{i-1}{i+1}, \frac{2i}{i+1}, \frac{2}{i+1}\right) G'(\beta) \frac{\partial \beta}{\partial t} \tag{3.9.32}$$

Equations (3.9.32) express the stated result that f, c, and v take stationary values with G. Moreover, the stationary values of reflected and transmitted signal functions f and g also occur at this time. If $g = g_1$ denotes a stationary value of $G(\beta)$ in the incident pulse, which corresponds to stationary values,

$$T = T(g_1), \qquad v = g_1 \tag{3.9.33}$$

and the corresponding stationary values of T and v at $X = 0$ in the transmitted wave are

$$T_{\mathrm{L}} = T(g_1 + L(g_1)), \qquad v = g_1 - L(g_1) \tag{3.9.34}$$

The corresponding stationary value of $G_{\mathrm{L}}(\beta_{\mathrm{L}})$ of the transmitted wave and the stationary value of $F(\alpha)$ of the reflected wave are

$$g_{\mathrm{L}1} = g_1 - L(g_1), \qquad f_1 = L(g_1) \tag{3.9.35}$$

The reflected α-wave in a region $0 \leq X \leq l_{\mathrm{L}}$ interacts with the incident wave, but beyond this region it is again a simple wave in which according to (3.9.11) and (3.9.12)

$$g = 0, \qquad c = -v = f \tag{3.9.36}$$

Again stationary values of the state variables occur simultaneously and can be found from the stationary value of any state variable in the incident pulse, for example,

$$T = T(f_1) = T(L(g_1)), \qquad v = -f_1 = -L(g_1) \tag{3.9.37}$$

Thus the amplitudes of reflected and transmitted waves are completely determined upon one reflection.

The reflected wave, upon arrival at $X = D$, will undergo the same process just described. If a subscript R denotes the state variables in $X = D+$, then we have

$$T_{\mathrm{R}} = T_{\mathrm{R}}(c_{\mathrm{R}}) \tag{3.9.38}$$

and with the same reasoning applied to the boundary $X = 0$, we have

$$g = R(f), \qquad \text{at} \quad X = D; \qquad r(f) = R'(f) = [j(f) - 1]/[j(f) + 1] \tag{3.9.39}$$

where $j(f)$ is the *impedance*:

$$j(f) = \rho_R a_R / \rho a \tag{3.9.40}$$

For simplicity, suppose that g_1 is the only stationary value of G in the incident wave. Then upon the first reflection of the incident wave at $X = 0$ the stationary values of f and g and consequently all other state variables occur simultaneously. They are

$$f_1 = L(g_1), \qquad g_2 = R(f_1) = R(L(g_1)) \tag{3.9.41}$$

g_2 is also the stationary value of g in the pulse during the second approach to $X = 0$. More generally, the stationary value of g in the pulse during the nth approach to the interface $X = 0$ is

$$g_n = R(L(g_{n-1})), \qquad \text{for all} \quad n \geq 2 \tag{3.9.42}$$

and this wave is a simple wave outside interaction regions I_L and I_R bordering $X = 0$ and $X = D$, respectively. The width of the interaction regions varies with n. Similarly, the stationary value of f in the pulse during its nth approach to the interface $X = D$ is

$$f_n = L(R(f_{n-1})), \qquad \text{for all} \quad n \geq 2 \tag{3.9.43}$$

In the case where elastic media obey Hooke's law, we have

$$L(g) = lg, \qquad R(f) = rf \tag{3.9.44}$$

where the reflection coefficients l and r are constants; therefore (3.9.42) and (3.9.43) take the simple forms

$$g_n = (lr)^{n-1} g_1, \qquad f_n = (lr)^{n-1} f_1 \tag{3.9.45}$$

where by (3.9.37) $f_1 = lg_1$. Since l and r are given by positive impedance i and j, it follows that

$$|l| \leq 1, \qquad |r| \leq 1$$

The equality sign holds when the interface is either perfectly free or perfectly rigid. Except for this case (3.9.45) indicates that both $|f_n|$ and $|g_n|$ and consequently $|v|$ and $|c|$ decrease as n increases.

Finally, we give the stationary values of the waves transmitted to the regions $X < 0$ and $X > D$ in the nth transmitted pulse. These are given by a direct generalization of (3.9.35):

$$g_{Ln} = g_n - L(g_n), \qquad f_{Ln} = f_n - R(f_n) \tag{3.9.46}$$

These results do not have simple extensions to the case where either a shock occurs or the width of the pulse exceeds the width of the medium D.

C. Decay of Pulse in Fully Saturated Soil

Following Çekirge and Varley [1973] we now illustrate the preceding theory with wave propagation in a layer between sea water and rock. The soil–rock interface is assumed to be rigid. A plane pulse is moving back and forth in the layer each time with reflection in the soil–rock interface and reflection and transmission in the soil–water interface.

According to Cole [1948] the dynamic pressure–density relation for sea water is well approximated over a wide range of pressures by

$$p_{\mathrm{L}} = p_{\mathrm{LO}} + (\rho_{\mathrm{LO}} A_{\mathrm{LO}}^2/\gamma)[(\rho_{\mathrm{L}}/\rho_{\mathrm{LO}})^{\gamma} - 1] \tag{3.9.47}$$

where ρ_{LO} and A_{LO} are, respectively, the density and sound speed in the water when it is at uniform pressure p_{LO}. This equation also relates pressure and density in an isentropic flow of inviscid, polytropic gas. For sea water $\gamma = 7.15$ typically, and for a gas $\gamma = 1.40$. In terms of T_{L} and e_{L} it reads

$$T_{\mathrm{L}} = T_{\mathrm{LO}} + (\rho_{\mathrm{LO}} A_{\mathrm{LO}}^2/\gamma)[1 - (1 + e_{\mathrm{L}})^{-\gamma}] \tag{3.9.48}$$

From (3.9.6) and (3.9.8) it follows that

(3.9.49)

$$1 + e_{\mathrm{L}} = \left(1 - \frac{\gamma - 1}{2}\frac{c_{\mathrm{L}}}{A_{\mathrm{LO}}}\right)^{-2/(\gamma-1)}, \qquad A_{\mathrm{L}} = A_{\mathrm{LO}}\left(1 - \frac{\gamma - 1}{2}\frac{c_{\mathrm{L}}}{A_{\mathrm{LO}}}\right)^{(\gamma+1)/(\gamma-1)}$$

$$T_{\mathrm{L}} = T_{\mathrm{LO}} + \frac{\rho_{\mathrm{LO}} A_{\mathrm{LO}}^2}{\gamma}\left[1 - \left(1 - \frac{\gamma - 1}{2}\frac{c_{\mathrm{L}}}{A_{\mathrm{LO}}}\right)^{2\gamma/(\gamma-1)}\right]$$

Fully saturated soil is a material that hardens under compression. In Fig. 3.9.2 the shape of the stress–strain relations is shown together with the experimental results taken from various sources. Also shown is the curve fitting based on

$$T = T_0 + 3e_1\rho_0 A_0^2\{[1 - (e/e_1)]^{-1/3} - 1\} \tag{3.9.50}$$

which is excellent in the range $0 \le e/e_1 \le 0.85$. For a typical *locking strain* $e_1 = -0.33\cdots\%$, this corresponds to the upper limit $e = 0.85e_1 = -0.28\,\%$.

To simplify the algebra we introduce the normalized variables

$$(\bar{v}, \bar{c}, \bar{f}, \bar{g}) = (3e_1 A_0)^{-1}(v, c, f, g) \tag{3.9.51}$$

Then when (3.9.50) holds,

$$\begin{gathered} \bar{e} = e/e_1 = 1 - (1 - \bar{c})^3, \qquad \bar{A} = A/A_0 = (1 - \bar{c})^{-2} \\ \bar{T} = (T - T_0)/3e_1\rho_0 A_0^2 = \bar{c}/(1 - \bar{c}) = (1 - \bar{e})^{-1/3} - 1 \end{gathered} \tag{3.9.52}$$

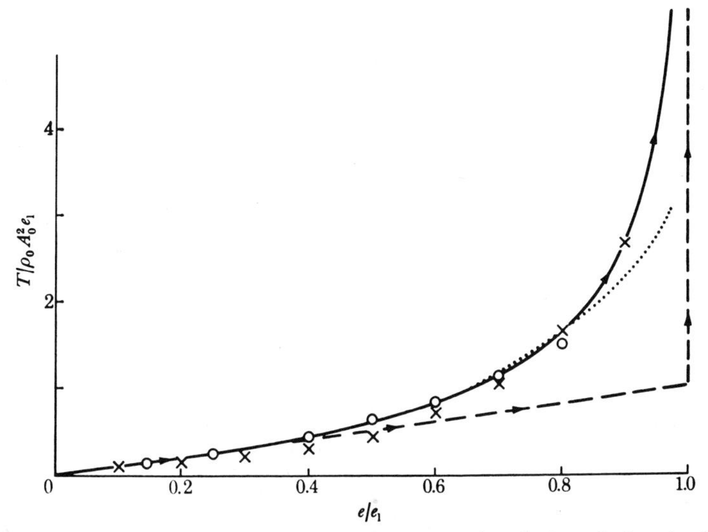

Fig. 3.9.2 Comparison of the experimental stress–strain relations during the dynamic compression of saturated soil, dry sand, and clay silt with the theoretical law (*after* Çekirge and Varley [1973]).

To determine various state variables s, we plot

(3.9.53) $$s = [-f/A_0, -g/A_0, 0.1A/A_0, (T_0 - T)/\rho A_0^2, -\tfrac{3}{2}e\,\%]$$

in Fig. 3.9.3. In the first quadrant the state variables are plotted as functions of $\bar{g}$ in the incident pulse where $\bar{f} = 0$. The variation of $\bar{g}$ is confined to the corresponding range of $\bar{e}$, $0 \leq \bar{e} \leq 0.875$, beyond which experimental agreement is not satisfactory. To obtain the relation between $\bar{f}$ and $\bar{g}$ at the soil–water interface, following the procedure previously described, we set $\bar{T} = \bar{T}_{\mathrm{L}}$ and $c_{\mathrm{L}} = v$ in (3.9.53) and (3.9.49):

(3.9.54)
$$\bar{T} = \bar{c}/(1 - \bar{c}) = (i_0/\delta)(\{1 + [(\gamma - 1)/2\gamma]\delta\bar{v}\}^{2\gamma/(\gamma-1)} - 1), \qquad \text{at} \quad X = 0$$
$$\bar{c} = \bar{g} + \bar{f}, \qquad \bar{v} = \bar{g} - \bar{f}, \qquad i_0 = \rho_{\mathrm{L0}} A_{\mathrm{L0}}/\rho_0 A_0, \qquad \delta = -3\gamma e_1 A_0/A_{\mathrm{L0}} > 0$$

When $|\delta\bar{v}| \ll 1$, $(3.9.54)_1$ can be approximated by the linear relationship

(3.9.55) $$\bar{c} = i_0\bar{v}, \qquad c = i_0 v$$

which together with $(3.9.54)_2$ gives

(3.9.56) $$\bar{f} = (i_0 - 1)\bar{g}/(i_0 + 1)$$

A typical example of the reflection function $\bar{f} = \bar{L}(\bar{g})$ which results from (3.9.54) is depicted in the upper left quadrant of Fig. 3.9.3. This corresponds

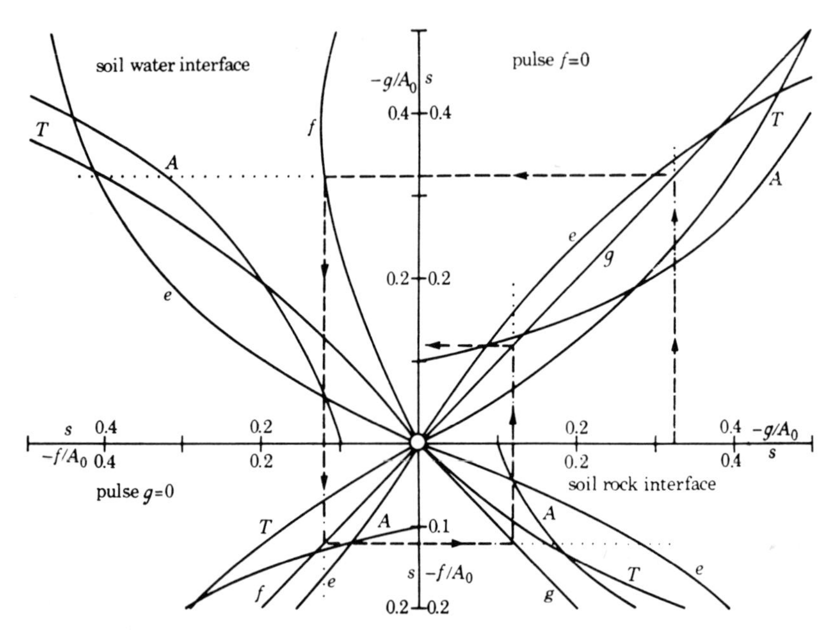

Fig. 3.9.3 Variations of the state variables of Eq. (3.9.53) with g in the incident pulse and at the soil–water interface, and with f in the reflected pulse and at the soil–rock interface. To calculate the maximum value of $-f/A_0$ at the soil–water interface given the maximum value of $-g/A_0$ in the incident pulse follow the broken lines. To calculate the corresponding maxima of the state variables s follow the dotted lines. To calculate the maxima of $-f/A_0$ in the reflected pulse and the maxima of $-f/A_0$ and $-g/A_0$ at the soil–rock interface continue to follow the broken lines. To calculate the corresponding values of the state variables follow the dotted lines. The maximum value of $-g/A_0$ at the soil–rock interface is identical with that in the pulse during its second approach to the soil–water interface. By making n similar circuits the reduction in the amplitude of the pulse after n reflections from both interfaces can be calculated (*after* Çekirge and Varley [1973]).

to values $(\rho_0, \rho_{L0}) = (2, 1)$ gr/cm^2, $(A_0, A_{L0}) = (0.2, 1.5)$ km/sec, respectively, and $e = -0.333\%$, taken from Cristescu [1967] and Hampton and Huck [1968]. With $\gamma = 7.15$ we obtain

$$i_0 = 3.750, \qquad \delta = 0.953 \times 10^{-2}$$

Since $i_0 > 1$, a small amplitude compression pulse in which $\bar{g} > 0$ is reflected as a compression pulse in which $\bar{f} > 0$, with its amplitude reduced by almost half and a large part of its energy transmitted to the water.

According to Fig. 3.9.3, at the soil–water interface, $\bar{f}$ first increases and then begins to decrease as $\bar{g}$ increases. $\bar{f}_{\max} = 0.126$ occurs when the local impedance $i = 1$ at $\bar{g} = 0.394$.

The state variables of the reflected pulse ($\bar{g} = 0$) are plotted as functions of $\bar{f}$ in the soil–water interface in the lower left quadrant and in the soil–rock interface in the lower right quadrant. Since here $\bar{v} = 0$,

$$\bar{g} = \bar{f} \equiv \bar{R}(\bar{f}), \qquad \text{at} \quad X = D$$

The figure can be used to obtain the decay of disturbance in the soil. To determine $f_{\max} = f_1$ at the soil–water interface and $g_{\max} = g_2$ at the soil–rock interface, we follow the broken line from the maximum value of g $(=g_1)$ in the incident pulse. After one complete circuit around the origin we obtain f_2 and g_2, and after n circuits we have f_{n+1} and g_{n+1}. The corresponding values of state variables at the interface outside the interaction regions follow from the dotted lines.

D. Determination of Signal Functions $F(t)$ and $G(t)$

When the values of g at the free surface $X = 0$ and of f at the other free surface $X = D$ of the slab are given then, as discussed before, we can determine the state variables at (X, t). However this task is formidable mathematically and can be achieved only for special types of media. Çekirge and Varley [1973] have studied a special type of medium whose stress–strain relations can be approximated by a polynomial in the strain, including up to third-order terms.

We now express (3.9.12) in the characteristic variables (α, β):

$$\partial X/\partial\alpha = -A(c)\,\partial t/\partial\alpha, \qquad \partial X/\partial\beta = A(c)\,\partial t/\partial\beta \tag{3.9.57}$$

where in accordance with (3.9.16)

$$c = F(\alpha) + G(\beta) \tag{3.9.58}$$

Eliminating X from (3.9.57) we obtain the *hodograph equation*

$$\frac{\partial^2 t}{\partial\alpha\,\partial\beta} + \frac{1}{2A}\frac{dA}{dc}\left(\frac{\partial c}{\partial\alpha}\frac{\partial t}{\partial\beta} + \frac{\partial c}{\partial\beta}\frac{\partial t}{\partial\alpha}\right) = 0 \tag{3.9.59}$$

which may be written in either of the forms

$$\begin{aligned} &\frac{\partial}{\partial\beta}\left(A^{1/2}\frac{\partial t}{\partial\alpha}\right) + \frac{1}{2}A^{-1/2}\frac{dA}{dc}\frac{\partial t}{\partial\beta}F'(\alpha) = 0 \\ &\frac{\partial}{\partial\alpha}\left(A^{1/2}\frac{\partial t}{\partial\beta}\right) + \frac{1}{2}A^{-1/2}\frac{dA}{dc}\frac{\partial t}{\partial\alpha}G'(\beta) = 0 \end{aligned} \tag{3.9.60}$$

The media to be studied are defined by the property that $A(c)$ satisfies the differential equation of the form

$$dA/dc = \mu A^{1/2} + \nu A^{3/2} \tag{3.9.61}$$

where μ and ν are material constants. Using (3.9.57) we can now write (3.9.60) in the forms

$$2\frac{\partial}{\partial\beta}\left(A^{1/2}\frac{\partial t}{\partial\alpha}\right) + F'(\alpha)\frac{\partial}{\partial\beta}(\mu t + \nu X) = 0$$
(3.9.62)
$$2\frac{\partial}{\partial\alpha}\left(A^{1/2}\frac{\partial t}{\partial\beta}\right) + G'(\beta)\frac{\partial}{\partial\alpha}(\mu t - \nu X) = 0$$

These equations integrate to give

(3.9.63)
$$2A^{1/2}(\partial t/\partial\alpha) + [\mu(t-\alpha) + \nu X]F'(\alpha) = m(\alpha)$$
$$2A^{1/2}(\partial t/\partial\beta) + [\mu(t-\beta) + \nu(D-X)]G'(\beta) = n(\beta)$$

where the integration constants $m(\alpha)$ and $n(\beta)$ can be evaluated at $X = 0$ and $X = D$:

(3.9.64)
$$m(t) = A^{1/2}, \quad \text{at} \quad X = 0, \quad \text{where} \quad t = \alpha$$
$$n(t) = A^{1/2}, \quad \text{at} \quad X = D, \quad \text{where} \quad t = \beta$$

since at constant X we have

$$\frac{Dt}{D\alpha} = \frac{\partial t}{\partial\alpha} + \frac{\partial t}{\partial\beta}\frac{d\beta}{d\alpha} = \frac{\partial t}{\partial\alpha} + A^{-1}\frac{\partial X}{\partial\beta}\frac{d\beta}{d\alpha}, \qquad \text{by } (3.9.57)_2$$

$$= \frac{\partial t}{\partial\alpha} - A^{-1}\frac{\partial X}{\partial\alpha} = 2\frac{\partial t}{\partial\alpha}, \qquad \text{by } (3.9.57)_1$$

and similarly $Dt/D\beta = 2\,\partial t/\partial\beta$. At $X = \text{const}$, (3.9.63) may be written as

(3.9.65)
$$A^{1/2}(Dt/D\alpha) + [\mu(t-\alpha) + \nu X]F'(\alpha) = m(\alpha)$$
$$A^{1/2}(Dt/D\beta) + [\mu(t-\beta) + \nu(D-X)]G'(\beta) = n(\beta)$$

When $F(\alpha)$, $G(\beta)$, $m(\alpha)$, and $n(\beta)$ are known, and A has been determined, then Eqs. (3.9.65) provide two nonlinear differential equations for the variations of α and β with t at any X. Once these are determined, F and G, and consequently all other state variables, can be calculated.

As an example, we discuss the reflection of a shockless centered simple wave (at a time interval $0 \le t \le t_R$) from an interface. The wave is assumed to be centered at $X = D$ and the surface $X = 0$ is free. To calculate the condition at $X = 0$ we employ $(3.9.65)_2$. In this wave $F \equiv 0$, and at $X = D$, $G(\beta)\,(=v=c)$ and $n(\beta) = A^{1/2}$ are known. A pulse moves towards $X = 0$, remaining a simple wave until it is affected by the reflected wave from $X = 0$. Equation $(3.9.65)_2$ by use of (3.9.61) integrates to give

(3.9.66)
$$t = \beta + (D - X)/A(G) = \beta + (D - X)/n^2(\beta)$$

This equation together with the condition that in the simple wave

$$v = c = G(\beta) \tag{3.9.67}$$

are sufficient to calculate the change in the state variables with t at any X. At the interaction region I_L neighboring $X = 0$, the signal function F is not identically zero. Consequently the interaction term containing $A^{1/2}$ in (3.9.65) is not, in general, a known function of (β, X), and we need additional information. But at $X = 0$, according to (3.9.31), we have

$$F = L(G), \qquad c = G + L(G) \tag{3.9.68}$$

so that $A^{1/2}$ is a known function of β, say $A^{1/2} = \hat{m}(\beta)$. When this is inserted into $(3.9.65)_2$ at $X = 0$, we obtain

$$\hat{m}(\beta)(d/d\beta)[t - \beta + (v/\mu)D] + \mu[t - \beta + (v/\mu)D]G'(\beta) = n(\beta) - \hat{m}(\beta) \tag{3.9.69}$$

The solution of this first-order differential equation under the initial condition:

$$t = D/A_0 \qquad \text{when} \quad \beta = 0 \tag{3.9.70}$$

gives G as a function of t at $X = 0$. Equation (3.9.68) then gives $F(t)$, the signal carried by the reflected wave, and therefore $G_L(t)$ $(= v$ at $X = 0)$, the signal carried by the transmitted wave.

In the case of a centered wave at $X = D$, Eq. (3.9.65) must be interpreted carefully, since at the singular point $X = D$, G changes a finite amount over a vanishingly small variation in β $(= t$ at $X = D)$. To obtain the limiting form (3.9.65), we take the incident wave at $X = X_0 < D$ at some time $t = t_0 > 0$, and then let $(X_0, t_0) \to (D, 0)$. In such a centered wave

$$A = (X_0 - X)/(t - t_0) \tag{3.9.71}$$

According to (3.9.66) then, $G(\beta)$ is determined from

$$A(G) = (X_0 - D)/(\beta - t_0) \tag{3.9.72}$$

When β is solved from this equation and inserted into $(3.9.65)_2$ it yields

$$A^{1/2}(dt/dG) + \mu(t - t_0) + v(X_0 - X) = 0, \qquad \text{at any} \quad X \tag{3.9.73}$$

where we used the fact that $n(\beta) = A^{1/2}(G)$ and $A(c)$ satisfy (3.9.61). At $X = 0$, where c is given in terms of G by (3.9.68), Eq. (3.9.73) predicts G as a function of t through

$$A^{1/2}(dt/dG) + \mu t + vD = 0 \tag{3.9.74}$$

which integrates to give

$$\frac{A_0 t}{D} = 1 + \frac{1 - M}{M}\left[1 - \exp\left(-\mu \int_0^G A^{-1/2}\, dG\right)\right] \tag{3.9.75}$$

where $M = -\mu/A_0 \nu$. The constant of integration has been determined by use of the condition that the front of the centered wave, at which $G = 0$, arrives at $X = 0$ at $t = D/A_0$. The function $F(t)$ is now determined from (3.9.68). Equation (3.9.75) determines $G(t)$ at $X = 0$. If the interface $X = 0$ is *perfectly free*, we have $c = 0$ and $A = A_0$ there. In this case (3.9.75) gives

(3.9.76)

$$v/A_0 = 2G/A_0 = -2F/A_0 = -2(\mu A_0^{1/2})^{-1} \log\{1 - [M/(1-M)][(A_0 t/D) - 1]\}$$

If the interface $X = 0$ is *perfectly rigid*, then

$$v = 0, \qquad G = F = \tfrac{1}{2}c, \qquad \text{at} \quad X = 0$$

Then (3.9.75) implies a variation of c with t:

$$\frac{A_0 t}{D} = 1 + \frac{1-M}{M}\left[1 - \exp\left(-\frac{1}{2}\mu \int_0^c A^{-1/2}\, dc\right)\right] \tag{3.9.77}$$

E. Constitutive Equations

The determination of the stress and velocity field requires the knowledge of the constitutive equations, e.g., the form of A which satisfies (3.9.61). Çekirge and Varley [1973][1] found some solutions of this equation, (3.9.10) and (3.9.8) appropriate to a class of elastic solids, gases, and saturated soil. Here we mention briefly only some of these results.

For *polycrystalline solids,*

$$\begin{aligned} A/A_0 &= \eta_2 \coth^2[\eta_0 + \eta_1(c/A_0)] \\ T/\rho_0 {A_0}^2 &= t_0 + \eta_2 \eta_1^{-1}\{\eta_1(c/A_0) - \coth[\eta_0 + \eta_1(c/A_0)]\} \\ e &= e_0 + (\eta_1\eta_2)^{-1}\{\eta_1(c/A_0) - \tanh[\eta_0 + \eta_1(c/A_0)]\} \end{aligned} \tag{3.9.78}$$

where

$$A_0^{1/2}\mu = 2\eta_1\eta_2^{1/2}, \qquad A_0^{3/2}\nu = -2\eta_1\eta_2^{-1/2} \tag{3.9.79}$$

Further, taking $A = A_0$, $T = e = 0$ at the natural state ($c = 0$),

$$\eta_2 = \tanh^2\eta_0, \qquad t_0 = \eta_1^{-1}\tanh\eta_0, \qquad e_0 = \eta_1^{-1}\coth\eta_0 \tag{3.9.80}$$

we see that we have only two parameters μ and ν to provide curve fitting. If T_M and e_M are, respectively, maximum compressive stress and strain, and

$$c_M = \tfrac{4}{3}e_M A_M, \qquad A_M = [(1/2\rho_0)T_M/e_M]^{1/2} \tag{3.9.81}$$

[1] A solution of (3.9.61) approximating the stress–strain relations up to fourth order in train has the form

$$T = \rho_0 {A_0}^2[e + pe^2 + qe^3 + 0(e^4)]$$

where

$$\mu = \tfrac{1}{2}p(7q-6)/qA_0^{1/2}, \qquad \nu = \tfrac{1}{2}p(6-5q)/qA_0^{3/2}, \qquad A_0 \equiv A(0)$$

excellent curve fitting is obtained in the range $0.1 \leq c/c_M \leq 1$, which corresponds to $0.046 \leq e/e_M \leq 1$ with Bell's [1968] parabolic stress–strain law:

$$T/T_M = (e/e_M)^{1/2} \tag{3.9.82}$$

When $-e_M = 4 \times 10^{-2}$, considered by Bell, the range corresponds to

$$1.8 \times 10^{-3} \leq -e \leq 4 \times 10^{-2}$$

For a best fitting, Eq. (3.9.78) has the form

(3.9.83)

$$\begin{aligned} A/A_M &= 0.9655 \coth^2[0.6785 + 1.4088(c/c_M)] \\ T/T_M &= 0.8274 + 0.6437(c/c_M) - 0.4569 \coth[0.6785 + 1.4088(c/c_M)] \\ e/e_M &= 0.5693 + 1.3809(c/c_M) - 0.9802 \tanh[0.6785 + 1.4088(c/c_M)] \end{aligned}$$

The curve fitting fails near $e = 0$; otherwise, as shown in Fig. 3.9.4, it is excellent.

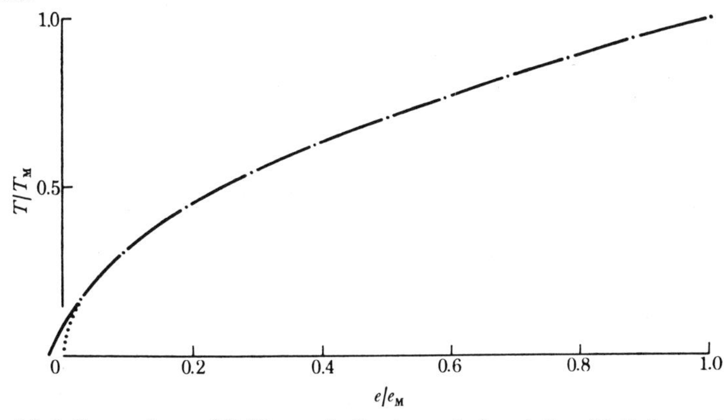

Fig. 3.9.4 Comparison of Bell's parabolic stress–strain relation (dotted curve) with a model stress–strain relation (solid curve) (*after* Çekirge and Varley [1973]).

By adjustment of the parameters μ, ν, and A_0, a variety of different kinds of stress–strain curves can be produced from (3.9.78) or from other representations of solutions of (3.9.61) for hard and soft materials. For example, the solution of (3.9.61), (3.9.10), and (3.9.8) may also be expressed as

$$\begin{aligned} \frac{T}{T_1} &= (1 + M)\left[1 + M^{1/2}\left(\bar{c} - \frac{1 + M^{1/2}\tanh\bar{c}}{M^{1/2} + \tanh\bar{c}}\right)\right] \\ \frac{e}{e_1} &= (1 + M^{-1})\left[1 + M^{-1/2}\left(c - \frac{M^{1/2} + \tanh\bar{c}}{1 + M^{1/2}\tanh\bar{c}}\right)\right] \\ \frac{A}{A_0} &= M\left(\frac{1 + M^{1/2}\tanh\bar{c}}{M^{1/2} + \tanh\bar{c}}\right)^2 \end{aligned} \tag{3.9.84}$$

where

$$\begin{aligned}
&\bar{c} = [M^{1/2}/(1+M)]c/e_1 A_0, && M = -\mu/\nu A_0 \equiv A_\infty/A_0 > 0 \\
(3.9.85)\quad &\mu = [2A_\infty/(A_0 + A_\infty)]A_0^{-1/2}/e_1, && \nu = -[2/(A_0 + A_\infty)]A_0^{-1/2}/e_1 \\
&T_1 = \rho_0 A_0{}^2 e_1, && 0 \le \bar{c} < \infty
\end{aligned}$$

Equations (3.9.84) can be shown to be identical to (3.9.78) in the range $0 \le M \le 1$. For $M = 1$ they give Hooke's law, and for $1 < M \le \infty$ they give *hardening* materials. Two such classes of stress–strain curves for nonideal materials are shown in Figs. 3.9.5 and 3.9.7.

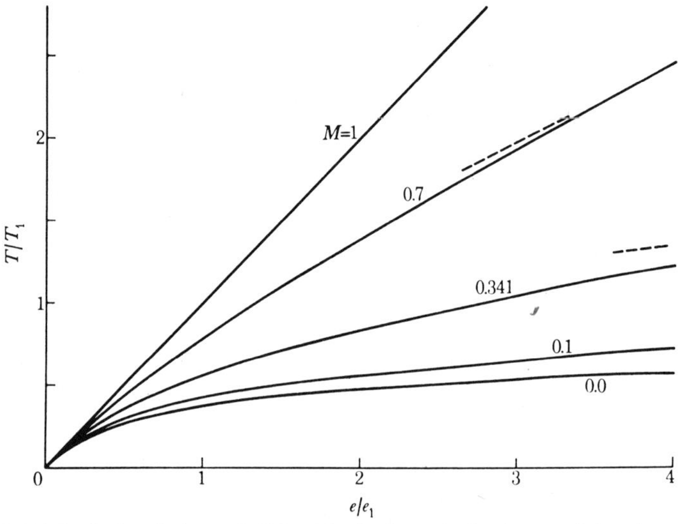

Fig. 3.9.5 Typical variations of T/T_1 with e/e_1 for soft elastic materials. The parameter $M = A_\infty/A_0$. The case $M = 0.341$ corresponds to materials which satisfy Bell's parabolic law (*after* Çekirge and Varley [1973]).

In different ranges of parameters M and $\bar{c}$, (3.9.84) give different classes of stress–strain curves. For example, in the ranges

$$(3.9.86)\qquad 1 \le M \le \infty, \qquad -\bar{\eta}_0 \le \bar{c} < 0, \qquad \text{where} \quad \coth \bar{\eta}_0 = M^{1/2}$$

we have one family of *ideally soft materials*. In this range (3.9.84) may be written as

(3.9.87)

$$\begin{aligned}
T/T_2 &= (\bar{\eta}_0 - \tanh \bar{\eta}_0)^{-1}[\tanh(\bar{\eta}_0 + \bar{c}) - \bar{c} - \tanh \bar{\eta}_0] \\
\rho_0 A_0{}^2 e/T_2 &= \tanh^4 \bar{\eta}_0 (\bar{\eta}_0 - \tanh \bar{\eta}_0)^{-1}[\coth(\bar{\eta}_0 + \bar{c}) - \bar{c} - \coth \bar{\eta}_0] \\
A/A_0 &= \coth^2 \bar{\eta}_0 \tanh^2(\bar{\eta}_0 + \bar{c})
\end{aligned}$$

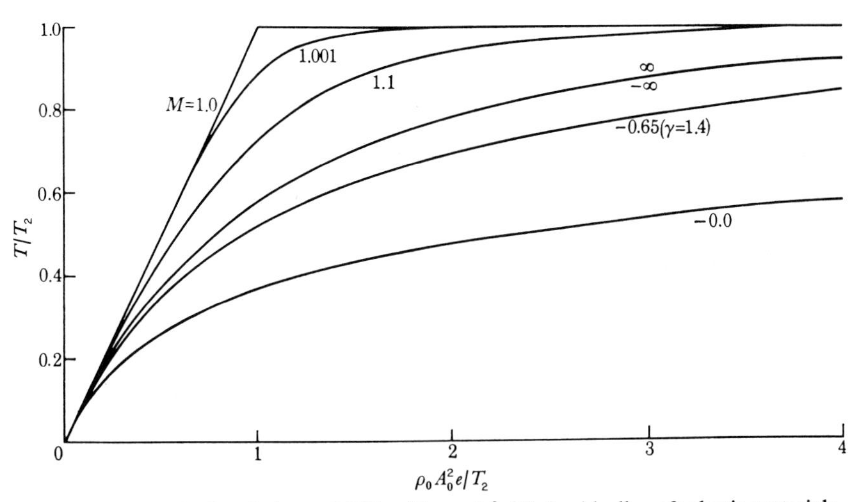

Fig. 3.9.6 Typical variations of T/T_2 with $\rho_0 A_0{}^2 e/T_2$ for ideally soft elastic materials. The case $M = -0.65$ corresponds to a gas with isentropic exponent $\gamma = 1.4$ (*after* Çekirge and Varley [1973]).

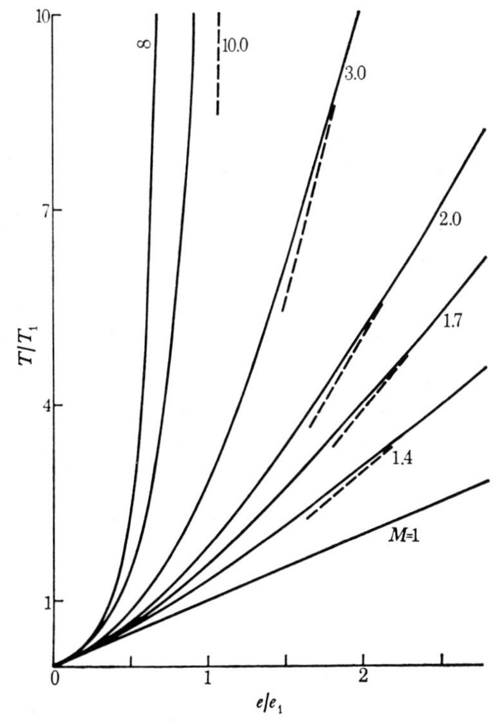

Fig. 3.9.7 Typical variations of T/T_1 with e/e_1 for hard elastic materials. The parameter $M = A_\infty/A_0$ (*after* Çekirge and Varley [1973]).

where

$$\bar{c} = -\coth^2 \bar{\eta}_0(\bar{\eta}_0 - \tanh \bar{\eta}_0)\rho_0 A_0 c/T_2$$
(3.9.88) $$\mu = -2 \coth^3 \bar{\eta}_0(\bar{\eta}_0 - \tanh \bar{\eta}_0)\rho_0 A_0^{3/2}/T_2$$
$$\nu = 2 \coth \bar{\eta}_0(\bar{\eta}_0 - \tanh \bar{\eta}_0)\rho_0 A_0^{1/2}/T_2$$

We observe that $T/T_2 \to 1$, $\rho_0 A_0{}^2 e/T_2 \to \infty$, and $A/A_0 \to 0$ as $\bar{c} \to -\bar{\eta}_0$. The parameters for curve fitting are T_2, A_0, and $\bar{\eta}_0$. Several curves characterized by (3.9.87) are shown in Fig. 3.9.6. They are parametrized by

(3.9.89) $$M = -\mu/A_0 \nu = \coth^2 \bar{\eta}_0$$

In the range

(3.9.90) $$0 \leq M \leq 1, \qquad -\bar{\eta}_0 \leq \bar{c} \leq 0, \qquad \text{where} \qquad \tanh \eta_0 = M^{1/2}$$

Eqs. (3.9.84) characterize *ideally hard materials*. In this range we may express these equations in the form

$$T/\rho_0 A_0{}^2 e_2 = \tanh^4 \eta_0(\eta_0 - \tanh \eta_0)^{-1}[\coth(\eta_0 + \bar{c}) - \bar{c} - \coth \eta_0]$$
(3.9.91) $$e/e_2 = (\eta_0 - \tanh \eta_0)^{-1}[\tanh(\eta_0 + \bar{c}) - \bar{c} - \tanh \eta_0]$$
$$A/A_0 = \tanh^2 \eta_0 \coth^2(\eta_0 + \bar{c})$$

where

$$\bar{c} = -\coth^2 \eta_0(\eta_0 - \tanh \eta_0)c/e_2 A_0 \qquad (=\eta_1 c/A_0)$$
(3.9.92) $$\mu = -2 \coth \eta_0(\eta_0 - \tanh \eta_0)/e_2 A_0^{1/2}$$
$$\nu = 2 \coth^3 \eta_0(\eta_0 - \tanh \eta_0)/e_2 A_0^{3/2}$$

F. Particle Velocity and Motion

The preceding constitutive equations can be used to determine the particle velocity v, Riemann invariant c, and other state variables as functions of time upon reflection. We compare the variation of state variables at $X = 0$ with those at any point X in the incident wave:

(3.9.93) $$v = c, \qquad A_0 t/(D - X) = A_0/A$$

For soft materials, employing (3.9.84) this gives

(3.9.94)

$$\frac{v}{A_0 e_1} = \frac{1 + M}{M^{1/2}} \bar{c}, \qquad \text{where} \quad \frac{A_0 t}{D - X} = \left(\frac{1 + M^{-1/2} \tanh \bar{c}}{1 + M^{1/2} \tanh \bar{c}}\right)^2, \quad M \leq 1$$

For a class of soft materials described by (3.9.86)–(3.9.89), (3.9.93) implies that

$$\begin{aligned} \rho_0 A_0 v/T_2 &= -\tanh^2 \bar{\eta}_0(\bar{\eta}_0 - \tanh \bar{\eta}_0)^{-1}\bar{c} \\ A_0 t/(D - X) &= \tanh^2 \bar{\eta}_0 \coth^2(\bar{\eta}_0 + \bar{c}) \end{aligned} \tag{3.9.95}$$

A centered simple wave cannot be generated in a Hookean material, since the Lagrangian sound of speed A_0 does not vary with c, so that any initial discontinuity in c cannot be smoothed by amplitude dispersion. The limit $M \to 1$ to a first approximation follows from (3.9.94):

$$\frac{v}{A_0 e_1} = -\log\left[1 - \frac{1}{1-M}\left(\frac{A_0 t}{D - X} - 1\right)\right] \quad \text{as} \quad M \to 1-$$

Accordingly, as $M \to 1$, $v/A_0 e_1$ changes by a finite amount in a layer near the front of pulse, where

$$[A_0 t/(D - X)] - 1 = 0(1 - M)$$

In the case of reflection from a perfectly free interface $X = 0$, by using (3.9.95) in (3.9.76) we calculate

$$\rho_0 A_0 v/T_2 = \tanh^3 \bar{\eta}_0(\bar{\eta}_0 - \tanh \bar{\eta}_0)^{-1} \log\{1 + \cosh^2 \bar{\eta}_0[(A_0 t/D) - 1]\} \tag{3.9.96}$$

The range of $A_0 t/D$ in this equation is determined by the range of v in the incident wave. If v varies in the range $[0, v_m]$ so that G is in the range $[0, v_m]$, then (3.9.76) predicts that at $X = 0$, v varies in the range $[0, 2v_m]$. $v = 0$ when $A_0 t/D = 1$, and $v = 2v_m$ when

$$A_0 t/D = 1 + [(1 + M)/M][1 - \exp(-\mu v_m A_0{}^{-1/2})]$$

Typical variations of v with $A_0 t/(D - X)$ are shown in Fig. 3.9.8. The displacement x of the interface can be calculated by integrating (3.9.76):

$$x/D = 2(\mu A_0^{1/2})^{-1}[(1 - M)/M][(1 + t^*) \log(1 + t^*) - t^*] \tag{3.9.97}$$

where

$$t^* = -[M/(1 - M)][(A_0 t/D) - 1] \tag{3.9.98}$$

In Fig. 3.9.9 we give the variation of x with time.

Similarly, we can calculate the variation of traction at a *rigid interface* by using (3.9.77) or (3.9.87). Upon employing the expression of A given by either $(3.9.84)_3$ or $(3.9.87)_3$ we obtain, respectively,

$$\begin{aligned} A_0 t/D &= 1 + [(1 - M)/M][1 - (\cosh \bar{c} + M^{1/2} \sinh \bar{c})^{-1}] \\ A_0 t/D &= 1 + \operatorname{sech}^2 \bar{\eta}_0\{[\sinh \bar{\eta}_0/\sinh(\bar{\eta}_0 + \bar{c})] - 1\} \end{aligned} \tag{3.9.99}$$

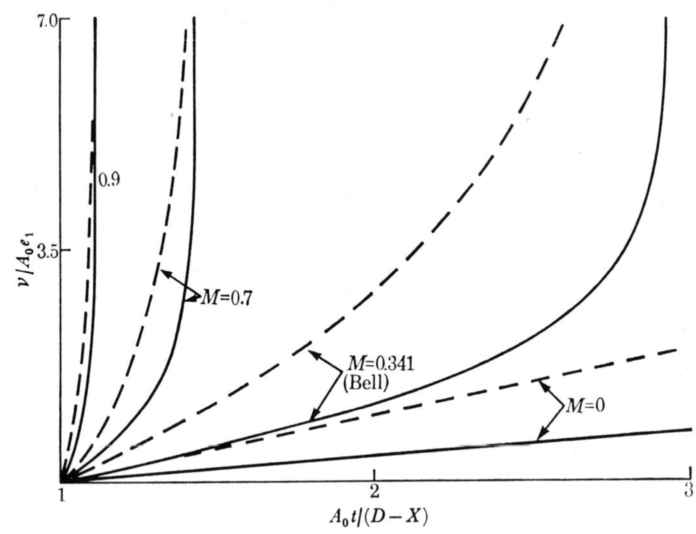

Fig. 3.9.8 v against $A_0\,t/(D-x)$ in an incident centered wave (solid lines) and v against $A_0\,t/D$ (dashed lines) during its reflection from a perfectly free interface: nonideal materials (*after* Çekirge and Varley [1973]).

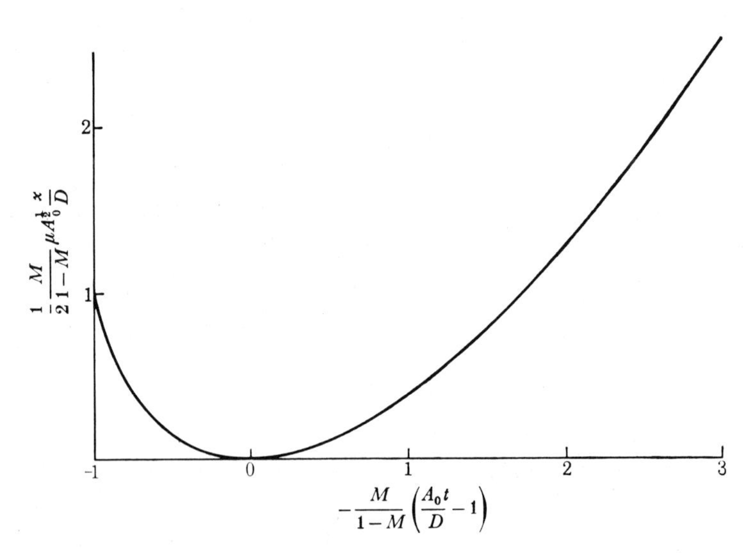

Fig. 3.9.9 The variation of the displacement x of a free interface during the reflection of a centered simple wave (*after* Çekirge and Varley [1973]).

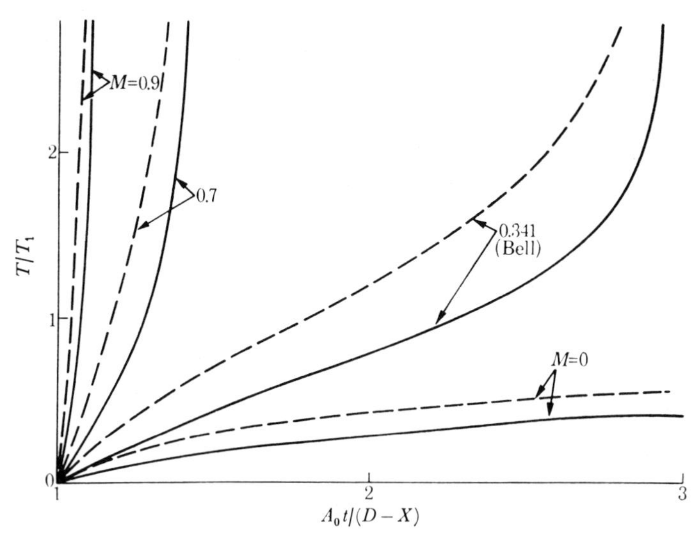

Fig. 3.9.10 T against $A_0 t/(D - X)$ in an incident centered wave (solid lines) and T against $A_0 t/D$ (dashed lines) during its reflection from a perfectly rigid interface: nonideal materials (*after* Çekirge and Varley [1973]).

Once $\bar{c}$ is calculated from these equations, then T and e can be found from the appropriate equations of (3.9.84) or (3.9.87). In Fig. (3.9.10) T/T_1 is given for soft materials. Calculations can be carried out for other types of materials. For these and other detailed discussions we refer the reader to Çekirge and Varley [1973].

3.10 SOME SOLUTIONS FOR SPECIAL MATERIALS

Here we present a brief discussion on special motions of an elastic solid having special material constitution. Although the results obtained are valid for isotropic solids, a similar treatment can be given for the transversely isotropic materials. A discussion of this topic is to be found in a report by Carroll [1966].

We consider a certain special class of finite motions in which kinematical nonlinearities are not present *exactly*. Therefore, only constitutive nonlinearities may take part in the differential equations which describe the motion of the material. This permits the discussion of some exact or approximate solutions for special types of materials. We study here only one class of such motions, represented by the equations

$$x_1 = \lambda_1 X_1, \qquad x_2 = \lambda_2 X_2, \qquad x_3 = \lambda_3 X_3 + w(x_1, x_2, t) \tag{3.10.1}$$

This is, in fact, a finite *generalized shear motion*, or *antiplane motion*, superimposed on a homogeneous deformation specified by three constant principal stretches λ_1, λ_2, and λ_3. The rectangular coordinate axes in both the reference and deformed states are selected to coincide with the principal directions of the initial deformation. Using (3.10.1) we calculate

(3.10.2)

$$\mathbf{F} = [x_{k,K}] = \begin{bmatrix} \lambda_1 & 0 & 0 \\ 0 & \lambda_2 & 0 \\ \lambda_1 w_{,1} & \lambda_2 w_{,2} & \lambda_3 \end{bmatrix}$$

$$\overset{-1}{\mathbf{c}} = \begin{bmatrix} {\lambda_1}^2 & 0 & {\lambda_1}^2 w_{,1} \\ 0 & {\lambda_2}^2 & {\lambda_2}^2 w_{,2} \\ {\lambda_1}^2 w_{,1} & {\lambda_2}^2 w_{,3} & {\lambda_3}^2 + {\lambda_1}^2 (w_{,1})^2 + {\lambda_2}^2 (w_{,2})^2 \end{bmatrix}$$

$$I = {\lambda_1}^2[1 + (w_{,1})^2] + {\lambda_2}^2[1 + (w_{,2})^2] + {\lambda_3}^2$$

$$II = {\lambda_3}^2({\lambda_1}^2 + {\lambda_2}^2) + {\lambda_1}^2{\lambda_2}^2[1 + (w_{,1})^2 + (w_{,2})^2]$$

$$III = {\lambda_1}^2{\lambda_2}^2{\lambda_3}^2$$

where

$$w_{,1} = \partial w/\partial x_1, \qquad w_{,2} = \partial w/\partial x_2$$

and consequently

$$\partial w/\partial X_1 = \lambda_1 w_{,1}, \qquad \partial w/\partial X_2 = \lambda_2 w_{,2}$$

For isotropic solids the stress–strain relations can be put in the form

$$t_{kl} = p\,\delta_{kl} + (\Phi + I\Psi)\overset{-1}{c}_{kl} - \Psi\overset{-2}{c}_{kl} \tag{3.10.3}$$

where

(3.10.4)

$$p = 2III^{1/2}\,\partial\Sigma/\partial III, \qquad \Phi = 2III^{-1/2}\,\partial\Sigma/\partial I, \qquad \Psi = 2III^{-1/2}\,\partial\Sigma/\partial II$$

Since $III = \text{const}$, the superposed deformation is isochoric, and the density ρ of the medium in motion remains unchanged as related to its density ρ_0 at the natural stress-free state by

$$\rho = \rho_0 III^{-1/2} = \rho_0/\lambda_1\lambda_2\lambda_3 \tag{3.10.5}$$

Substituting (3.10.2) into (3.10.3) we find the stress components

(3.10.6)

$$t_{11} = p + \Phi{\lambda_1}^2 + \Psi{\lambda_1}^2\{{\lambda_3}^2 + {\lambda_2}^2[1 + (w_{,2})^2]\}$$

$$t_{22} = p + \Phi{\lambda_2}^2 + \Psi{\lambda_2}^2\{{\lambda_3}^2 + {\lambda_1}^2[1 + (w_{,1})^2]\}$$

$$t_{33} = p + \Phi[{\lambda_3}^2 + {\lambda_1}^2(w_{,1})^2 + {\lambda_2}^2(w_{,2})^2] + \Psi\{{\lambda_3}^2({\lambda_1}^2 + {\lambda_2}^2) + {\lambda_1}^2{\lambda_2}^2[(w_{,1})^2 + (w_{,2})^2]\}$$

$$t_{12} = -\Psi{\lambda_1}^2{\lambda_2}^2 w_{,1} w_{,2}$$

$$t_{13} = (\Phi + \Psi{\lambda_2}^2){\lambda_1}^2 w_{,1}$$

$$t_{23} = (\Phi + \Psi{\lambda_1}^2){\lambda_2}^2 w_{,2}$$

where Φ, Ψ, and p are prescribed functions of unknowns $(w_{,1})^2$ and $(w_{,2})^2$ through the invariants. Therefore, stresses are also functions of x_1, x_2, and t only. The acceleration components reduce to ($v_1 = v_2 = 0$, $v_3 = \partial w/\partial t$)

$$a_1 = a_2 = 0, \qquad a_3 = \partial^2 w/\partial t^2 \tag{3.10.7}$$

Thus the problem is kinematically linear and the equations of motion are independent of x_3, i.e.,

$$t_{11,1} + t_{12,2} = 0, \qquad t_{12,1} + t_{22,2} = 0, \qquad t_{13,1} + t_{23,2} = \rho\, \partial^2 w/\partial t^2 \tag{3.10.8}$$

where ρ is given by (3.10.5). Inserting (3.10.6) into (3.10.8) we get

$$\begin{aligned}
&[p + \lambda_1{}^2\Phi + \lambda_1{}^2(\lambda_2{}^2 + \lambda_3{}^2)\Psi]_{,1} \\
&\qquad + \lambda_1{}^2\lambda_2{}^2[(\Psi w_{,2})_{,1} w_{,2} - (\Psi w_{,2})_{,2}\, w_{,1}] = 0 \\
&[p + \lambda_2{}^2\Phi + \lambda_2{}^2(\lambda_1{}^2 + \lambda_3{}^2)\Psi]_{,2} \\
&\qquad + \lambda_1{}^2\lambda_2{}^2[(\Psi w_{,1})_{,2}\, w_{,1} - (\Psi w_{,1})_{,1} w_{,2}] = 0 \\
&\lambda_1{}^2[(\Phi + \lambda_2{}^2\Psi) w_{,1}]_{,1} + \lambda_2{}^2[(\Phi + \lambda_1{}^2\Psi) w_{,2}]_{,2} = \rho\, \partial^2 w/\partial t^2
\end{aligned} \tag{3.10.9}$$

In principle, the last equation of (3.10.9) determines w. If the solution so obtained is consistent with the first two equations, the deformation is controllable, i.e., it can be maintained by application of surface tractions only. However, in general it is very unlikely that the solution for w is a consistent one. The nonvanishing left-hand side of the first two equations of (3.10.9) with the opposite sign may be identified as the components of some body force distribution acting in the plane $x_1 x_2$. But this scheme results in highly unrealistic body forces. Therefore, we must conclude that, in general, a generalized shear motion described by (3.10.1) is improbable unless a special constitution of material which renders the first two of (3.10.9) identically satisfied.

The situation becomes somewhat more hopeful in the case of incompressible materials. In such solids, p is to be regarded as an unknown function of x_1, x_2, and t. Hence the first two equations of (3.10.9) can be employed to determine this function, provided that the following compatibility condition is satisfied:

$$\begin{aligned}
&(\lambda_1{}^2 - \lambda_2{}^2)(\Phi + \lambda_3{}^2\Psi)_{,12} - \lambda_1{}^2\lambda_2{}^2\{[(\Psi w_{,2})_{,1} w_{,2} - (\Psi w_{,2})_{,2}\, w_{,1}]_{,2} \\
&\qquad - [(\Psi w_{,1})_{,2}\, w_{,1} - (\Psi w_{,1})_{,1} w_{,2}]_{,1}\} = 0
\end{aligned} \tag{3.10.10}$$

This condition for incompressible solids may, of course, be considered less restrictive in comparison with the corresponding ones in compressible solids, since the number of redundancies is reduced by one. If the material is initially stress free, or more generally if it is subjected only to a simple extension in the direction of motion x_3 ($\lambda_1 = \lambda_2$), then the compatibility condition (3.10.10) involves only one material function Ψ. If this vanishes by any chance, i.e., if the material constitution is such that the strain energy function of the incompressible solid is also independent of II, as in the case of neo-Hookean materials, then the generalized shear motion becomes entirely realizable. We are then left with only one nonlinear partial differential equation for the determination of w:

$$\lambda_1{}^2[(\Phi w_{,1})_{,1} + (\Phi w_{,2})_{,2}] = \rho_0\, \partial^2 w/\partial t^2 \tag{3.10.11}$$

Compatibility condition (3.10.10) is satisfied identically if w depends on only one spatial variable, say x_1. Then $(3.10.9)_3$ provides an exact nonlinear differential equation for w:[1]

$$\lambda_1{}^2[(\Phi + \lambda_2{}^2\Psi)w_{,1}]_{,1} = \rho_0\, \partial^2 w/\partial t^2 \tag{3.10.12}$$

The unknown function p can now be chosen in such a way that the stress component t_{11} is equal to zero everywhere, i.e.,

$$p = -\lambda_1{}^2\Phi - \lambda_1{}^2(\lambda_2{}^2 + \lambda_3{}^2)\Psi$$

subject to the restriction $\lambda_1\lambda_2\lambda_3 = 1$. Hence the stress components corresponding to this special case are determined from (3.10.6) as

$$\begin{aligned} t_{11} &= t_{12} = t_{23} \equiv 0 \\ t_{22} &= (\lambda_2{}^2 - \lambda_1{}^2)(\Phi + \lambda_3{}^2\Psi) + \lambda_1{}^2\lambda_2{}^2\Psi(w_{,1})^2 \\ t_{33} &= (\Phi + \lambda_2{}^2\Psi)[\lambda_3{}^2 - \lambda_1{}^2 + \lambda_1{}^2(w_{,1})^2] \\ t_{13} &= \lambda_1{}^2(\Phi + \lambda_2{}^2\Psi)w_{,1} \end{aligned} \tag{3.10.13}$$

The tractions on planes $x_1 = \text{const}$ are given by

$$t_1 = t_2 = 0, \qquad t_3 = \lambda_1{}^2(\Phi + \lambda_2{}^2\Psi)w_{,1}$$

Therefore, this formalism may demonstrate the solution of the problem of the generalized shear motion of an initially strained incompressible elastic half-space loaded on its boundary $x_1 = 0$ by a spatially uniform but time-dependent, tangential force (Fig. 3.10.1). The simple wave solution of this problem was considered in Section 3.6.

We now consider a special incompressible material, namely, Mooney material. In these materials, the strain energy is given by

$$\Sigma = \alpha(I - 3) + \beta(II - 3)$$

where $\Phi = 2\alpha$ and $\Psi = 2\beta$ are constant. Equation (3.10.12) now reduces to a linear equation:

$$2\lambda_1{}^2(\alpha + \lambda_2{}^2\beta)\, \partial^2 w/\partial x_1{}^2 = \rho_0\, \partial^2 w/\partial t^2$$

[1] Equation (3.10.12) was first obtained by Green [1963].

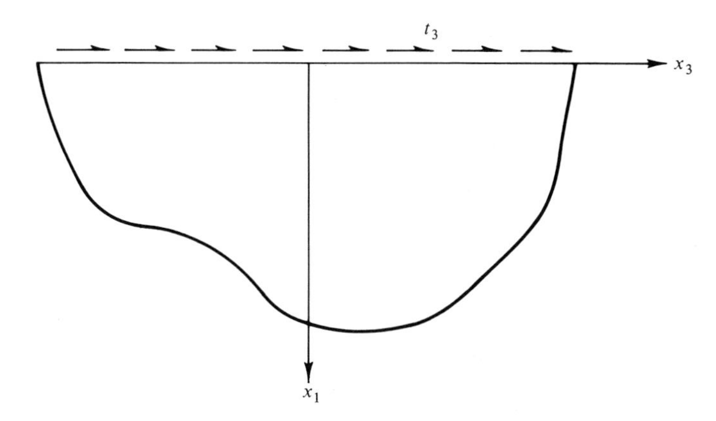

Fig. 3.10.1 Generalized shear motion in half-space.

which is the usual linear wave equation in one spatial variable with the constant propagation velocity

$$U^2 = 2\lambda_1{}^2(\alpha + \lambda_2{}^2\beta)/\rho_0$$

Therefore, the waves of finite amplitude propagate without change of form in such a material, and the displacement w can be determined at once by utilizing the corresponding solution in the linear elasticity. However, the state of stress in the nonlinear case differs considerably due to the nonlinear terms in (3.10.13).

The field equations (3.10.9) and compatibility equation (3.10.10) are simplified to some extent for Mooney materials. Equation (3.10.9) reduces to

$$\begin{aligned} p_{,1} + 2\lambda_1{}^2\lambda_2{}^2\beta(w_{,12}w_{,2} - w_{,22}w_{,1}) &= 0 \\ p_{,2} + 2\lambda_1{}^2\lambda_2{}^2\beta(w_{,12}w_{,1} - w_{,11}w_{,2}) &= 0 \\ 2\lambda_1{}^2(\alpha + \lambda_2{}^2\beta)w_{,11} + 2\lambda_2{}^2(\alpha + \lambda_1{}^2\beta)w_{,22} &= \rho_0\,\partial^2 w/\partial t^2 \end{aligned} \tag{3.10.14}$$

and compatibility condition (3.10.10) reduces to

$$(\nabla^2 w)_{,1}w_{,2} - (\nabla^2 w)_{,2}w_{,1} = 0 \tag{3.10.15}$$

where ∇^2 is a two-dimensional Laplace operator in x_1 and x_2. One can easily show that a motion in the form

$$w = w(\xi, t) \tag{3.10.16}$$

where $\xi = n_1x_1 + n_2x_2$ (n_1, n_2 are constants) satisfies (3.10.15) and the first two of (3.10.14) with $p = \text{const}$. We can take, without loss of generality, $n_1{}^2 + n_2{}^2 = 1$. The last equation of (3.10.14) becomes

$$2[(\lambda_1{}^2n_1{}^2 + \lambda_2{}^2n_2{}^2)\alpha + \lambda_1{}^2\lambda_2{}^2\beta]\,\partial^2 w/\partial\xi^2 = \rho_0\,\partial^2 w/\partial t^2 \tag{3.10.17}$$

Clearly (3.10.16) represents a plane shear wave propagating in the direction of the unit vector $\mathbf{n}(n_1, n_2, 0)$ in the x_1x_2-plane. Therefore, in Mooney materials transverse waves of finite amplitude, polarized in a principal direction of the initial deformation (e.g., x_3-direction), propagate without change of form in any direction lying in a plane parallel to the other two principal directions, with a velocity U:

$$U^2 = 2[(\lambda_1{}^2 n_1{}^2 + \lambda_2{}^2 n_2{}^2)\alpha + \lambda_1{}^2\lambda_2{}^2\beta]/\rho_0$$

Consider now a solution of (3.10.17) in the form $A \cos(k\xi - \omega t)$. This represents a harmonic plane wave propagating in the given direction $\mathbf{n}$ with an angular frequency ω and a wave number k related to ω by

$$2k^2[(\lambda_1{}^2 n_1{}^2 + \lambda_2{}^2 n_2{}^2)\alpha + \lambda_1{}^2\lambda_2{}^2\beta] = \rho_0\,\omega^2$$

Combining two solutions of this kind with $\mathbf{n}_1(n_1, n_2, 0)$ and $\mathbf{n}_2(-n_1, n_2, 0)$, we obtain

(3.10.18)

$$w(x_1, x_2, t) = \tfrac{1}{2}A[\cos(k_1x_1 + k_2x_2 - \omega t) + \cos(k_1x_1 - k_2x_2 + \omega t)]$$

where A is a known constant designating the amplitude of the wave, and $k_1 \equiv kn_1$, $k_2 \equiv kn_2$. As we shall see in Volume II, Chapter VII, (3.10.18) represents waves of the form (3.10.18) incident on, and reflected from, the boundary $x_1 = 0$ of a half-space. One can easily verify that this motion can be maintained only by the application of normal tractions:

$$t_1 = 2\beta\lambda_1{}^2\lambda_2{}^2k^2n_2{}^2A^2\sin^2(k_2x_2 - \omega t), \qquad n_2 \neq 0 \tag{3.10.19}$$

where we selected the constant p in $(3.10.6)_1$ as $-2\lambda_1{}^2[\alpha + (\lambda_2{}^2 + \lambda_3{}^2)\beta]$ to reduce the traction on the boundary $x_1 = 0$ to the form $t_1 = 2\beta\lambda_1{}^2\lambda_2{}^2(w_{,2})^2$. Therefore unless $\beta = 0$, such a motion cannot be supported by a free boundary, thus indicating that finite amplitude, horizontally polarized shear waves, incident obliquely on the free boundary of a half-space, cannot be reflected as pure waves but give rise to other kinds of waves as well. However, for small amplitudes t_1 is of second order and may be discarded. Hence in a linear solid, such shear waves are reflected from a free boundary as pure shear waves. If $n_2 = 0$, i.e., for normal incidence, one can easily verify that $t_1 = 0$, and the solution is in complete agreement with its classical counterpart.

If the reference configuration for the superposed motion is assumed to be the natural state, the restrictions can be relaxed to some extent. In this case $\lambda_1 = \lambda_2 = \lambda_3 = 1$ and (3.10.6) and (3.10.9) reduce to

$$\begin{aligned}
t_{11} &= p + \Phi + \Psi[2 + (w_{,2})^2], & t_{12} &= -\Psi w_{,1}w_{,2} \\
t_{22} &= p + \Phi + \Psi[2 + (w_{,1})^2], & t_{13} &= (\Phi + \Psi)w_{,1} \\
t_{33} &= p + \Phi[1 + (w_{,1})^2 + (w_{,2})^2] & t_{23} &= (\Phi + \Psi)w_{,2} \\
&\quad + \Psi[2 + (w_{,1})^2 + (w_{,2})^2], &&
\end{aligned} \tag{3.10.20}$$

and

$$
\begin{aligned}
&(p + \Phi + 2\Psi)_{,1} + (\Psi w_{,2})_{,1} w_{,2} - (\Psi w_{,2})_{,2} w_{,1} = 0 \\
(3.10.21)\quad &(p + \Phi + 2\Psi)_{,2} + (\Psi w_{,1})_{,2} w_{,1} - (\Psi w_{,1})_{,1} w_{,2} = 0 \\
&[(\Phi + \Psi) w_{,1}]_{,1} + [(\Phi + \Psi) w_{,2}]_{,2} = \rho_0\, \partial^2 w / \partial t^2
\end{aligned}
$$

with invariants

$$I = II = 3 + (w_{,1})^2 + (w_{,2})^2, \qquad III = 1$$

It is still impossible, in general, to satisfy all the equations of (3.10.21) by a single function $w(x_1, x_2, t)$. However, some special classes of materials, which we now explore, admit consistent solutions.

Consider a compressible material whose response functions satisfy

$$(3.10.22)\qquad p + \Phi = \Psi = 0$$

whenever

$$(3.10.23)\qquad I = II, \qquad III = 1$$

Then for all motions satisfying (3.10.23), the stress–strain relations (3.10.3) take the form

$$t_{kl} = \Phi\left(\overset{-1}{c}_{kl} - \delta_{kl}\right)$$

Here Φ is, of course, a function of invariants in general. Thus (3.10.20) and (3.10.21), for such materials, reduce to

$$
(3.10.24)\qquad
\begin{aligned}
&t_{11} = t_{12} = t_{22} = 0, && t_{33} = \Phi[(w_{,1})^2 + (w_{,2})^2] \\
&t_{13} = \Phi w_{,1}, && t_{23} = \Phi w_{,2}
\end{aligned}
$$

and

$$(3.10.25)\qquad (\Phi w_{,1})_{,1} + (\Phi w_{,2})_{,2} = \rho_0\, \partial^2 w / \partial t^2$$

If we suppose that the material is bounded by a cylindrical surface $f(x_1, x_2) = 0$, this surface keeps its shape unchanged during the motion, and the stress boundary conditions on this surface become

$$(3.10.26)\qquad t_1 = t_2 = 0, \qquad t_3 = n_1 t_{13} + n_2 t_{23} = \Phi(w_{,1} n_1 + w_{,2} n_2) = \Phi\, \partial w / \partial n$$

where $\mathbf{n} = (n_1, n_2, 0)$ is the exterior unit normal to the boundary. Equations (3.10.25) and (3.10.26) determine w under given compatible initial conditions.

Materials for which Φ is independent of invariants whenever (3.10.23) holds are called *effectively linear*. For such materials, the stress tensor depends *linearly* on $\overset{-1}{\mathbf{c}}$, so that (3.10.25) becomes a linear equation

$$\nabla^2 w = (\rho_0 / \Phi)\, \partial^2 w / \partial t^2$$

and the classical solutions for antiplane motion can be applied immediately. However, all stress components are *not* linear functions of displacement gradients, and t_{33} is not present in the classical solutions.

For incompressible materials, p is an unknown function of position and time, and the first two equations of (3.10.21) on w are replaced by a compatibility condition

$$[(\Psi w_{,2})_{,1} w_{,2} - (\Psi w_{,2})_{,2} w_{,1}]_{,2} - [(\Psi w_{,1})_{,2} w_{,1} - (\Psi w_{,1})_{,1} w_{,2}]_{,1} = 0 \tag{3.10.27}$$

Now if

$$\Psi = 0 \tag{3.10.28}$$

whenever

$$I = II \tag{3.10.29}$$

then (3.10.27) is identically satisfied. Further, if we choose p in such a way that $t_{11} = 0$, the problem reduces to (3.10.24)–(3.10.26).

The materials for which Φ is independent of invariants whenever (3.10.29) and (3.10.28) are satisfied will be called *effectively neo-Hookean* materials. The field equation is linear in such materials.

For one-dimensional motions, from (3.10.13) and (3.10.12) it follows that

$$t_{11} = t_{12} = t_{23} = 0, \qquad t_{22} = \Psi(w_{,1})^2,$$
$$t_{33} = (\Phi + \Psi)(w_{,1})^2, \qquad t_{13} = (\Phi + \Psi)w_{,1}$$

and

$$[(\Phi + \Psi)w_{,1}]_{,1} = \rho_0\, \partial^2 w/\partial t^2$$

subject to boundary conditions, on a plane $x_1 = \text{const}$,

$$t_1 = t_2 = 0, \qquad t_3 = (\Phi + \Psi)w_{,1}$$

If $\Phi + \Psi$ is independent of invariants whenever (3.10.29) is satisfied, this material will be called *effectively linear incompressible material.* If, in addition, Φ and Ψ are constant under condition (3.10.29), it will be called *effectively Mooney material.* In these materials, one-dimensional motion w is determined from a linear equation, and the boundary conditions depend linearly on $w_{,1}$. Therefore, in such solids the classical solution administers the propagation of a finite amplitude wave of the type under consideration. However, it should be emphasized that the stresses associated with finite motion differ from the corresponding expressions of the infinitesimal motions.

Other types of special motions can be investigated. The results, including those corresponding to transversely isotropic solids, can be found in the report by Carroll [1966].

3.11 OTHER SOLUTIONS

The one-dimensional dynamic problems, being the most accessible to analytical treatments, have attracted the attention of many investigators. Bland [1967] derived the general equations for cylindrically and spherically radial motions. He also obtained [1964b, Sections 4–7] similarity solutions for longitudinal and spherically symmetric motions and showed that the displacement can be represented in both cases by $u = Xf(X/t)$, X being the radial variable in the case of spherical motion. The function f can be determined, at least in principle, as a solution of a nonlinear ordinary differential equation. In addition, he gave a numerical solution of the problem of a spherical cavity subject to explosion, in an infinite quadratic solid. Comparison with the classical solution indicated significant difference in the field quantities at the vicinity of the cavity.

The method of characteristics was used by Valanis and Sun [1968] to investigate the propagation of finite longitudinal waves in semiinfinite thin rods. Their treatment and results differ very little in essence from those obtained by Chu [1964] for shear waves in incompressible solids.

John [1960] studied the propagation of plane waves in a material in plane strain for a special strain energy function. He later investigated [1966] the general plane waves extensively and rigorously for what he calls *Hadamard* and *harmonic* materials. He proved that Hadamard materials, defined by

$$\Sigma = aI + bII + F(III)$$

where a and b are arbitrary constants and F is an arbitrary function of III, are the only materials for which plane waves in arbitrary directions can be decomposed into three polarized waves, one longitudinal and the other two transverse. He defined harmonic materials by the strain energy function

$$\Sigma = as + bt + f(r)$$

where

$$r^2 - 2s = I, \qquad s^2 - 2rt = II, \qquad t^2 = III$$

with a, b arbitrary constants, and f an arbitrary function of r. He showed that the plane wave deformation in harmonic materials is pseudoirrotational, i.e., the curl of $\mathbf{x}$ with respect to material coordinates $\mathbf{X}$ vanishes everywhere, and these materials are capable of transmitting certain kinds of polarized waves. In this context we also refer the reader to a recent paper by Currie and Hayes [1969].

Collins [1966] studied simple wave solutions in isotropic and transversely isotropic incompressible materials and discussed in great detail the shock formation and shock stability for quadratic solids. In a later paper, [1967], he investigated, again for a quadratic solid, the motion of an isotropic, incompressible elastic half-space subject to time-dependent uniform shear stresses at the boundary. A similar problem was treated by Howard [1966], and more recently by Waterstone [1968] for compressible transversely isotropic solids.

An ordinary perturbation analysis is employed by Fine and Shield [1966] to obtain approximate solutions of the general equations of motion of a compressible isotropic quadratic solid. The method gives the same wave speed, corresponding to the linear approximation, at every stage of the computation, thus generating secular terms which may grow indefinitely with time. Therefore, the results can be valid only for very short time intervals. An improved perturbation technique, proposed by Davison [1968] for plane waves to remove this difficulty, incorporates a corrected estimate of the wave speed at every stage of the computation. The method is somewhat similar to the conventional Poincaré–Lighthill–Kuo technique in that both dependent and independent variables are expanded into perturbation series.

Another technique, devised by Achenbach and Reddy [1967], was employed by Achenbach [1967] to investigate plane shear waves in an incompressible, isotropic half-space. The essence of the method is to develop the displacement function into a Taylor series about the time of arrival of the wave front to a certain point in the material. The time derivatives of the displacement evaluated at the wave front are calculated from the compatibility conditions of the discontinuities across the wave front and from the given velocity function at the boundary, under the assumption that the shear stress can be expressed as a polynomial in the shear deformation. This method along with the method of characteristics is employed by Reddy and Achenbach [1968] to study simple wave propagation, shock formation, and shock growth in semiinfinite thin elastic rods loaded at one end.

Regarding solutions for special classes of motion, there exist a few attempts to give some explicit results for incompressible elastic materials. Nowinski [1966] studied the axial shearing motion of rectangular tubes made of neo-Hookean materials. Recently Wang [1969] considered quasi-free oscillations[1] of bodies with particular geometries, such as the plane shearing of a slab of finite uniform thickness and the gyratory shearing of a thick-walled cylindrical tube specified by

$$r = R, \qquad \theta = \Theta + \phi(r, t), \qquad z = Z$$

where R, Θ, and Z are cylindrical coordinates of a material point in the undeformed state, and r, θ, and z are the coordinates of the same point at

[1] Since all the boundaries cannot be entirely stress free, due to the higher-order effects, we use the term quasi-free.

time t. The longitudinal shearing of a cylindrical tube[1] is specified by

$$r = r(R, t), \qquad \theta = \Theta, \qquad z = Z + w(r, t)$$

in cylindrical coordinates. The solutions corresponding to the first two cases can be obtained from linear differential equations for either Mooney or neo-Hookean materials. In the last case, the longitudinal and radial motions become uncoupled for neo-Hookean materials, yielding a linear equation for the determination of the longitudinal motion and the solution of (3.3.19). Wang also considered the strain energy function of the form[2]

$$\Sigma = \alpha(I - 3) + \beta(II - 3) + \gamma(I - 3)^2 \tag{3.11.1}$$

and gave a perturbative solution to the problems mentioned earlier.

The propagation of acceleration waves in a spherical, elastic earth has been investigated by Sawyers and Rivlin [1969], who assumed that the earth has spherical symmetry and is transversely isotropic, with density depending on the radial variable. They discussed the seismic aspects of this interesting problem.

Finally, a short monograph by Bland [1969], which compiles mainly the author's own works, should be mentioned.

[1] This problem has also been studied by Nowinski and Schultz [1964] for neo-Hookean materials.

[2] Proposed by Ishihara *et al.* [1951].

CHAPTER IV

Small Motions Superimposed on Large Static Deformations

4.1 SCOPE OF THE CHAPTER

In some mechanical problems the final deformed state of the material body can be reached by the composition of two deformations: an initial finite static deformation and a superimposed infinitesimal deformation, characterized by a small displacement field that may be time dependent. If the response of the material is known for the given initial deformation, then it is possible to derive a set of *linear* differential equations to determine the superimposed infinitesimal displacements. The coefficients of these differential equations are expressible in terms of the functions describing the initial deformation. This approach provides an invaluable tool in investigating whether or not a given initial deformation is stable. It is, therefore, not surprising that this subject first attracted the attention of investigators interested in the question of stability, e.g., the theory of elastic stability. The importance of the theory of stability in engineering problems needs no further comment.

Unfortunately, the linear equations obtained for the description of the superimposed motion, in general, are complicated, and often one must be content with the more modest information that can be drawn from these rather than the complete solution. Nevertheless, it is worthwhile to treat this problem here briefly, but as completely as possible within the space limitations.

To maintain clarity and simplicity, we leave all thermodynamic considerations aside. We are interested in the general problem and not primarily with the problem of stability, although this aspect of the theory may be considered the real justification of this chapter. We confine ourselves to the simple problem of determining all motions superimposed on some simple known static deformations in special geometries.

The general problem of infinitesimal displacements superimposed on a large elastic deformation has been tackled by many investigators. Here we mention only a few: Prager [1946]; Green *et al.* [1952]; Pearson [1956]; Ericksen and Toupin [1956]; Hill [1957]; Toupin and Bernstein [1961]; Urbanowski [1959, 1961]; Guo [1962a]; Guo and Urbanowski [1963]; and Zorski [1964]. An elaborate and clear exposition by Truesdell and Noll [1965, Sections 68–70] should also be cited.

In Section 4.2 we derive the fundamental equations, i.e., the equations of motion, boundary conditions, and constitutive relations for both compressible and incompressible solids. Section 4.3 is devoted to the investigation of the propagation of plane waves in homogeneously deformed infinite media. Surface waves in a homogeneously prestressed half-space are the subject of Section 4.4. In Section 4.5 we study the small torsional and longitudinal oscillations of a circular cylinder subject to a uniform extension. We discuss, in Section 4.6, the oscillations of a finitely twisted circular cylinder made of Mooney material. A brief review of other solutions of dynamical problems is given in Section 4.7.

4.2 FUNDAMENTAL EQUATIONS

Consider an elastic body occupying a spatial region R in the configuration B, which we henceforth call the reference configuration. Suppose that this reference configuration does not correspond to a stress-free state and that the body is brought into the reference configuration B by a *finite static* deformation from a configuration B_0 corresponding to the natural state of the body occupying a region R_0. If the coordinates of a material particle in the natural state are denoted by (X_1, X_2, X_3) and in the reference state by (x_1, x_2, x_3), this deformation is specified by functional relations

$$\mathbf{x} = \mathbf{x}(\mathbf{X}) \tag{4.2.1}$$

The stresses induced by the deformation (4.2.1) are, of course, determined throughout the body via constitutive relations and satisfy the equilibrium equations

$$t^0_{kl,k} + \rho f_l^{\,0} = 0 \qquad \text{in} \quad R \tag{4.2.2}$$

and the boundary conditions

$$t^0_{kl} n_k = t_l^{\,0} \qquad \text{on} \quad S \tag{4.2.3}$$

In these expressions t^0_{kl} is the Cauchy stress tensor, ρ the known mass density of the material, $f_l^{\,0}$ the body force per unit mass in B, and $t_l^{\,0}$ the prescribed traction vector on the boundary S of R. We now consider a *small* time and position dependent displacement vector superimposed on the reference configuration B. In order to carry out properly the order-of-magnitude analysis crucial to the linearization, we denote this displacement by $\varepsilon\mathbf{u}(\mathbf{x}, t)$, where ε is a small parameter. If one would like to absorb ε into the displacement field and the quantities induced by it, all one need do is take $\varepsilon = 1$.

The superimposed displacement field deforms the body into another *time-dependent* configuration $B'(t)$, which occupies a region $R'(t)$. But the region R' is very close to the region R on account of the assumed smallness of the superimposed displacement field. The coordinates of the material particles in the current configuration B' may be written as (Fig. 4.2.1):

$$x_k' = x_k + \varepsilon u_k(\mathbf{x}, t) \tag{4.2.4}$$

in the same fixed frame of reference of B. Consider now the state of stress in the configuration B'. The stress vector in B' referred to the area element in

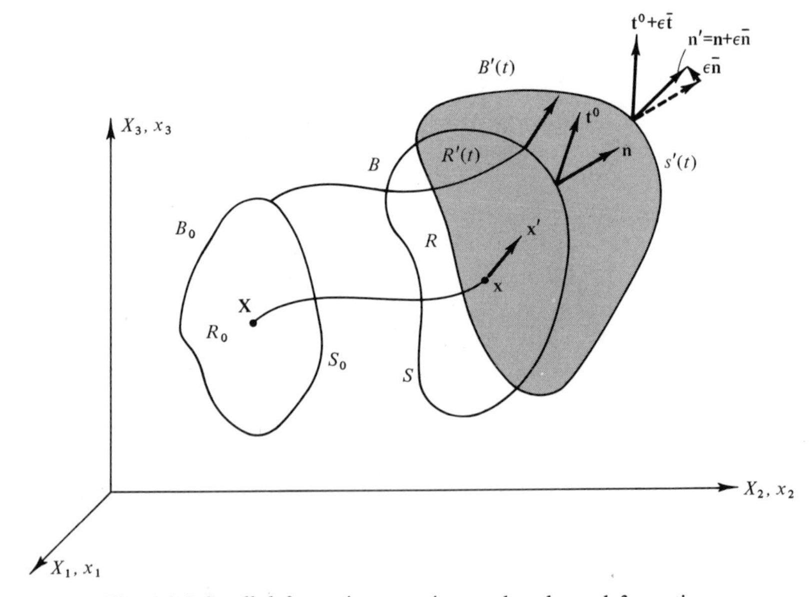

Fig. 4.2.1 Small deformation superimposed on large deformations.

the reference configuration B is characterized by the Piola–Kirchhoff pseudo-stress tensor T_{kl} (cf. Section 1.12). Within an order of magnitude linear in ε, this may be written as

(4.2.5) $$T_{kl}(\mathbf{x}, t) = t_{kl}^0(\mathbf{x}) + \varepsilon \overline{T}_{kl}(\mathbf{x}, t)$$

where $\varepsilon \overline{T}_{kl}$ stands for the increment in the Piola–Kirchhoff tensor referred to the configuration B, caused by the infinitesimal displacement $\varepsilon \mathbf{u}$. The Cauchy stress tensor $\mathbf{t}$ referred to B' is related to $\mathbf{T}$ by

(4.2.6) $$t_{kl} = j^{-1} (\partial x_k'/\partial x_m) T_{ml}$$

where j is the Jacobian of the infinitesimal deformation, i.e.,

(4.2.7) $$j = \det(\partial x_k'/\partial x_l)$$

If we use (4.2.4) in (4.2.7) and neglect the powers of ε higher than one, we get

(4.2.8) $$j = 1 + \varepsilon u_{k,k}, \qquad j^{-1} = 1 - \varepsilon u_{k,k}$$

recalling that $\partial x_k'/\partial x_l = \delta_{kl} + \varepsilon u_{k,l}$. Hence (4.2.6) may be approximated by

(4.2.9) $$t_{kl} = t_{kl}^0 + \varepsilon \bar{t}_{kl}$$

where

(4.2.10) $$\bar{t}_{kl} = \overline{T}_{kl} + t_{ml}^0 u_{k,m} - t_{kl}^0 u_{m,m}$$

is the increment in the actual stress tensor in the deformed body due to the superimposed displacements. If we define a displacement gradient tensor $\mathbf{H}$ by

$$H_{kl} = u_{k,l}$$

we can rewrite (4.2.10) in a compact form:

(4.2.10)′ $$\bar{\mathbf{t}} = \overline{\mathbf{T}} + \mathbf{H}\mathbf{t}^0 - \mathbf{t}^0 \operatorname{tr} \mathbf{H}$$

Therefore, the knowledge of displacement gradients, the initial stress distribution, and the tensor $\overline{\mathbf{T}}$ completely determine the increment in the Cauchy stress tensor $\bar{\mathbf{t}}$. As we shall soon see, $\overline{\mathbf{T}}$ is found from the constitutive relations for the material.

Next we investigate the boundary conditions: The unit normal $\mathbf{n}$ to the deformed boundary S', within our framework of approximation, may be written as (Fig. 4.2.1)

$$\mathbf{n}' = \mathbf{n} + \varepsilon \bar{\mathbf{n}}$$

Since the surface element in S' is given by [cf. (1.7.2)]

$$da_k' = j\,(\partial x_l/\partial x_k')\, da_l, \qquad da'/da = j\left(\overset{-1}{C}_{kl} n_k n_l\right)^{1/2}$$

we have

$$n_k' = j\,(\partial x_l/\partial x_k')n_l\,(da/da') = n_l\,(\partial x_l/\partial x_k')\left(\overset{-1}{C}_{mn} n_m n_n\right)^{-1/2} \tag{4.2.11}$$

Within the present approximation,

$$\begin{aligned} \partial x_l/\partial x_k' &= \delta_{kl} - \varepsilon u_{l,k} \\ \overset{-1}{C}_{kl} &= (\partial x_k/\partial x_m')(\partial x_l/\partial x_m') = \delta_{kl} - \varepsilon(u_{k,l} + u_{l,k}) = \delta_{kl} - 2\varepsilon\tilde{e}_{kl} \end{aligned} \tag{4.2.12}$$

where $\tilde{e}_{kl}$ is the infinitesimal strain tensor for the displacement $\mathbf{u}$. Thus

$$\overset{-1}{C}_{mn} n_m n_n = 1 - 2\varepsilon\tilde{e}_{mn} n_m n_n = 1 - 2\varepsilon\tilde{e}_{(\mathbf{n})}$$

where $\tilde{e}_{(\mathbf{n})}$ is the infinitesimal extension in the direction $\mathbf{n}$. It now follows from (4.2.11) that

$$n_k' \simeq n_l(\delta_{lk} - \varepsilon u_{l,k})(1 + \varepsilon\tilde{e}_{(\mathbf{n})}) = n_k + \varepsilon(\tilde{e}_{(\mathbf{n})}\,\delta_{lk} - u_{l,k})n_l \tag{4.2.13}$$

Therefore, we obtain

$$\bar{n}_k = \tilde{e}_{(\mathbf{n})} n_k - u_{l,k} n_l, \qquad \bar{\mathbf{n}} = \tilde{e}_{(\mathbf{n})}\mathbf{n} - \mathbf{H}^{\mathrm{T}}\mathbf{n} \tag{4.2.14}$$

The boundary conditions on tractions now take the form

$$(t_{kl}^0 + \varepsilon\bar{t}_{kl})(n_k + \varepsilon\bar{n}_k) = t_l^{\,0} + \varepsilon\bar{t}_l \qquad \text{on} \quad S$$

where $\bar{t}_l$ are the components of the additional tractions on the boundary induced by the displacement $\mathbf{u}$. This expression approximates the boundary conditions on S' with those on the surface S of the body in the strained reference configuration. Considering (4.2.3) from the above expression, we get

$$\bar{t}_{kl} n_k + t_{kl}^0 \bar{n}_k = \bar{t}_l \qquad \text{on} \quad S \tag{4.2.15}$$

Substituting (4.2.14) into (4.2.15) and considering (4.2.3) we finally find the following boundary conditions:

$$\bar{t}_{kl} n_k = \bar{t}_l - t_l^{\,0}\,\tilde{e}_{(\mathbf{n})} + t_{kl}^0 u_{m,k} n_m \tag{4.2.16}$$

The right-hand side of (4.2.16) does not vanish generally even in the case of free boundaries. If we introduce (4.2.10) into (4.2.16) we obtain the boundary conditions on the Piola–Kirchhoff stress tensor:

$$\bar{T}_{kl} n_k = \bar{t}_l + (u_{k,k} - \tilde{e}_{(\mathbf{n})})t_l^{\,0} \tag{4.2.17}$$

Since

$$da'/da \simeq 1 + \varepsilon(u_{k,k} - \tilde{e}_{(\mathbf{n})})$$

we can deduce from the relation

$$t_k\, da = t_k'\, da'$$

where t_k' and t_k are traction vectors referred to the surfaces S' and S, respectively, that the right-hand side of (4.2.17) is none other than the increment in the boundary traction $t_k^0 + \varepsilon \bar{t}_k$ referred to an area in the surface of the reference configuration.

The equations of motion in the reference state are

$$T_{kl,k} + \rho f_l = \rho \varepsilon \, \partial^2 u_l / \partial t^2$$

where the higher-order terms are neglected in the acceleration. If we write the body force $\mathbf{f}$ as

$$\mathbf{f} = \mathbf{f}^0 + \varepsilon \bar{\mathbf{f}}$$

replace T_{kl} by $t_{kl}^0 + \varepsilon \overline{T}_{kl}$, and use (4.2.2), we arrive at the equations of motion in incremental stresses:

$$\overline{T}_{kl,k} + \rho \bar{f}_l = \rho \, \partial^2 u_l / \partial t^2 \tag{4.2.18}$$

Substituting from (4.2.10) into (4.2.18) and again taking (4.2.2) into consideration, we obtain the equations of motion in terms of the increments in Cauchy stresses as follows:

$$\bar{t}_{kl,k} - t_{ml,k}^0 u_{k,m} + \rho \bar{f}_l - \rho f_l^0 u_{k,k} = \rho \, \partial^2 u_l / \partial t^2 \tag{4.2.19}$$

We now define a stress tensor $\tilde{\mathbf{t}}$ by the relation

$$\bar{t}_{kl} = \tilde{t}_{kl} + t_{km}^0 u_{l,m} + t_{lm}^0 u_{k,m} \tag{4.2.20}$$

or in compact form

$$\bar{\mathbf{t}} = \tilde{\mathbf{t}} + \mathbf{t}^0 \mathbf{H}^{\mathrm{T}} + \mathbf{H} \mathbf{t}^0 \tag{4.2.20$'$}$$

If we substitute (4.2.20) into the equation of motion (4.2.19) and the boundary conditions (4.2.16), we find that

$$\tilde{t}_{kl,k} + t_{km}^0 u_{l,km} + t_{lm}^0 u_{k,km} + \rho \bar{f}_l - \rho (f_l^0 u_{k,k} + f_k^0 u_{l,k}) = \rho \, \partial^2 u_l / \partial t^2 \tag{4.2.21}$$

and

$$\tilde{t}_{kl} n_k = \bar{t}_l - [\tilde{e}_{(\mathbf{n})} \, \delta_{lk} + u_{l,k}] t_k^0 \qquad \text{on } S \tag{4.2.22}$$

Equation (4.2.21) is essentially the same as that given by Green *et al.* [1952].

The reader who is familiar with tensor analysis will at once identify $\tilde{t}^{kl}$ as the contravariant component of the incremental Cauchy stress tensor in *convective* or *dragged* curvilinear coordinates x^k, which are connected to the local Cartesian coordinates x'^k by $x'^k = x^k + \varepsilon u^k(\mathbf{x}, t)$. In effect, since t^{kl} are the components of the Cauchy stress tensor in Cartesian coordinates x'^k, we can immediately write the usual transformation formulas, denoting the components of the stress tensor in the curvilinear system by $\tau^{mn} = \tau^{0mn} + \varepsilon \tilde{t}^{mn}$, as

$$t^{kl} = \tau^{mn} \, (\partial x'^k / \partial x^m) \, \partial x'^l / \partial x^n$$

or approximately

$$t^{0kl} + \varepsilon \bar{t}^{kl} = (\tau^{0mn} + \varepsilon \tilde{t}^{mn})(\delta_m{}^k + \varepsilon u^k{}_{;m})(\delta_n{}^l + \varepsilon u^l{}_{;n})$$

where semicolons represent covariant differentiation. Equating the coefficients of zeroth and first powers of ε, we get

$$t^{0kl} = \tau^{0kl}, \qquad \tilde{t}^{kl} = \bar{t}^{kl} + t^{0km}\, u^{l}{}_{;m} + t^{0ml}\, u^{k}{}_{;m}$$

Similarly, employing the transformations

$$f^{k} = \mathfrak{f}^{m}\, \partial x'^{k}/\partial x^{m}, \qquad \rho = \rho' \det(\partial x'^{k}/\partial x^{l})$$

and writing $f^{k} = f^{0k} + \varepsilon \tilde{f}^{k}$, $\mathfrak{f}^{m} = \mathfrak{f}^{0m} + \varepsilon \bar{f}^{m}$, and $\partial x'^{k}/\partial x^{l} = \delta_{l}{}^{k} + \varepsilon u^{k}{}_{;l}$ gives

$$\tilde{f}^{k} = \bar{f}^{k} - f^{0k}\, u^{m}{}_{;m} - f^{0m}\, u^{k}{}_{;m}$$

The above expressions of $\tilde{t}$ and $\tilde{f}$ when substituted into Cauchy's equations of motion give (4.2.21) in curvilinear coordinates.

The equations that we have derived so far are valid regardless of the constitution of the material. The incremental stresses $\bar{\mathbf{T}}$, $\bar{\mathbf{t}}$, or $\tilde{\mathbf{t}}$, however, can only be determined after the constitutive equations of the material are known. In the hyperelastic materials dealt with here, these are expressed as

$$T_{Kk} = \partial\Sigma/\partial F_{kK} \tag{4.2.23}$$

where $F_{kK} = \partial x_{k}'/\partial X_{K}$ are the components of the current deformation gradient tensor in the material. The Piola–Kirchhoff stress tensor referred to the natural configuration B_0 is connected to the Cauchy stress tensor, referred to the instantaneous configuration B', by

$$T_{Kk} = \det(\partial x_{m}'/\partial X_{M})\,(\partial X_{K}/\partial x_{l}')t_{lk}$$

or

$$T_{Kk} = \det\left(\frac{\partial x_{m}'}{\partial x_{n}}\frac{\partial x_{n}}{\partial X_{M}}\right)\frac{\partial X_{K}}{\partial x_{p}}\frac{\partial x_{p}}{\partial x_{l}'}\, t_{lk} = j^{0}\, j\, \frac{\partial X_{K}}{\partial x_{p}}\frac{\partial x_{p}}{\partial x_{l}'}\, t_{lk} \tag{4.2.24}$$

where j is given by (4.2.7) and

$$j^{0} = \det(\partial x_{k}/\partial X_{K})$$

The Piola–Kirchhoff stress tensor referred to the reference configuration B is similarly given by

$$T_{pk} = j\,(\partial x_{p}/\partial x_{l}')t_{lk} \tag{4.2.25}$$

Thus introducing (4.2.25) into (4.2.24) we have

$$T_{Kk} = j^{0} X_{K,l}\, T_{lk}$$

or inverting this expression for T_{lk},

$$T_{kl} = (j^{0})^{-1} x_{k,K}\, T_{Kl} \tag{4.2.26}$$

The linear approximation to the constitutive relations, in the infinitesimal superimposed displacement gradients, yields

$$T_{Kk} = T^0_{Kk} + \varepsilon A^0_{KkLl} u_{l,L} = T^0_{Kk} + \varepsilon A^0_{KkLl} x_{m,L} u_{l,m} \tag{4.2.27}$$

where

$$T^0_{Kk} = \partial\Sigma/\partial F_{kK}|_{F_{kK}=x_{k,K}}, \qquad A^0_{KkLl} = \partial^2\Sigma/\partial F_{kK}\,\partial F_{lL}|_{F_{mM}=x_{m,M}}$$

are evaluated at the reference configuration. It now follows from (4.2.26), (4.2.27), and (4.2.5) that

$$t^0_{kl} + \varepsilon \overline{T}_{kl} = (j^0)^{-1} x_{k,K} T^0_{Kl} + \varepsilon (j^0)^{-1} A^0_{KlLn} x_{k,K} x_{m,L} u_{n,m}$$

If we recall that the initial Cauchy stress tensor is given by

$$t^0_{kl} = (j^0)^{-1} x_{k,K} T^0_{Kl}$$

then from the above expression we deduce that

$$\overline{T}_{kl} = B^0_{klmn} u_{n,m} \tag{4.2.28}$$

where

$$B^0_{klmn} = (j^0)^{-1} A^0_{KlLn} x_{k,K} x_{m,L} \tag{4.2.29}$$

All the components of the tensor $\mathbf{B}^0$ can be calculated for a given material and deformation $\mathbf{x} = \mathbf{x}(\mathbf{X})$. Evidently B^0_{klmn} are functions of the position $\mathbf{x}$, except for *homogeneous deformations*.

Substitution of (4.2.28) into (4.2.18) and (4.2.17) results in the equations of motion and boundary conditions expressed in terms of the displacement vector $\mathbf{u}$:

$$\begin{aligned} (B^0_{klmn} u_{n,m})_{,k} + \rho \bar{f}_l &= \rho\, \partial^2 u_l/\partial t^2 \qquad \text{in } R \\ B^0_{klmn} u_{n,m} n_k &= \bar{t}_l + (u_{k,k} - \tilde{e}_{(\mathbf{n})}) t_l^{\,0}, \qquad \text{on } S \end{aligned} \tag{4.2.30}$$

Equations $(4.2.30)_1$, which can be rewritten as

$$B^0_{klmn}(\mathbf{x}) u_{n,mk} + B^0_{klmn,k}(\mathbf{x}) u_{n,m} + \rho \bar{f}_l = \rho\, \partial^2 u_l/\partial t^2$$

constitute a system of three linear second-order partial differential equations with variable coefficients. This system is totally hyperbolic if the characteristic equation

$$\det(Q_{kl} - \rho U^2\, \delta_{kl}) = 0$$

has distinct and positive roots U^2. In this equation Q_{kl} is given by

$$Q_{kl} = B^0_{mknl} n_m n_n$$

where $\mathbf{n}$ is the unit normal to the characteristic surface, or more precisely to the wave front propagating with velocity U into the prestressed medium.

The conditions of total hyperbolicity are satisfied whenever $\mathbf{B}^0$ is strongly elliptic, i.e.,

$$B^0_{klmn}\lambda_k\lambda_m\mu_l\mu_n > 0$$

where $\boldsymbol{\lambda}$ and $\boldsymbol{\mu}$ are arbitrary nonzero vectors. It can be shown that this condition of strong ellipticity is a sufficient but not necessary condition for the uniqueness of the solution to the boundary value problem (4.2.30) (Hayes [1963], [1964]).

Presently, the questions of existence, uniqueness, and stability of the solution to the system (4.2.30) are still open, except for certain special cases primarily confined to homogeneous initial deformations. These questions are of the utmost importance in dealing with the stability of elastic systems. For a concise treatment of this topic the reader is referred to Truesdell and Noll [1965, Sections 68–69b].

We now return to the equations of motion for the superimposed infinitesimal displacements. If we substitute $(1.17.7)_1$ into (4.2.29), take into account $(1.13.22)_1$, and define

$$C^0_{klmn} = 4(j^0)^{-1}(\partial^2\Sigma/\partial C_{KL}\,\partial C_{MN})x_{k,K}\,x_{l,L}\,x_{m,M}\,x_{n,N}$$

we obtain the following representation:

$$B^0_{klmn} = \delta_{ln}t^0_{km} + C^0_{klmn} \tag{4.2.31}$$

The tensor C^0_{klmn} enjoys the symmetry relations

$$C^0_{klmn} = C^0_{lkmn} = C^0_{klnm} = C^0_{mnkl}$$

hence the number of its independent components is only 21.

If we employ (4.2.31) in (4.2.28) and insert the resulting expression into (4.2.10), we obtain

$$\bar{t}_{kl} = t^0_{km}u_{l,m} + t^0_{lm}u_{k,m} - t^0_{kl}u_{m,m} + C^0_{klmn}u_{n,m}$$

or considering the symmetries of C^0_{klmn},

$$\bar{t}_{kl} = t^0_{km}u_{l,m} + t^0_{lm}u_{k,m} - t^0_{kl}u_{m,m} + C^0_{klmn}\tilde{e}_{mn} \tag{4.2.32}$$

Comparing (4.2.32) with (4.2.20), we identify $\tilde{t}_{kl}$ as

$$\tilde{t}_{kl} = -\tilde{e}_{mm}t^0_{kl} + C^0_{klmn}\tilde{e}_{mn} \tag{4.2.33}$$

Finally, for the equations of motion and boundary conditions we have from (4.2.19) and (4.2.16)

$$\begin{aligned} &(C^0_{klmn}u_{m,n})_{,k} + t^0_{km}u_{l,km} - \rho f_k{}^0 u_{l,k} + \rho\bar{f}_l = \rho\,\partial^2 u_l/\partial t^2 && \text{in } R \\ &C^0_{klmn}u_{m,n}n_k = \bar{t}_l + t_l{}^0(u_{k,k} - \tilde{e}_{(\mathbf{n})}) - t_k{}^0 u_{l,k} && \text{on } S \end{aligned} \tag{4.2.34}$$

In passing, we note that the actual stress-deformation relations between the tensor $\mathbf{t} = \mathbf{t}^0 + \varepsilon\bar{\mathbf{t}}$ and the displacement gradients $\varepsilon\mathbf{H}$ (with $\varepsilon = 1$) may be written as[1]

$$t_{kl} = t^0_{kl}(1 - u_{m,m}) + t^0_{km}u_{l,m} + t^0_{lm}u_{k,m} + C^0_{klmn}u_{m,n}$$

or in another form,

$$t_{kl} = t^0_{kl} + \bar{C}^0_{klmn}u_{n,m}$$

where

$$\bar{C}^0_{klmn} = C^0_{klmn} + \delta_{ln}t^0_{km} + \delta_{km}t^0_{ln} - \delta_{mn}t^0_{kl}$$

Therefore, the linear stress-deformation relations for infinitesimal deformations of a prestressed, initially anisotropic, elastic solid are specified by prescribing 27 coefficients: 21 independent components of $\mathbf{C}^0$ and 6 components of the initial stress tensor—in contrast to the 21 coefficients that appear in the linear constitutive relations of an anisotropic solid in its stress-free natural state—and they are, in general, point functions.

In the same fashion, as above, we can write

$$\tilde{t}_{kl} = \tilde{C}^0_{klmn}\tilde{e}_{mn}$$

where

$$\tilde{C}^0_{klmn} = C^0_{klmn} - \delta_{mn}t^0_{kl}$$

In isotropic materials we are capable of finding a more explicit expression for the incremental stress tensor $\tilde{\mathbf{t}}$. The tensor $\mathbf{C}^0$ is expressed in this case as

$$C^0_{klmn} = \delta_{mn}t^0_{kl} - \delta_{lm}t^0_{kn} - \delta_{km}t^0_{ln} + 2F^0_{klmp}\overset{-1}{c}_{np}$$

where the tensor F^0_{klmp}, given in (1.17.16), is evaluated at the initial deformation, and $\overset{-1}{\mathbf{c}}$ is the Finger tensor of the initial deformation. Carrying (1.17.15) into (4.2.29) and (4.2.31), we obtain

$$\begin{aligned}
(4.2.35)\qquad C^0_{klmn} &= h_0(\delta_{kl}\delta_{mn} - \delta_{km}\delta_{ln} - \delta_{kn}\delta_{lm}) + h_1\delta_{mn}\overset{-1}{c}_{kl} \\
&+ h_2\left(\delta_{mn}\overset{-2}{c}_{kl} + \overset{-1}{c}_{km}\overset{-1}{c}_{ln} + \overset{-1}{c}_{lm}\overset{-1}{c}_{kn}\right) \\
&+ 2\,\delta_{kl}\left[\frac{\partial h_0}{\partial I}\overset{-1}{c}_{mn} + \frac{\partial h_0}{\partial II}\left(I\overset{-1}{c}_{mn} - \overset{-2}{c}_{mn}\right) + III\frac{\partial h_0}{\partial III}\delta_{mn}\right] \\
&+ 2\overset{-1}{c}_{kl}\left[\frac{\partial h_1}{\partial I}\overset{-1}{c}_{mn} + \frac{\partial h_1}{\partial II}\left(I\overset{-1}{c}_{mn} - \overset{-2}{c}_{mn}\right) + III\frac{\partial h_1}{\partial III}\delta_{mn}\right] \\
&+ 2\overset{-2}{c}_{kl}\left[\frac{\partial h_2}{\partial I}\overset{-1}{c}_{mn} + \frac{\partial h_2}{\partial II}\left(I\overset{-1}{c}_{mn} - \overset{-2}{c}_{mn}\right) + III\frac{\partial h_2}{\partial III}\delta_{mn}\right]
\end{aligned}$$

[1] These equations were first derived by Cauchy [1829, p. 342]. However, he had 36 components for the tensor C^0_{klmn}.

Substituting (4.2.35) into (4.2.32), the expression for the incremental Cauchy stress tensor in naturally isotropic solids follows:

(4.2.36)

$$\bar{t}_{kl} = t^0_{ml} u_{k,m} + t^0_{km} u_{l,m} - 2h_0 \tilde{e}_{kl} + 2h_2 \overset{-1}{c}_{km} \tilde{e}_{mn} \overset{-1}{c}_{nl} + 2 \sum_{a=0}^{2} \left[\frac{\partial h_a}{\partial I} \overset{-1}{c}_{mn} \tilde{e}_{mn} + \frac{\partial h_a}{\partial II} \left(I \overset{-1}{c}_{mn} \tilde{e}_{mn} - \overset{-2}{c}_{mn} \tilde{e}_{mn} \right) + III \frac{\partial h_a}{\partial III} \tilde{e}_{mm} \right] \overset{-a}{c}_{kl}$$

In compact notation (in invariant form) we have

(4.2.37)

$$\bar{\mathbf{t}} = \mathbf{t}^0 \mathbf{H}^T + \mathbf{H} \mathbf{t}^0 - 2h_0 \tilde{\mathbf{e}} + 2h_2 \overset{-1}{\mathbf{c}} \tilde{\mathbf{e}} \overset{-1}{\mathbf{c}} + 2 \sum_{a=0}^{2} \left[\frac{\partial h_a}{\partial I} \operatorname{tr} \overset{-1}{\mathbf{c}} \tilde{\mathbf{e}} + \frac{\partial h_a}{\partial II} \left(I \operatorname{tr} \overset{-1}{\mathbf{c}} \tilde{\mathbf{e}} - \operatorname{tr} \overset{-2}{\mathbf{c}} \tilde{\mathbf{e}} \right) + III \frac{\partial h_a}{\partial III} \operatorname{tr} \tilde{\mathbf{e}} \right] \overset{-a}{\mathbf{c}}$$

On the other hand, (4.2.33) results immediately in

(4.2.38)

$$\tilde{\mathbf{t}} = -2h_0 \tilde{\mathbf{e}} + 2h_2 \overset{-1}{\mathbf{c}} \tilde{\mathbf{e}} \overset{-1}{\mathbf{c}} + 2 \sum_{a=0}^{2} \left[\frac{\partial h_a}{\partial I} \operatorname{tr} \overset{-1}{\mathbf{c}} \tilde{\mathbf{e}} + \frac{\partial h_a}{\partial II} \left(I \operatorname{tr} \overset{-1}{\mathbf{c}} \tilde{\mathbf{e}} - \operatorname{tr} \overset{-2}{\mathbf{c}} \tilde{\mathbf{e}} \right) + III \frac{\partial h_a}{\partial III} \operatorname{tr} \tilde{\mathbf{e}} \right] \overset{-a}{\mathbf{c}}$$

It is a simple exercise to show that (4.2.38) can be made identical to the form given by Green *et al.* [1952] if we recall the relations (1.14.14) defining the response functions h_0, h_1, and h_2:

(4.2.39)
$$\tilde{\mathbf{t}} = \bar{\Phi} \overset{-1}{\mathbf{c}} + \bar{\Psi} \mathbf{B} + \Psi \bar{\mathbf{B}} + \bar{p} \mathbf{I} - 2p \tilde{\mathbf{e}}$$

where

(4.2.40)

$$\mathbf{B} = I \overset{-1}{\mathbf{c}} - \overset{-2}{\mathbf{c}}, \qquad \bar{\mathbf{B}} = \bar{I} \overset{-1}{\mathbf{c}} - 2 \overset{-1}{\mathbf{c}} \tilde{\mathbf{e}} \overset{-1}{\mathbf{c}}$$

$$\bar{p} = III(E\bar{I} + D\overline{II} + C\overline{III}) + (p/2III)\overline{III}$$

$$\bar{\Phi} = A\bar{I} + F\overline{II} + E\overline{III} - (\Phi/2III)\overline{III}$$

$$\bar{\Psi} = F\bar{I} + B\overline{II} + D\overline{III} - (\Psi/2III)\overline{III}$$

$$\bar{I} = 2 \operatorname{tr} \overset{-1}{\mathbf{c}} \tilde{\mathbf{e}}, \qquad \overline{II} = 2\left(I \operatorname{tr} \overset{-1}{\mathbf{c}} \tilde{\mathbf{e}} - \operatorname{tr} \overset{-2}{\mathbf{c}} \tilde{\mathbf{e}}\right), \qquad \overline{III} = 2III \operatorname{tr} \tilde{\mathbf{e}}$$

and

(4.2.41)

$$p = 2III^{1/2}\, \partial\Sigma/\partial III, \qquad \Phi = 2III^{-1/2}\, \partial\Sigma/\partial I, \qquad \Psi = 2III^{-1/2}\, \partial\Sigma/\partial II$$

$$A = 2III^{-1/2}\, \partial^2\Sigma/\partial I^2, \qquad B = 2III^{-1/2}\, \partial^2\Sigma/\partial II^2, \qquad C = 2III^{-1/2}\, \partial^2\Sigma/\partial III^2$$

$$D = 2III^{-1/2}\, \partial^2\Sigma/\partial II\, \partial III, \qquad E = 2III^{-1/2}\, \partial^2\Sigma/\partial I\, \partial III, \qquad F = 2III^{-1/2}\, \partial^2\Sigma/\partial I\, \partial II$$

Equations (4.2.37) or (4.2.38) clearly indicate that the response of an elastic solid, which is isotropic in its undistorted state, to an infinitesimal deformation is no longer so if it has undergone finite deformation before the infinitesimal deformation. Therefore, in general, an isotropic material may apparently display some features of anisotropy if it is prestressed. Alternatively, an isotropic material may lose its isotropy if it is finitely strained prior to the observed deformation. However, it should be pointed out that this loss of isotropy is only apparent since the material behaves isotropically if the final state of deformation is known to be evaluated with reference to its natural state. Hence it is impossible to distinguish the material, at a first glance, from an anisotropic one without a knowledge of the history of deformation. Through a series of elaborate experiments, however, it should be possible to reveal this fact, since initial stresses appear as additional material functions in the stress–strain relations.

As may be anticipated, a compressible elastic material preserves its apparent isotropy if it is subject to a hydrostatic initial stress. In this case, we have

$$\mathbf{t}^0 = -p^0\mathbf{I} = (h_0 + \lambda_0{}^2 h_1 + \lambda_0{}^4 h_2)\Big|_{\lambda_\alpha = \lambda_0} \mathbf{I}$$

where λ_0 is the dilatational stretch and $\overset{-1}{\mathbf{c}} = \lambda_0{}^2\mathbf{I}$. Substitution of hydrostatic stress into (4.2.37) results in

$$\bar{\mathbf{t}} = \bar{\lambda}(\lambda_0)(\operatorname{tr}\tilde{\mathbf{e}})\mathbf{I} + 2\bar{\mu}(\lambda_0)\tilde{\mathbf{e}} \tag{4.2.42}$$

which is identical to the *linear, isotropic stress–strain relations* with *apparent* Lamé constants $\bar{\lambda}$ and $\bar{\mu}$ given by

$$\bar{\lambda} = 2\lambda_0{}^2 \sum_{a=0}^{2} \lambda_0^{2a}\left(\frac{\partial h_a}{\partial I} + 2\lambda_0{}^2\frac{\partial h_a}{\partial II} + \lambda_0{}^4\frac{\partial h_a}{\partial III}\right)\Bigg|_{\overset{-1}{\mathbf{c}} = \lambda_0{}^2\mathbf{I}}$$

$$\bar{\mu} = \lambda_0{}^2(h_1 + 2\lambda_0{}^2 h_2)\Big|_{\overset{-1}{\mathbf{c}} = \lambda_0{}^2\mathbf{I}}$$

Using the formulas derived on p. 124, one can show that

$$\bar{\lambda} = \rho\,(dp_0/d\rho) - \tfrac{2}{3}\bar{\mu} = (dp_0/d\log\rho) - \tfrac{2}{3}\bar{\mu}$$

where $\rho = \rho_0\lambda_0^{-3}$, with ρ_0 the density of the material in the undistorted state.

Incompressible materials. If we are dealing with incompressible elastic solids, some simplifications can be achieved. In this case, the superimposed displacements are subject to the constraint

$$u_{k,k} = 0 \qquad \text{or} \qquad \operatorname{tr}\tilde{\mathbf{e}} = \operatorname{tr}\mathbf{H} = 0 \tag{4.2.43}$$

The stress tensor in such a solid is decomposed into the form

$$t_{kl} = -p\,\delta_{kl} + {}_E t_{kl}, \qquad T_{Kk} = -pX_{K,k} + {}_E T_{Kk}$$

where ${}_E\mathbf{t}$ or ${}_E\mathbf{T}$ can be determined by the constitutive relations. The analysis in terms of infinitesimal superimposed deformations is entirely similar to the one in compressible materials that we have just finished discussing. Therefore, we abstain from repeating the detailed calculations here and only give a list of the results.

$\overline{T}_{kl}$ in (4.2.5) now takes the form

$$\overline{T}_{kl} = -\bar{p}\,\delta_{kl} + p^0 u_{k,l} + {}_E\overline{T}_{kl}$$

and $\bar{t}_{kl}$ in (4.2.10) has the form

$$\bar{t}_{kl} = -\bar{p}\,\delta_{kl} + {}_E t^0_{ml}\, u_{k,m} + {}_E\overline{T}_{kl}$$

Next we find, as in (4.2.32),

$$\bar{t}_{kl} = -\bar{p}\,\delta_{kl} + {}_E\bar{t}_{kl} = -\bar{p}\,\delta_{kl} + {}_E t^0_{km}\, u_{l,m} + {}_E t^0_{lm}\, u_{k,m} + C^0_{klmn}\,\tilde{e}_{mn} \tag{4.2.44}$$

and if we define $\tilde{t}_{kl}$ by

$$\bar{t}_{kl} = -\bar{p}\,\delta_{kl} + t^0_{km}\, u_{l,m} + t^0_{lm}\, u_{k,m} + \tilde{t}_{kl}$$

we have

$$\tilde{t}_{kl} = -\bar{p}\,\delta_{kl} + 2p^0\tilde{e}_{kl} + C^0_{klmn}\,\tilde{e}_{mn} = -\bar{p}\,\delta_{kl} + {}_E\tilde{t}_{kl} \tag{4.2.45}$$

The equations of motion and boundary conditions

(4.2.46)

$$\begin{aligned} -\bar{p}_{,l} + {}_E\bar{t}_{kl,k} + p^0{}_{,k}\, u_{k,l} - {}_E t^0_{ml,k}\, u_{k,m} + \rho_0 \bar{f}_l &= \rho_0\, \partial^2 u_l/\partial t^2 && \text{in } R \\ -\bar{p}n_l + {}_E\bar{t}_{kl}\, n_k &= \bar{t}_l - \bar{t}_l{}^0 \tilde{e}_{(\mathbf{n})} + t^0_{kl}\, u_{m,k}\, n_m && \text{on } S \end{aligned}$$

become

(4.2.47)

$$\begin{aligned} -\bar{p}_{,l} + (C^0_{klmn}\, u_{m,n})_{,k} + p^0{}_{,k}(u_{k,l} + u_{l,k}) + {}_E t^0_{km}\, u_{l,km} & \\ - \rho_0 f_k^0\, u_{l,k} + \rho_0 \bar{f}_l = \rho_0\, \partial^2 u_l/\partial t^2 & \quad \text{in } R \\ -\bar{p}n_l + C^0_{klmn}\, u_{m,n}\, n_k = \bar{t}_l - (\tilde{e}_{(\mathbf{n})}\,\delta_{lk} + u_{l,k})t_k{}^0 - 2p^0\tilde{e}_{kl}\, n_k & \quad \text{on } S \end{aligned}$$

or

$$-\bar{p}n_l + {}_E\tilde{t}_{kl}\, n_k = \bar{t}_l - (\tilde{e}_{(\mathbf{n})}\,\delta_{kl} + u_{l,k})t^0_k \qquad \text{on } S \tag{4.2.48}$$

Equations (4.2.43) and $(4.2.47)_1$ constitute a linear system of four partial differential equations to determine the four unknowns $\mathbf{u}(\mathbf{x}, t)$ and $\bar{p}(\mathbf{x}, t)$.

If the material is isotropic, then the initial stress satisfies

$$\mathbf{t}^0 = -p^0\mathbf{I} + h_1 \overset{-1}{\mathbf{c}} + h_2 \overset{-2}{\mathbf{c}}$$

where

$$h_1 = 2\left(\frac{\partial \Sigma}{\partial I} + \mathrm{I}\,\frac{\partial \Sigma}{\partial II}\right), \qquad h_2 = -\frac{\partial \Sigma}{\partial II}$$

and the tensor $\mathbf{C}^0$ in (4.2.44) becomes

$$C^0_{klmn} = \delta_{mn}\, {}_E t^0_{kl} - \delta_{lm}\, {}_E t^0_{kn} - \delta_{km}\, {}_E t^0_{ln} + 2F^0_{klmp}\, \overset{-1}{c}_{pn}$$

where F^0_{klmp} is given by (1.17.16) with terms containing derivatives with respect to III and h_0 omitted. Thus

$$\bar{t}_{kl} = -\bar{p}\,\delta_{kl} + {}_E t^0_{km} u_{l,m} + {}_E t^0_{lm} u_{k,m} + 2h_2\, \overset{-1}{c}_{km}\tilde{e}_{mn}\overset{-1}{c}_{nl} + 2\sum_{a=1}^{2}\left[\frac{\partial h_a}{\partial I}\overset{-1}{c}_{mn}\tilde{e}_{mn} + \frac{\partial h_a}{\partial II}\left(I\,\overset{-1}{c}_{mn}\tilde{e}_{mn} - \overset{-2}{c}_{mn}\tilde{e}_{mn}\right)\right]\overset{-a}{c}_{kl} \tag{4.2.49}$$

or in compact notation,

$$\bar{\mathbf{t}} = -\bar{p}\mathbf{I} + {}_E\mathbf{t}^0\mathbf{H}^{\mathrm{T}} + \mathbf{H}_E\mathbf{t}^0 + 2h_2\, \overset{-1}{\mathbf{c}}\tilde{\mathbf{e}}\overset{-1}{\mathbf{c}} + 2\sum_{a=1}^{2}\left[\frac{\partial h_a}{\partial I}\,\mathrm{tr}\,\overset{-1}{\mathbf{c}}\tilde{\mathbf{e}} + \frac{\partial h_a}{\partial II}\left(I\,\mathrm{tr}\,\overset{-1}{\mathbf{c}}\tilde{\mathbf{e}} - \mathrm{tr}\,\overset{-2}{\mathbf{c}}\tilde{\mathbf{e}}\right)\right]\overset{-a}{\mathbf{c}} \tag{4.2.50}$$

If desired, the expression for $\bar{\mathbf{t}}$ can be written by use of (4.2.45).

4.3 PLANE WAVES IN HOMOGENEOUSLY DEFORMED ELASTIC MATERIALS[1]

The coefficients of displacement gradients in the field equations (4.2.34) are, in general, functions of the position in the body, since the initial stress or deformation fields are not usually homogeneous. Therefore, these equations with vanishing body forces cannot admit plane wave solutions under an arbitrary initial state of stress. If, however, the initial deformation is homogeneous, both tensors $\mathbf{C}^0$ and $\mathbf{t}^0$ are constants everywhere in the material, and (4.2.34) with zero body forces reduces to

$$(C^0_{klmn} + \delta_{ln}t^0_{km})u_{n,km} = \rho\,\partial^2 u_l/\partial t^2$$

or, recalling (4.2.31), we have

$$B^0_{klmn}u_{n,km} = \rho\,\partial^2 u_l/\partial t^2 \tag{4.3.1}$$

where $\mathbf{B}^0$ is a constant tensor. A plane wave solution to (4.3.1) can now be sought by assuming

$$u_k = a_k \exp i(\mathbf{k}\cdot\mathbf{x} - \omega t) \tag{4.3.2}$$

where $\mathbf{k} = k\mathbf{n}$, the unit vector $\mathbf{n}$ is the direction of propagation of the plane wave, k its wave number, and $\mathbf{a}$ a constant amplitude vector. If we substitute

[1] This problem has been investigated in detail by Hayes and Rivlin [1961a] for isotropic materials.

(4.3.2) into (4.3.1) and define the wave velocity by $c = \omega/k$, we get

$$(Q^0_{kl} - \rho c^2 \delta_{kl})a_l = 0 \tag{4.3.3}$$

where

$$Q^0_{kl} = B^0_{mknl} n_m n_n \tag{4.3.4}$$

as before. Equation (4.3.3) shows that the plane waves propagate in the direction $\mathbf{n}$ in the homogeneously deformed medium with the velocity c of an acceleration front to be determined from

$$\det(Q^0_{kl} - \rho c^2 \delta_{kl}) = 0$$

[cf. (2.12.14) and (2.12.15)].

Consider now another *Cartesian* coordinate system $\boldsymbol{\xi}$ related to the system $\mathbf{x}$ by an orthogonal transformation $\boldsymbol{\alpha}$:

$$\xi_i = \alpha_{ij} x_j, \qquad \alpha_{ij}\alpha_{kj} = \delta_{ik} \tag{4.3.5}$$

The components of the displacement vector in the $\boldsymbol{\xi}$-system read

$$U_k = \alpha_{kl} u_l = b_k \exp i(\mathbf{k} \cdot \boldsymbol{\xi} - \omega t) \tag{4.3.6}$$

where $b_k = \alpha_{kl} a_l$. Since the components of the unit vector $\mathbf{n}$ in the $\boldsymbol{\xi}$-system can be written as

$$\nu_k = \alpha_{kl} n_l$$

by use of (4.3.4), Eq. (4.3.3) now takes the form

$$(\boldsymbol{\alpha}\mathbf{Q}\boldsymbol{\alpha}^{\mathrm{T}} - \rho c^2 \mathbf{I})\mathbf{b} = \mathbf{0} \tag{4.3.7}$$

Since $\mathbf{Q}$ is a symmetric matrix, there exists an orthogonal transformation which diagonalizes it. If we choose $\boldsymbol{\alpha}$ as the orthogonal matrix representing this transformation, then writing

$$\mathbf{R} = \boldsymbol{\alpha}\mathbf{Q}\boldsymbol{\alpha}^{\mathrm{T}}$$

in (4.3.7) we obtain its diagonal form[1]

$$(R_{\mathfrak{a}\mathfrak{a}} - \rho c_{\mathfrak{a}}^2) b_{\mathfrak{a}} = 0, \qquad \mathfrak{a} = 1, 2, 3 \tag{4.3.8}$$

Hence the wave velocities are

$$c_{\mathfrak{a}}^2 = R_{\mathfrak{a}\mathfrak{a}}/\rho \tag{4.3.9}$$

[1] Henceforth, the summation convention is suspended on German (fraktur) indices.

The amplitude of the wave propagating with the velocity $c_\mathfrak{a}$ is necessarily in the direction of $\xi_\mathfrak{a}$-axis, according to (4.3.8). Therefore, the directions of the axes ξ_i so chosen are three mutually perpendicular directions of polarizations for three mutually uncoupled waves which may propagate in the direction $\mathbf{n}$.

If $\mathbf{Q}$ is a positive definite matrix, all $R_{\mathfrak{aa}}$ are positive, and consequently all $c_\mathfrak{a}{}^2$ are real. However, if one or more of the solutions for $c_\mathfrak{a}{}^2$ is negative for any direction of propagation of the plane wave, one must deduce that the initial state of homogeneous deformation is inherently *unstable*. For, let us assume that one of $R_{\mathfrak{aa}}$, say R_{11}, is negative. Then we have, in general,

$$U_1 = b_1 \exp(-\omega_1 t + i\mathbf{k} \cdot \boldsymbol{\xi}) + \bar{b}_1 \exp(\omega_1 t + i\mathbf{k} \cdot \boldsymbol{\xi})$$

where $\omega_1 = k(-c_1{}^2)^{1/2}$. It is evident that the displacement parallel to the ξ_1-axis builds up exponentially with time in this case. Therefore, any infinitesimal disturbance tends to grow indefinitely with the passage of time, showing that the initial state of deformation is not stable. The necessary and sufficient condition for stability in this case is the positive definiteness of $\mathbf{Q}$, which may conveniently be expressed in terms of its invariants as

$$\operatorname{tr} \mathbf{Q} > 0, \qquad (\operatorname{tr} \mathbf{Q})^2 > \operatorname{tr} \mathbf{Q}^2, \qquad \det \mathbf{Q} > 0$$

This simple example clearly illustrates the power of the technique of infinitesimal displacements superimposed on large ones in investigating the stability of an existing deformation. Unfortunately, Eqs. (4.2.34) rarely lend themselves to simple mathematical treatment when the initial state is not homogeneous.

More explicit results are obtained in the case of isotropic materials. If we choose the axes x_i oriented so as to coincide with the principal directions of the initial large deformation, we have

(4.3.10)

$$t^0_{\mathfrak{aa}} = h_0 + \lambda_\mathfrak{a}{}^2 h_1 + \lambda_\mathfrak{a}{}^4 h_2 = 2III^{-1/2}\left\{III\frac{\partial\Sigma}{\partial III} + \lambda_\mathfrak{a}{}^2\left[\frac{\partial\Sigma}{\partial I} + (I - \lambda_\mathfrak{a}{}^2)\frac{\partial\Sigma}{\partial II}\right]\right\}$$

$$t^0_{\mathfrak{ab}} = 0, \qquad \text{if} \quad \mathfrak{a} \neq \mathfrak{b}$$

where $\lambda_\mathfrak{a}$ ($\mathfrak{a} = 1, 2, 3$) are principal stretches. We can then show from (4.2.35), (4.2.36), and (4.3.4), after some lengthy manipulations, that

(4.3.11)

$$\bar{t}_{\mathfrak{aa}} = \sum_{\mathfrak{b}=1}^{3} a_{\mathfrak{ab}}\, \partial u_\mathfrak{b}/\partial x_\mathfrak{b}$$

$$\bar{t}_{\mathfrak{ab}} = 2III^{-1/2}\left[\frac{\partial\Sigma}{\partial I} + (I - \lambda_\mathfrak{a}{}^2 - \lambda_\mathfrak{b}{}^2)\frac{\partial\Sigma}{\partial II}\right]\left(\lambda_\mathfrak{a}{}^2\frac{\partial u_\mathfrak{b}}{\partial x_\mathfrak{a}} + \lambda_\mathfrak{b}{}^2\frac{\partial u_\mathfrak{a}}{\partial x_\mathfrak{b}}\right), \qquad \mathfrak{a} \neq \mathfrak{b}$$

with

(4.3.12)

$$
\begin{aligned}
a_{ab} = 2III^{-1/2}\Bigg\{&\left[\frac{\partial\Sigma}{\partial I} + (I - \lambda_a{}^2 - \lambda_b{}^2)\frac{\partial\Sigma}{\partial II}\right]\lambda_a{}^2(2\,\delta_{ab} - 1) \\
&+ \lambda_a{}^2\lambda_b{}^2\frac{\partial\Sigma}{\partial II} + III\frac{\partial\Sigma}{\partial III} + 2\Bigg\{\lambda_a{}^2\left[\frac{\partial^2\Sigma}{\partial I^2}\right. \\
&\left. + (I - \lambda_a{}^2)\frac{\partial^2\Sigma}{\partial I\,\partial II}\right] + III\frac{\partial^2\Sigma}{\partial I\,\partial III}\Bigg\}\lambda_b{}^2 \\
&+ 2\Bigg\{\lambda_a{}^2\left[\frac{\partial^2\Sigma}{\partial I\,\partial II} + (I - \lambda_a{}^2)\frac{\partial^2\Sigma}{\partial II^2}\right] + III\frac{\partial^2\Sigma}{\partial II\,\partial III}\Bigg\}\lambda_b{}^2(I - \lambda_b{}^2) \\
&+ 2\Bigg\{\lambda_a{}^2\left[\frac{\partial^2\Sigma}{\partial I\,\partial III} + (I - \lambda_a{}^2)\frac{\partial^2\Sigma}{\partial II\,\partial III}\right] + III\frac{\partial^2\Sigma}{\partial III^2}\Bigg\}III\Bigg\}
\end{aligned}
$$

and

(4.3.13)

$$
\begin{aligned}
Q_{ab} = 2III^{-1/2}\Bigg\{&\left(\lambda_a{}^2\lambda_b{}^2\frac{\partial\Sigma}{\partial II} + III\frac{\partial\Sigma}{\partial III}\right)n_a n_b + \delta_{ab}\sum_{c=1}^{3}\lambda_c{}^2 n_c{}^2\left[\frac{\partial\Sigma}{\partial I}\right. \\
&\left. + (I - \lambda_a{}^2 - \lambda_c{}^2)\frac{\partial\Sigma}{\partial II}\right] + 2\Bigg\{\left[\lambda_a{}^2\left[\frac{\partial^2\Sigma}{\partial I^2} + (I - \lambda_a{}^2)\frac{\partial^2\Sigma}{\partial I\,\partial II}\right]\right. \\
&\left. + III\frac{\partial^2\Sigma}{\partial I\,\partial III}\right]\lambda_b{}^2 + \left[\lambda_a{}^2\left[\frac{\partial^2\Sigma}{\partial I\,\partial II} + (I - \lambda_a{}^2)\frac{\partial^2\Sigma}{\partial II^2}\right]\right. \\
&\left. + III\frac{\partial^2\Sigma}{\partial II\,\partial III}\right]\lambda_b{}^2(I - \lambda_b{}^2) + \left[\lambda_a{}^2\left[\frac{\partial^2\Sigma}{\partial I\,\partial III} + (I - \lambda_a{}^2)\frac{\partial^2\Sigma}{\partial II\,\partial III}\right]\right. \\
&\left. + III\frac{\partial^2\Sigma}{\partial III^2}\right]III\Bigg\}n_a n_b\Bigg\}
\end{aligned}
$$

For an arbitrary direction of propagation $\mathbf{n}$, the subsequent analysis differs very little from the one corresponding to an anisotropic solid. However, some preferred directions for $\mathbf{n}$ lead to tangible results in isotropic materials. Let us assume that $\mathbf{n}$ is in the direction of one of the principal axes, say x_1. Following our previous terminology we call such a wave a *principal plane wave.* It then follows from (4.3.13) that

(4.3.14)

$$Q_{\mathfrak{ab}} = 0, \qquad \mathfrak{a} \neq \mathfrak{b}$$

$$Q_{11} = \frac{2\lambda_1}{\lambda_2 \lambda_3}\left\{\frac{\partial \Sigma}{\partial I} + (\lambda_2^2 + \lambda_3^2)\frac{\partial \Sigma}{\partial II} + \lambda_2^2\lambda_3^2 \frac{\partial \Sigma}{\partial III} + 2\lambda_1^2\left[\frac{\partial^2 \Sigma}{\partial I^2}\right.\right.$$

$$+ 2(\lambda_2^2 + \lambda_3^2)\frac{\partial^2 \Sigma}{\partial I\,\partial II} + 2\lambda_2^2\lambda_3^2 \frac{\partial^2 \Sigma}{\partial I\,\partial III} + 2\lambda_2^2\lambda_3^2(\lambda_2^2 + \lambda_3^2)\frac{\partial^2 \Sigma}{\partial II\,\partial III}$$

$$\left.\left. + (\lambda_2^2 + \lambda_3^2)^2 \frac{\partial^2 \Sigma}{\partial II^2} + \lambda_2^4\lambda_3^4 \frac{\partial^2 \Sigma}{\partial III^2}\right]\right\}$$

$$Q_{22} = \frac{2\lambda_1}{\lambda_2 \lambda_3}\left(\frac{\partial \Sigma}{\partial I} + \lambda_3^2 \frac{\partial \Sigma}{\partial II}\right), \qquad Q_{33} = \frac{2\lambda_1}{\lambda_2 \lambda_3}\left(\frac{\partial \Sigma}{\partial I} + \lambda_2^2 \frac{\partial \Sigma}{\partial II}\right)$$

Since $\mathbf{Q}$ is already in a diagonal form, we see that there will be three uncoupled principal plane waves polarized in the principal directions. One of these waves is longitudinal and the other two transverse. The velocity of the longitudinal waves is

$$c_1^2 = Q_{11}/\rho$$

while the velocities of transverse waves polarized in the x_2- and x_3-directions are, respectively,[1]

$$c_2^2 = Q_{22}/\rho, \qquad c_3^2 = Q_{33}/\rho$$

These are exactly the same wave velocities we found in (2.13.8) and (2.13.9). Therefore, *for a linear solid*, we again obtain formula (2.13.32) for the differences of the velocities of two shear waves polarized in principal directions perpendicular to the principal direction coinciding with the direction of propagation of plane waves:

(4.3.15) $$c_2 - c_3 = (c_0/2\mu)(t_2 - t_3)$$

where t_2 and t_3 are the principal stresses in the directions of polarization and

$$c_0 = (\mu/\rho_0)^{1/2}$$

is the velocity of shear waves in the undistorted medium. Equation (4.3.15) expresses the fact that a transverse wave propagating in a principal direction in a *homogeneously deformed linear elastic material* has two components

[1] The velocities of the principal waves propagating in the x_2- and x_3-directions may be found from (4.3.14) by a cyclic permutation of indices.

that travel with different velocities, their differences being proportional to the differences between principal stresses in the directions perpendicular to the direction of propagation. In short, the deformed elastic material acquires the property of *birefringence* as far as the transverse waves are concerned. This is called the *acoustoelastic effect*, which bears a strong similarity to the very familiar photoelastic effect and, just as in the photoelastic effect, may be regarded as a physical basis for a nondestructive experimental technique to investigate the stress distribution within the material.[1]

The phase difference between two components of a transverse wave that cause acoustical fringes after having traversed a specimen of thickness h may be found from (4.3.15) as

$$\phi = \omega h\left(\frac{1}{c_3} - \frac{1}{c_2}\right) \simeq \alpha h(t_2 - t_3) \tag{4.3.16}$$

where

$$\alpha = \omega/2\rho_0 c_0{}^3$$

This is called the *acoustoelastic sensitivity* of the material and is directly proportional to the frequency of the waves. This demonstrates one of the benefits of the use of ultrasonic waves in the experimental setup.

The *acoustoelastic law* (4.3.16) is exact only if the material undergoes a homogeneous deformation. The underlying advantage of employing ultrasonic waves in experiments is that it permits us to suppose that this law is locally valid, thus allowing us to determine a nonhomogeneous stress distribution in a specimen. It is important to recall that a sound wave in a nonhomogeneously deformed material cannot be a plane wave, since the coefficients in Eqs. (4.2.34) are usually functions of the coordinates x_i. However, if the initial deformation field and consequently the initial stresses and the coefficients C^0_{klmn} are but slowly varying functions of position, then the sound wave may be considered locally plane. In addition, in the ultrasonic frequency range, i.e., for frequencies larger than 5 MHz, the wave number is very large, especially in metals. Therefore, the second gradients of the displacement

[1] The technique that was first described by Benson and Raelson [1959] is entirely similar to the technique used in photoelastic measurements. Two crystal transducers, usually Y-cut quartz, take the place of the polarizer and the analyzer in the photoelastic apparatus, one connected with a radio frequency pulse generator, generating transversally polarized ultrasonic waves, and the other picking up the signals traversing the specimen. The advantage of the acoustoelastic method over the photoelastic method is that there is no need to build transparent models of the elements to be analyzed. The experiment can be performed on the actual element and makes it possible to determine the residual stresses in a structural element without destroying it. The only requirement is that sound waves must be able to penetrate the specimen, and the best results are obtained for metallic materials that absorb the sound least.

which are proportional to k^2 in view of (4.3.2), produced by the propagating ultrasonic wave, are much larger than the strain gradients that exist initially in the material. We can thus neglect in (4.2.34) the terms $C^0_{klmn,k} u_{m,n}$ in comparison with the remaining terms in $u_{k,lm}$. Hence in the ultrasonic frequency range the displacement field may still be assumed approximately to satisfy (4.3.1), the coefficients of which are now slowly varying functions of position. Therefore, the acoustoelastic law (4.3.16) holds locally for ultrasonic waves but is now position dependent.

It can be shown that an acoustoelastic law similar to (4.3.16) holds for quadratic solids [cf. Eqs. (2.13.31)]. Detailed discussions of this subject comprising anisotropic solids and crystalline materials may be found in the works of Toupin and Bernstein [1961], Tokuoka and Iwashimizu [1968], Tokuoka and Saito [1969a,b], and Tokuoka [1969].

Ultrasonic wave propagation in prestressed elastic materials also provides an effective method for the experimental determination of elastic coefficients. For this, we refer the reader to Hughes and Kelly [1953], Toupin and Bernstein [1961], Thurston and Brugger [1964], and Thurston [1965].

We now turn our attention to the case of incompressible materials undergoing homogeneous deformations. The constraint of incompressibility requires that

$$\mathbf{u} \cdot \mathbf{k} = 0 \qquad \text{or} \qquad \mathbf{u} \cdot \mathbf{n} = 0 \tag{4.3.17}$$

which means that all plane waves in incompressible solids are necessarily transverse. This result is, of course, in complete agreement with that stated in Section 2.14.

The incremental stresses are still given by (4.3.11), with $III = 1$ and $\bar{t}_{\mathfrak{aa}}$ being replaced by ${}_E\bar{t}_{\mathfrak{aa}}$, and the $a_{\mathfrak{ab}}$ in (4.3.12) are simplified to

$$\begin{aligned} a_{\mathfrak{ab}} = 2\lambda_{\mathfrak{a}}^2 \bigg\{ & \left[\frac{\partial \Sigma}{\partial I} + (I - \lambda_{\mathfrak{a}}^2 - \lambda_{\mathfrak{b}}^2) \frac{\partial \Sigma}{\partial II}\right](2\,\delta_{\mathfrak{ab}} - 1) + \lambda_{\mathfrak{b}}^2 \frac{\partial \Sigma}{\partial II} \\ & + 2\lambda_{\mathfrak{b}}^2 \left[\frac{\partial^2 \Sigma}{\partial I^2} + (I - \lambda_{\mathfrak{a}}^2) \frac{\partial^2 \Sigma}{\partial I\, \partial II}\right] + 2\lambda_{\mathfrak{b}}^2 (I - \lambda_{\mathfrak{b}}^2) \left[\frac{\partial^2 \Sigma}{\partial I\, \partial II} \right. \\ & \left. + (I - \lambda_{\mathfrak{a}}^2) \frac{\partial^2 \Sigma}{\partial II^2}\right] \bigg\} \end{aligned} \tag{4.3.18}$$

and the initial stresses are given by

$$\begin{aligned} t^0_{\mathfrak{aa}} &= -p^0 + 2\lambda_{\mathfrak{a}}^2 \left[\frac{\partial \Sigma}{\partial I} + (I - \lambda_{\mathfrak{a}}^2) \frac{\partial \Sigma}{\partial II}\right] \\ t^0_{\mathfrak{ab}} &= 0, \qquad \mathfrak{a} \neq \mathfrak{b} \end{aligned} \tag{4.3.19}$$

For given principal stresses $t^0_{\mathfrak{aa}}$ ($\mathfrak{a} = 1, 2, 3$), Eqs. (4.3.19) determine the principal stretches $\lambda_\mathfrak{a}$ ($\lambda_1 \lambda_2 \lambda_3 = 1$) and the hydrostatic pressure p^0.

The propagation velocity of the transverse plane waves can, of course, be found just as in the case of compressible material. But the secular equation can now be written explicitly as

$$(4.3.20)\quad \begin{aligned}&(\rho_0 c^2)^2 - \rho_0 c^2[(b+d)n_1{}^2 + (a+d)n_2{}^2 + (a+b)n_3{}^2 - en_1 n_3\\ &\quad - fn_1 n_2 - gn_2 n_3] + bdn_1{}^2 + adn_2{}^2 + abn_3{}^2 - dfn_1 n_2\\ &\quad - agn_2 n_3 - ben_1 n_3 + hn_1 n_2 n_3 = 0\end{aligned}$$

where

$$(4.3.21)\quad \begin{aligned}a &= a_{11}n_1{}^2 + a_3\lambda_2{}^2 n_2{}^2 + a_2\lambda_3{}^2 n_3{}^2\\ b &= a_3\lambda_1{}^2 n_1{}^2 + a_{22}n_2{}^2 + a_1\lambda_3{}^2 n_3{}^2\\ d &= a_2\lambda_1{}^2 n_1{}^2 + a_1\lambda_2{}^2 n_2{}^2 + a_{33}n_3{}^2\\ e &= a_{13}n_1{}^2 + a_{31}n_3{}^2 + a_2(\lambda_1{}^2 + \lambda_3{}^2)n_1 n_3 = a_{31}n_3{}^2 + e_1 n_1\\ f &= a_{12}n_1{}^2 + a_{21}n_2{}^2 + a_3(\lambda_1{}^2 + \lambda_2{}^2)n_1 n_2 = a_{21}n_2{}^2 + f_1 n_1\\ g &= a_{23}n_2{}^2 + a_{32}n_3{}^2 + a_1(\lambda_2{}^2 + \lambda_3{}^2)n_2 n_3\\ h &= a_{31}(f_1 n_3 - h_1 n_2)n_3 + a_{21}(e_1 n_2 - h_2 n_3)n_2 + (e_1 f_1 - h_1 h_2)n_2\\ h_1 &= a_{13}n_1 - a_{23}n_2 + (a_2\lambda_1{}^2 + a_3\lambda_2{}^2 - a_1\lambda_2{}^2)n_3\\ h_2 &= a_{12}n_1 - a_{32}n_3 + (a_2\lambda_3{}^2 + a_3\lambda_1{}^2 - a_1\lambda_3{}^2)n_2\end{aligned}$$

In these expressions $a_{\mathfrak{ab}}$ were defined in (4.3.18), and $a_\mathfrak{a}$ are given by

$$(4.3.22)\quad a_\mathfrak{a} = 2\left(\frac{\partial\Sigma}{\partial I} + \lambda_\mathfrak{a}{}^2\frac{\partial\Sigma}{\partial II}\right)$$

One can show that if the waves propagate in one of the principal directions, say x_1, then the velocities of shear waves are given by (2.14.15). In the case of Mooney materials, (4.3.20) is somewhat simplified:

$$(4.3.23)\quad \begin{aligned}&[\rho_0 c^2 - 2(\lambda_1{}^2 n_1{}^2 + \lambda_3{}^2 n_3{}^2)(\alpha + \lambda_2{}^2\beta) - 2\lambda_2{}^2 n_2{}^2(\alpha + \lambda_3{}^2\beta)]\\ &\quad\cdot[\rho_0 c^2 - 2\lambda_1{}^2 n_1{}^2(\alpha + \lambda_3{}^2\beta) - 2(\lambda_2{}^2 n_2{}^2 + \lambda_3{}^2 n_3{}^2)(\alpha + \lambda_1{}^2\beta)]\\ &\quad - 2\beta\lambda_1{}^2\lambda_2{}^2(\lambda_1{}^2 - \lambda_3{}^2)(\lambda_2{}^2 - \lambda_3{}^2)n_1{}^2 n_2{}^2 = 0\end{aligned}$$

For neo-Hookean solids, $\beta = 0$, and we have only one wave velocity for plane waves polarized arbitrarily:

$$(4.3.24)\quad \rho_0 c^2 = 2(\lambda_1{}^2 n_1{}^2 + \lambda_2{}^2 n_2{}^2 + \lambda_3{}^2 n_3{}^2)\alpha$$

4.4 SURFACE WAVES IN PRESTRESSED MATERIALS

Consider a half space $x_2 \geq 0$ made of an isotropic elastic material (Fig. 4.4.1) which has undergone a homogeneous deformation in such a way that one principal direction is perpendicular to the deformed boundary $x_2 = 0$ which is assumed to be free of tractions. If the principal stretches are denoted

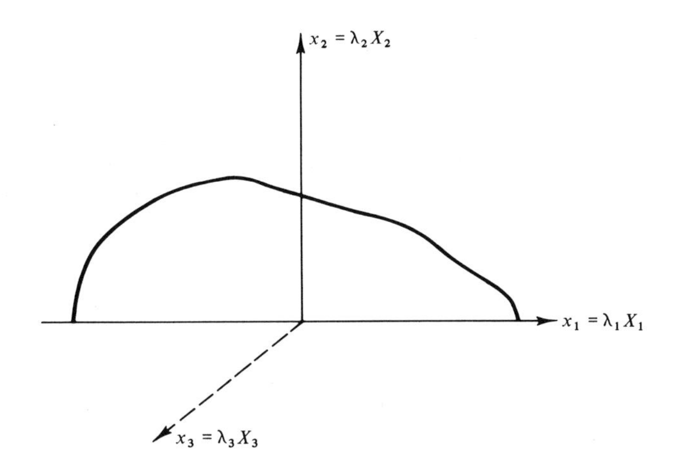

Fig. 4.4.1 Initially stretched half-space.

by λ_1, λ_2, and λ_3, the stresses are specified by (4.3.10), and $t_{ij}^0 = 0$, if $i \neq j$. Since t_{22}^0 vanishes at the boundary $x_2 = 0$, the principal stretches must satisfy

$$\frac{\partial \Sigma}{\partial I} + (\lambda_1{}^2 + \lambda_3{}^2)\frac{\partial \Sigma}{\partial II} + \lambda_1{}^2\lambda_3{}^2\frac{\partial \Sigma}{\partial III} = 0 \tag{4.4.1}$$

By using (4.4.1) in the expression for other stress components, we find the initial stresses:

$$\begin{aligned} t_{11}^0 &= 2\frac{\lambda_1{}^2 - \lambda_2{}^2}{\lambda_1\lambda_2\lambda_3}\left(\frac{\partial \Sigma}{\partial I} + \lambda_3{}^2\frac{\partial \Sigma}{\partial II}\right) \\ t_{33}^0 &= 2\frac{\lambda_3{}^2 - \lambda_2{}^2}{\lambda_1\lambda_2\lambda_3}\left(\frac{\partial \Sigma}{\partial I} + \lambda_1{}^2\frac{\partial \Sigma}{\partial II}\right) \\ t_{22}^0 &= t_{12}^0 = t_{13}^0 = t_{23}^0 = 0 \end{aligned} \tag{4.4.2}$$

Now superimpose a time-dependent infinitesimal displacement $\mathbf{u}$. The incremental stresses $\bar{t}_{kl}$ are given by the relations (4.3.11), which can be explicitly written as

(4.4.3)

$$\begin{aligned} \bar{t}_{11} &= a_{11}u_{1,1} + a_{12}u_{2,2} + a_{13}u_{3,3}, & \bar{t}_{12} &= a_3({\lambda_1}^2 u_{2,1} + {\lambda_2}^2 u_{1,2}) \\ \bar{t}_{22} &= a_{21}u_{1,1} + a_{22}u_{2,2} + a_{23}u_{3,3}, & \bar{t}_{13} &= a_2({\lambda_1}^2 u_{3,1} + {\lambda_3}^2 u_{1,3}) \\ \bar{t}_{33} &= a_{31}u_{1,1} + a_{32}u_{2,2} + a_{33}u_{3,3}, & \bar{t}_{23} &= a_1({\lambda_2}^2 u_{3,2} + {\lambda_3}^2 u_{2,3}) \end{aligned}$$

where

$$a_i = \frac{2}{\lambda_1 \lambda_2 \lambda_3}\left(\frac{\partial \Sigma}{\partial I} + {\lambda_i}^2 \frac{\partial \Sigma}{\partial II}\right), \qquad i = 1, 2, 3 \tag{4.4.4}$$

The equations of motion (4.2.19) with vanishing body forces then become

$$\begin{aligned} &a_{11}u_{1,11} + a_3{\lambda_2}^2 u_{1,22} + a_2{\lambda_3}^2 u_{1,33} + (a_{12} + a_3{\lambda_1}^2)u_{2,12} \\ &\qquad + (a_{13} + a_2{\lambda_1}^2)u_{3,13} = \rho\ddot{u}_1 \\ &(a_{21} + a_3{\lambda_2}^2)u_{1,12} + a_3{\lambda_1}^2 u_{2,11} + a_{22}u_{2,22} + a_1{\lambda_3}^2 u_{2,33} \\ &\qquad + (a_{23} + a_1{\lambda_2}^2)u_{3,23} = \rho\ddot{u}_2 \\ &(a_{31} + a_2{\lambda_3}^2)u_{1,13} + (a_{32} + a_1{\lambda_3}^2)u_{2,23} + a_2{\lambda_1}^2 u_{3,11} \\ &\qquad + a_1{\lambda_2}^2 u_{3,22} + a_{33}u_{3,33} = \rho\ddot{u}_3 \end{aligned} \tag{4.4.5}$$

We suppose that the boundary $x_2 = 0$ is still free from tractions. Hence the boundary conditions are

$$\bar{t}_{21} - t^0_{11}u_{2,1} = 0, \qquad \bar{t}_{22} = 0, \qquad \bar{t}_{23} - t^0_{33}u_{2,3} = 0 \tag{4.4.6}$$

We now look for a solution representing a plane wave propagation in a direction parallel to the surface of the half-space. This solution, if it exists, is known as the surface wave and was first discovered by Rayleigh (see Volume II, Section 7.5,) for linear elastic solids. Since (4.4.5) is a system with constant coefficients, the plane wave solution propagating in the direction of a unit vector $\mathbf{n}(n_1, 0, n_3)$ is known to be dependent exponentially on the depth variable x_2. Thus the desired solution for waves whose amplitudes decrease with depth may be represented by

$$u_i = B_i \exp k[-\sigma x_2 + i(n_1 x_1 + n_3 x_3 - ct)] \tag{4.4.7}$$

where $\mathbf{B}$ is an unknown constant amplitude vector, σ an unknown constant, and $c = \omega/k$ is the velocity of propagation of the waves. Substituting (4.4.7) into (4.4.5), we arrive at the following determinantal equation to determine σ:

(4.4.8) $$\det \Delta_{ij} = 0$$

$$\Delta_{11} = a_3 \lambda_2{}^2 k^2 \sigma^2 + \rho\omega^2 - k^2(a_{11} n_1{}^2 + a_2 \lambda_3{}^2 n_3{}^2) \qquad \Delta_{12} = -ik^2(a_{12} + a_3 \lambda_1{}^2)\sigma n_1$$

$$\Delta_{13} = -k^2(a_{13} + a_2 \lambda_1{}^2) n_1 n_3 \qquad \Delta_{21} = -ik^2(a_{21} + a_3 \lambda_2{}^2)\sigma n_1$$

$$\Delta_{22} = a_{22} k^2 \sigma^2 + \rho\omega^2 - k^2(a_3 \lambda_1{}^2 n_1{}^2 + a_1 \lambda_3{}^2 n_3{}^2) \qquad \Delta_{23} = -ik^2(a_{23} + a_3 \lambda_2{}^2)\sigma n_3$$

$$\Delta_{31} = -k^2(a_{31} + a_2 \lambda_3{}^2) n_1 n_3 \qquad \Delta_{32} = -ik^2(a_{32} + a_1 \lambda_3{}^2)\sigma n_3$$

$$\Delta_{33} = a_1 \lambda_2{}^2 k^2 \sigma^2 + \rho\omega^2 - k^2(a_2 \lambda_1{}^2 n_1{}^2 + a_{33} n_3{}^2)$$

Clearly (4.4.8) is a bicubic equation in σ having six roots of the form $\pm \sigma_j$, which depend on the yet unknown wave velocity c. The σ_j may, of course, be complex numbers. But we must demand that Re $\sigma > 0$, for physically significant solutions. This eliminates three roots out of six, and (4.4.7) can then be expressed as

(4.4.9)

$$u_i = \sum_{j=1}^{3} A_{ij} \exp k[-\sigma_j x_2 + i(n_1 x_2 + n_3 x_3 - ct)], \qquad \text{Re}\, \sigma_j > 0, \quad k > 0$$

The nine constant coefficients A_{ij} are dependent on only three constants, selected arbitrarily by using (4.4.5). For instance,

(4.4.10)

$$-i\sigma_j(a_{12} + a_3\lambda_1{}^2) n_1 A_{2j} + (a_{13} + a_2\lambda_1{}^2) n_1 n_3 A_{3j} + (a_3 \lambda_2{}^2 \sigma_j{}^2 + \rho c^2 - a_{11} n_1{}^2 - a_2 \lambda_3{}^2 n_3{}^2) A_{1j} = 0$$

$$i\sigma_j(a_{32} + a_1\lambda_3{}^2) A_{2j} - (a_1 \lambda_2{}^2 \sigma_j{}^2 + \rho c^2 - a_2 \lambda_1{}^2 n_1{}^2 - a_{33} n_3{}^2) A_{3j} + (a_{31} + a_2 \lambda_3{}^2) n_1 n_3 A_{1j} = 0$$

Now if we use (4.4.9) in the boundary condition (4.4.6) with $x_2 = 0$, we obtain

(4.4.11)

$$\lambda_2{}^2 a_3 \sum_{j=1}^{3} (i n_1 A_{2j} - \sigma_j A_{1j}) = 0$$

$$\sum_{j=1}^{3} (i a_{21} n_1 A_{1j} - a_{22} \sigma_j A_{2j} + i a_{23} n_3 A_{3j}) = 0$$

$$\lambda_2{}^2 a_1 \sum_{j=1}^{3} (i n_3 A_{2j} - \sigma_j A_{3j}) = 0$$

Equations (4.4.11) together with (4.4.10) constitute a system of homogeneous equations for the determination of A_{ij}. For a nontrivial solution, the determinant of the coefficients should be equal to zero. This yields a bicubic algebraic equation in c. A close scrutiny of (4.4.10) and (4.4.11) reveals that the coefficients are independent of the wave number k. Therefore, the velocity of propagation c does not depend on the wave number, and we conclude that *Rayleigh surface waves in a prestressed half-space are not dispersive*, just as in the case of a linear elastic solid. However, there is a fundamental difference between the two solutions: We will show in Volume II, Chapter 7, that Rayleigh waves in classical elasticity are plane polarized in the direction of propagation, viz., they do not have a transverse horizontal displacement component. However, one can easily see from the foregoing analysis that the horizontal transverse component $-n_3u_1 + n_1u_3$ does not vanish, in general. But we can show that if the wave propagates in one of the principal directions,[1] say x_1, then and only then does the transverse component disappear. Taking $n_1 = 1$, $n_3 = 0$ in the relevant expressions above, we see that the u_3-component uncouples from the others, and the boundary conditions $(4.4.6)_3$, which are now reduced to $\bar{t}_{23} = 0$, can be satisfied only if $u_3 \equiv 0$.[2] Equation (4.4.8), which determines σ, now becomes

$$
\begin{aligned}
(4.4.12)\quad & \lambda_2{}^2\, a\, a_{22}\, \sigma^4 + [a_{22}(\rho c^2 - a_{11}) + \lambda_2{}^2 a(\rho c^2 - \lambda_1{}^2 a) \\
& + (a_{12} + \lambda_1{}^2 a)(a_{21} + \lambda_2{}^2 a)]\sigma^2 + (\rho c^2 - a_{11})(\rho c^2 - \lambda_1{}^2 a) = 0
\end{aligned}
$$

where we wrote $a \equiv a_3$. The displacement components can thus be written as

$$
(4.4.13)\quad
\begin{aligned}
u_1 &= i(a_{12} + \lambda_1{}^2 a)k^2[\sigma_1 A \exp(-k\sigma_1 x_2) \\
&\quad + \sigma_2 B \exp(-k\sigma_2 x_2)] \exp i(kx_1 - \omega t) \\
u_2 &= k^2[(\rho c^2 - a_{11} + \sigma_1 \lambda_2{}^2 a)A \exp(-k\sigma_1 x_2) \\
&\quad + (\rho c^2 - a_{11} + \sigma_2{}^2 \lambda_2{}^2 a)B \exp(-k\sigma_2 x_2)] \exp i(kx_1 - \omega t)
\end{aligned}
$$

[1] This case has been studied by Hayes and Rivlin [1961b].

[2] However, we can show that such waves may exist for some materials. In this case (4.4.8) yields

$$\lambda_2{}^2 a_1 \sigma^2 = \lambda_1{}^2 a_2 - \rho c^2$$

from which we can write

$$u_3 = A \exp k[-\sigma x_2 + i(x_3 - ct)]$$

From the boundary conditions, we then have

$$\lambda_2{}^2 a_1 \sigma A = 0$$

It is easy to see that the only consistent possibility is $a_1 = 0$ or the severely restrictive condition on the material constitution

$$\frac{\partial \Sigma}{\partial I} = -\lambda_1{}^2 \frac{\partial \Sigma}{\partial II} = \lambda_1{}^4 \frac{\partial \Sigma}{\partial III}$$

obtained by having recourse to (4.4.1).

where σ_1 and σ_2 are roots of (4.4.12) which satisfy $\operatorname{Re}(\sigma_1, \sigma_2) > 0$, and A and B are arbitrary constants. The boundary conditions (4.4.6) are

$$\left.\begin{aligned} u_{2,1} + u_{1,2} &= 0 \\ a_{21}u_{1,1} + a_{22}u_{2,2} &= 0 \end{aligned}\right\} \quad \text{for} \quad x_2 = 0 \tag{4.4.14}$$

provided $a \neq 0$. Substituting u_1 and u_2 from (4.4.13) into (4.4.14), we get

$$\begin{aligned} &k^3\{\rho c^2 - a_{11} + \sigma_1{}^2[(\lambda_2{}^2 - \lambda_1{}^2)a - a_{12}]\}A \\ &\quad + k^3\{\rho c^2 - a_{11} + \sigma_2{}^2[(\lambda_2{}^2 - \lambda_1{}^2)a - a_{12}]\}B = 0 \\ &k^3\sigma_1(a_{12}a_{21} - a_{11}a_{22} + a_{21}a\lambda_1{}^2 + a_{22}\rho c^2 + a_{22}\sigma_1{}^2 a\lambda_2{}^2)A \\ &\quad + k^3\sigma_2(a_{12}a_{21} - a_{11}a_{22} + a_{21}a\lambda_1{}^2 + a_{22}\rho c^2 + a_{22}\sigma_2{}^2 a\lambda_2{}^2)B = 0 \end{aligned} \tag{4.4.15}$$

So that (4.4.15) has a nontrivial solution for constants A and B, the determinant of the coefficients must vanish:

$$\begin{aligned} &k^6(\sigma_1 - \sigma_2)\{(\rho c^2 - a_{11})[a_{12}a_{21} - a_{11}a_{22} + a_{21}a\lambda_1{}^2 + a_{22}\rho c^2 \\ &\quad + a_{22}a\lambda_2{}^2(\sigma_1{}^2 + \sigma_2{}^2)] - a_{22}a\lambda_2{}^2[a_{12} + (\lambda_1{}^2 - \lambda_2{}^2)a]\sigma_1{}^2\sigma_2{}^2 \\ &\quad - \sigma_1\sigma_2(a_{12} + \lambda_1{}^2 a)[a_{12}a_{21} - a_{11}a_{22} + a_{21}a(\lambda_1{}^2 - \lambda_2{}^2) + a_{22}\rho c^2]\} = 0 \end{aligned} \tag{4.4.16}$$

Inserting the values of $\sigma_1{}^2 + \sigma_2{}^2$ and $\sigma_1{}^2\sigma_2{}^2$ from (4.4.12) as

$$\begin{aligned} \sigma_1{}^2 + \sigma_2{}^2 &= -(1/\lambda_2{}^2 a a_{22})[a_{22}(\rho c^2 - a_{11}) + \lambda_2{}^2 a(\rho c^2 - \lambda_1{}^2 a) \\ &\qquad + (a_{12} + \lambda_1{}^2 a)(a_{21} + \lambda_2{}^2 a)] \\ \sigma_1{}^2\sigma_2{}^2 &= (1/\lambda_2{}^2 a a_{22})(\rho c^2 - a_{11})(\rho c^2 - \lambda_1{}^2 a) \end{aligned} \tag{4.4.17}$$

into (4.4.16) we find either

$$k = 0, \qquad \sigma_1 = \sigma_2 \tag{4.4.18}$$

or

$$\begin{aligned} (a_{12} + \lambda_1{}^2 a)(\rho c^2 - a_{11})[\rho c^2 - (\lambda_1{}^2 - \lambda_2{}^2)a] = (a_{12} + \lambda_1{}^2 a)[a_{11}a_{22} - a_{12}a_{21} \\ - a_{21}(\lambda_1{}^2 - \lambda_2{}^2)a - a_{22}\rho c^2]\sigma_1\sigma_2 \end{aligned}$$

Assuming $a_{12} + \lambda_1{}^2 a \neq 0$, squaring both sides of the above expression, again using $(4.4.17)_2$, and defining

$$\Omega = 1/\rho c^2 = k^2/\rho\omega^2$$

we obtain either

$$\Omega = a_{11}^{-1} \tag{4.4.19}$$

or

$$(1 - a_{11}\Omega)[1 - (\lambda_1^{\ 2} - \lambda_2^{\ 2})a\Omega]^2\lambda_2^{\ 2}aa_{22} = (1 - \lambda_1^{\ 2}a\Omega)\{[a_{11}a_{22} - a_{12}a_{21} - a_{21}(\lambda_1^{\ 2} - \lambda_2^{\ 2})a]\Omega - a_{22}\}^2 \tag{4.4.20}$$

to determine Ω. It has been shown by Hayes and Rivlin [1961b] that the solutions corresponding to (4.4.18) and (4.4.19) are degenerate, and we shall not be concerned with them here. Thus we call (4.4.20) the *frequency equation* determining the velocity of surface waves propagating in a principal direction in a homogeneously stressed elastic half-space.

As an illustrative example, Hayes and Rivlin considered a quadratic solid whose strain energy function Σ is assumed to take the following form:

$$\Sigma = \alpha_1 J_2 + \alpha_2 J_1^{\ 2} + \alpha_3 J_1 J_2 + \alpha_4 J_1^{\ 3} + \alpha_5 J_3 \tag{4.4.21}$$

$$J_1 = I - 3 = 2I_E, \quad J_2 = II - 2I + 3 = 4II_E, \quad J_3 = III - II + I - 1 = 8III_E$$

where α_1 to α_5 are constants. One can now identify the Lamé constants of a linear solid as

$$\lambda = 4(\alpha_1 + 2\alpha_2), \qquad \mu = -2\alpha_1$$

For a linear approximation in principal extensions $e_\alpha = \lambda_\alpha - 1$, we find that

$$\partial\Sigma/\partial I = \alpha_5 - 2\alpha_1 + 4(\alpha_2 - \alpha_3)(e_1 + e_2 + e_3)$$

$$\partial\Sigma/\partial II = \alpha_1 - \alpha_5 + 2\alpha_3(e_1 + e_2 + e_3), \qquad \partial\Sigma/\partial III = \alpha_5$$

$$\partial^2\Sigma/\partial I^2 = 2(\alpha_2 - 2\alpha_3) + 12\alpha_4(e_1 + e_2 + e_3), \qquad \partial^2\Sigma/\partial I\,\partial II = \alpha_3$$

and all the other derivatives of Σ vanish. Condition (4.4.1) then becomes

$$e_3 = -e_1 - [2\alpha_2/(\alpha_1 + 2\alpha_2)]e_2$$

which is the same as in the linear solid. An approximation to the roots Ω of (4.4.20) may now be written as

$$\Omega = \Omega_0 + \Omega_1 e_1 + \Omega_2 e_2$$

i.e., we specify Ω by small perturbations to the solution Ω_0 for a linear solid. Defining

(4.4.22)

$$\begin{aligned}
\bar{a} &= -512\alpha_1{}^3(\alpha_1 + 4\alpha_2)^2 = 16\mu^3(\lambda + \mu)^2 \\
\bar{b} &= -256\alpha_1{}^2(\alpha_1 + 4\alpha_2)(\alpha_1 + 6\alpha_2) = -8\mu^2(\lambda + \mu)(3\lambda + 4\mu) \\
\bar{c} &= -256\alpha_1\alpha_2(\alpha_1 + 4\alpha_2) = 8\mu(\lambda + \mu)(\lambda + 2\mu) \\
\bar{d} &= -16\alpha_2(\alpha_1 + 4\alpha_2) = -(\lambda + \mu)(\lambda + 2\mu) \\
\bar{a}_1 &= 512\alpha_1{}^2(\alpha_1 + 4\alpha_2)(-12\alpha_1{}^2 - 52\alpha_1\alpha_2 + 8\alpha_1\alpha_3 + 10\alpha_1\alpha_5 + 24\alpha_2\alpha_5) \\
\bar{b}_1 &= 256[4\alpha_1\alpha_2(-10\alpha_1{}^2 - 52\alpha_1\alpha_2 + 10\alpha_1\alpha_3 + 10\alpha_1\alpha_5 + 24\alpha_2\alpha_5) \\
&\quad + \alpha_1{}^2(-8\alpha_1{}^2 - 44\alpha_1\alpha_2 + 8\alpha_1\alpha_3 + 8\alpha_1\alpha_5 + 20\alpha_2\alpha_5)] \\
\bar{c}_1 &= -64\alpha_2(19\alpha_1{}^2 + 68\alpha_1\alpha_2 - 32\alpha_2\alpha_5 - 12\alpha_1\alpha_3 - 16\alpha_1\alpha_5) \\
\bar{d}_1 &= -32(\alpha_1\alpha_2 - \alpha_2\alpha_5) \\
\bar{a}_2 &= 512\alpha_1{}^2(\alpha_1 + 4\alpha_2)(-3\alpha_1{}^3 - 24\alpha_1{}^2\alpha_2 - 18\alpha_1{}^2\alpha_3 - 48\alpha_1{}^2\alpha_4 - 56\alpha_1\alpha_2{}^2 \\
&\quad + 48\alpha_2{}^2\alpha_5 - 8\alpha_1\alpha_2\alpha_3 + 20\alpha_1\alpha_2\alpha_5)(\alpha_1 + 2\alpha_2)^{-1} \\
\bar{b}_2 &= -512\alpha_1(2\alpha_1{}^4 + 26\alpha_1{}^3\alpha_2 + 8\alpha_1{}^3\alpha_3 + 30\alpha_1{}^3\alpha_4 + 176\alpha_1\alpha_2{}^3 - 96\alpha_2{}^3\alpha_5 \\
&\quad + 116\alpha_1{}^2\alpha_2{}^2 + 38\alpha_1{}^2\alpha_2\alpha_3 + 144\alpha_1{}^2\alpha_2\alpha_4 - 8\alpha_1\alpha_2{}^2\alpha_3 - 8\alpha_1{}^2\alpha_2\alpha_5 \\
&\quad - 60\alpha_1\alpha_2{}^2\alpha_5)(\alpha_1 + 2\alpha_2)^{-1} \\
\bar{c}_2 &= -64(17\alpha_1{}^3\alpha_2 + 24\alpha_1{}^3\alpha_4 + 248\alpha_1\alpha_2{}^3 - 64\alpha_2{}^3\alpha_5 + 134\alpha_1{}^2\alpha_2{}^2 \\
&\quad + 12\alpha_1{}^2\alpha_2\alpha_3 + 192\alpha_1{}^2\alpha_2\alpha_4 - 72\alpha_1\alpha_2{}^2\alpha_3 - 32\alpha_1\alpha_2{}^2\alpha_5)(\alpha_1 + 2\alpha_2)^{-1} \\
\bar{d}_2 &= -8(128\alpha_2{}^3 + 8\alpha_1{}^2\alpha_2 + 12\alpha_1{}^2\alpha_4 + 72\alpha_1\alpha_2{}^2 - 64\alpha_2{}^2\alpha_3 \\
&\quad - 8\alpha_2{}^2\alpha_5 - 4\alpha_1\alpha_2\alpha_3 + 96\alpha_1\alpha_2\alpha_4)(\alpha_1 + 2\alpha_2)^{-1}
\end{aligned}$$

after some tedious computations, we obtain

$$\bar{a}\Omega_0{}^3 + \bar{b}\Omega_0{}^2 + \bar{c}\Omega_0 + \bar{d} = 0 \tag{4.4.23}$$

or

$$8\bar{\Omega}_0{}^3 - 8(2 - \nu)\bar{\Omega}_0{}^2 + 8(1 - \nu)\bar{\Omega}_0 - (1 - \nu) = 0$$

where ν is the Poisson's ratio, $\bar{\Omega}_0 = \mu\Omega_0$, for the zeroth-order approximation, and

(4.4.24)

$$\Omega_\alpha = -(\bar{a}_\alpha\Omega_0{}^3 + \bar{b}_\alpha\Omega_0{}^2 + \bar{c}_\alpha\Omega_0 + \bar{d}_\alpha)/(3\bar{a}\Omega_0{}^2 + 2\bar{b}\Omega_0 + \bar{c}), \qquad \alpha = 1, 2$$

for the first-order corrections. After having solved (4.4.20) for Ω we can go back to (4.4.12) and determine σ_1 and σ_2. Define first $\gamma^2 = \sigma^2/\rho c^2$ and

(4.4.25)

$$A_0 = -16\alpha_1\alpha_2, \qquad A_1 = -32\alpha_2(\alpha_1 - \alpha_5)$$
$$B_0 = -2\alpha_1 + 8\alpha_2, \qquad B_1 = -4(\alpha_1 - \alpha_5)$$
$$C_1 = 8(\alpha_1{}^2 - 20\alpha_1\alpha_2 + 4\alpha_1\alpha_3 + 8\alpha_2\alpha_5)$$
$$D_1 = 8(\alpha_1{}^2 - 16\alpha_1\alpha_2 + 4\alpha_1\alpha_3 + 4\alpha_2\alpha_5)$$
$$E_1 = 4(-3\alpha_1 + 8\alpha_2 - 4\alpha_3 + \alpha_5)$$
$$A_2 = 32(-2\alpha_1{}^2\alpha_2 - 3\alpha_1{}^2\alpha_4 - 6\alpha_1\alpha_2{}^2 + \alpha_1\alpha_2\alpha_3 + 2\alpha_2{}^2\alpha_5)(\alpha_1 + 2\alpha_2)^{-1}$$
$$B_2 = 2(-\alpha_1{}^2 + 32\alpha_2{}^2 + 8\alpha_1\alpha_2 - 2\alpha_1\alpha_3 + 24\alpha_1\alpha_4 - 16\alpha_2\alpha_3 + 4\alpha_2\alpha_5)(\alpha_1 + 2\alpha_2)^{-1}$$
$$C_2 = 8(-\alpha_1{}^3 - 6\alpha_1{}^2\alpha_2 - 4\alpha_1{}^2\alpha_3 - 24\alpha_1{}^2\alpha_4 - 24\alpha_1\alpha_2{}^2 + 16\alpha_2{}^2\alpha_5)(\alpha_1 + 2\alpha_2)^{-1}$$
$$D_2 = 8(-\alpha_1{}^3 + 2\alpha_1{}^2\alpha_2 - 4\alpha_1{}^2\alpha_3 - 12\alpha_1{}^2\alpha_4 + 8\alpha_2{}^2\alpha_5 - 4\alpha_1\alpha_2\alpha_3)(\alpha_1 + 2\alpha_2)^{-1}$$
$$E_2 = 2(3\alpha_1{}^2 + 6\alpha_1\alpha_3 + 24\alpha_1\alpha_4 + 4\alpha_2\alpha_5)(\alpha_1 + 2\alpha_2)^{-1}$$

Now write

$$\gamma = \gamma_0 + \gamma_1 e_1 + \gamma_2 e_2$$

Then γ_0 satisfies

(4.4.26) $$A_0\gamma_0{}^4 + (B_0 - 2A_0\Omega_0)\gamma_0{}^2 + 1 - B_0\Omega_0 + A_0\Omega_0{}^2 = 0$$

which, after some manipulations, gives the two roots γ_{01}^2 and γ_{02}^2, as in the classical linear theory:

$$\gamma_{01}^2 = \Omega_0 - (1/\mu), \qquad \gamma_{02}^2 = \Omega_0 - [1/(\lambda + 2\mu)]$$

Therefore, the first-order corrections are found to be

(4.4.27) $$\gamma_\alpha = -[A_\alpha\gamma_0{}^4 + (B_\alpha - C_\alpha\Omega_0 - 2A_0\Omega_\alpha)\gamma_0{}^2 + D_\alpha\Omega_0{}^2 + 2A_0\Omega_0\Omega_\alpha - B_0\Omega_\alpha - E_\alpha\Omega_0]/[4A_0\gamma_0{}^3 + 2(B_0 - 2A_0\Omega_0)\gamma_0]$$

The corrections γ_{11}, γ_{21} to γ_{01}, and γ_{12}, γ_{22} to γ_{02} are found by replacing γ_0 by γ_{01} and γ_{02}, respectively, in expression (4.4.27). In the aforementioned paper, Hayes and Rivlin specialized the foregoing analysis to an incompressible body.[1]

The treatment of Rayleigh waves in general incompressible materials usually entails no greater simplicity compared to compressible solids. Some simplifications, however, can be achieved in a special incompressible solid, namely, Mooney materials.[2] First we can replace (4.4.5) by a simpler system. To this end, let us observe from (4.2.46) that

$$\nabla^2 \bar{p} = {}_E\bar{t}_{kl,kl}$$

[1] They also treated Love waves in a prestressed half-space with a superficial layer.
[2] This problem has been considered by Flavin [1963].

since

$$u_{l,l} = 0 \tag{4.4.28}$$

It is not difficult to see that in the case of Mooney materials ${}_E\bar{t}_{kl,kl} = 0$ because of (4.4.28). Thus by eliminating u_2 we can put the field equations into the following form:

$$\begin{aligned}
&-\bar{p}_{,1} + 2(\lambda_1{}^2 u_{1,11} + \lambda_2{}^2 u_{1,22})(\alpha + \lambda_3{}^2\beta) + 2\lambda_3{}^2(\alpha + \lambda_2{}^2\beta)u_{1,33} \\
&\qquad + 2\beta\lambda_1{}^2(\lambda_3{}^2 - \lambda_2{}^2)u_{3,13} = \rho_0 \ddot{u}_1 \\
&-\bar{p}_{,3} + 2\lambda_1{}^2(\alpha + \lambda_2{}^2\beta)u_{3,11} + 2(\lambda_3{}^2 u_{3,33} + \lambda_2{}^2 u_{3,22})(\alpha + \lambda_1{}^2\beta) \\
&\qquad + 2\beta\lambda_3{}^2(\lambda_1{}^2 - \lambda_2{}^2)u_{1,13} = \rho_0 \ddot{u}_3 \\
&\qquad \bar{p}_{,11} + \bar{p}_{,22} + \bar{p}_{,33} = 0 \\
&\qquad u_{2,2} = -u_{1,1} - u_{3,3}
\end{aligned} \tag{4.4.29}$$

We now seek a solution to these equations in the form

$$\text{const}\cdot\exp k[-\sigma x_2 + i(n_1 x_1 + n_3 x_3 - ct)]$$

It can readily be seen that σ satisfies the equation

$$\begin{aligned}
(\sigma^2 - 1)\{&[(\lambda_2{}^2\sigma^2 - \lambda_1{}^2 n_1{}^2)(1 + \lambda_3{}^2\gamma) - \lambda_3{}^2 n_3{}^2(1 + \lambda_2{}^2\gamma) + \kappa] \\
&\cdot[(\lambda_2{}^2\sigma^2 - \lambda_3{}^2 n_3{}^2)(1 + \lambda_1{}^2\gamma) - \lambda_1{}^2 n_1{}^2(1 + \lambda_2{}^2\gamma) + \kappa] \\
&- \lambda_1{}^2\lambda_3{}^2(\lambda_1{}^2 - \lambda_2{}^2)(\lambda_3{}^2 - \lambda_2{}^2)n_1{}^2 n_3{}^2\gamma\} = 0
\end{aligned} \tag{4.4.30}$$

where

$$\gamma \equiv \beta/\alpha, \qquad \kappa = \rho\omega^2/2\alpha k^2 = \rho c^2/2\alpha \tag{4.4.31}$$

The roots of (4.4.30) are $\sigma = 1$ and $\sigma = \sigma_1, \sigma_2$ (Re $\sigma_1, \sigma_2 > 0$), which are the roots of the expression within the brackets. We write the solution as

(4.4.32)

$$\begin{aligned}
\bar{p} &= -2ik\alpha\{\lambda_2{}^2(1 - \sigma_0{}^2) + \gamma[\lambda_2{}^2(\lambda_1{}^2 n_3{}^2 + \lambda_3{}^2 n_1{}^2) - \lambda_1{}^2\lambda_2{}^2]\}A \\
&\quad \cdot\exp k[-x_2 + i(\mathbf{n}\cdot\mathbf{x} - ct)] \\
u_1 &= \left[n_1 A\exp(-kx_2) + n_1 n_3(\lambda_3{}^2 - \lambda_2{}^2)\lambda_1{}^2\gamma\sum_{j=1}^{2} A_j\sigma_j\exp(-k\sigma_j x_2)\right] \\
&\quad \cdot\exp ik(\mathbf{n}\cdot\mathbf{x} - ct) \\
u_2 &= i\left\{A\exp(-kx_2) + \sum_{j=1}^{2} n_3[n_1{}^2(\lambda_3{}^2 - \lambda_2{}^2)\lambda_1{}^2\gamma + f(\sigma_j)]A_j\exp(-k\sigma_j x_2)\right\} \\
&\quad \cdot\exp ik(\mathbf{n}\cdot\mathbf{x} - ct) \\
u_3 &= \left[n_3 A\exp(-kx_2) + \sum_{j=1}^{2}\sigma_j f(\sigma_j)A_j\exp(-k\sigma_j x_2)\right]\cdot\exp ik(\mathbf{n}\cdot\mathbf{x} - ct)
\end{aligned}$$

where A, A_1, and A_2 are arbitrary constants,

$$f(\sigma_j) = (\lambda_2{}^2\sigma_j{}^2 - \lambda_1{}^2 n_1{}^2)(1 + \lambda_3{}^2\gamma) - \lambda_3{}^2(1 + \lambda_2{}^2\gamma)n_3{}^2 + \kappa$$

and

$$\sigma_0{}^2 = \lambda_2^{-2}(\lambda_1{}^2 n_1{}^2 + \lambda_3{}^2 n_3{}^2 - \kappa)$$

is the common value of $\sigma_1{}^2$ and $\sigma_2{}^2$ in (4.4.30) when γ is set equal to zero.

The boundary conditions at $x_2 = 0$ are

$$\begin{aligned} &u_{1,2} + u_{2,1} = 0, \qquad u_{3,2} + u_{2,3} = 0 \\ &-\bar{p} - 4\beta\lambda_2{}^2\lambda_3{}^2 u_{1,1} + 4\alpha\lambda_2{}^2 u_{2,2} - 4\beta\lambda_1{}^2\lambda_2{}^2 u_{3,3} = 0^1 \end{aligned} \tag{4.4.33}$$

The substitution of (4.4.32) into (4.4.33) results in the secular equation to determine the wave velocity c. We shall not evaluate c in the general case, but rather consider some special cases which yield manageable expressions as far as the algebra is concerned.

We first deal with the case in which waves propagate in one of the principal directions, say x_1. It can be seen that u_3 can be taken to be zero, and the solution reduces to

$$\begin{aligned} \bar{p} &= -2ik(\alpha + \lambda_3{}^2\beta)\lambda_2{}^2(1 - \sigma_0{}^2)A \exp k[-x_2 + i(x_1 - ct)] \\ u_1 &= A[\exp(-kx_2) - \tfrac{1}{2}(1 + \sigma_0{}^2)\exp(-k\sigma_0 x_2)] \exp ik(x_1 - ct) \\ u_2 &= iA[\exp(-kx_2) - \tfrac{1}{2}\sigma_0^{-1}(1 + \sigma_0{}^2)\exp(-k\sigma_0 x_2)] \exp ik(x_1 - ct) \end{aligned} \tag{4.4.34}$$

One can also find that the only possible wave velocity is

$$c^2 = \frac{2\lambda_1{}^2(\alpha + \lambda_3{}^2\beta)}{\rho_0}\left(1 - \frac{\lambda_2{}^2}{\lambda_1{}^2}\sigma_0{}^2\right) \tag{4.4.35}$$

and the boundary conditions yield the equation to determine σ_0 by excluding the factor $\sigma - 1$:

$$\sigma^3 + \sigma^2 + 3\sigma - 1 = 0 \tag{4.4.36}$$

which gives $\sigma_0 \simeq 0.2956$. The other two roots are complex with negative real parts; therefore, they should be rejected. If $c^2 < 0$, then the initial disturbances grow indefinitely, which means that the initial deformation is inherently unstable. The stability condition follows from (4.4.35). In order that $c^2 > 0$, we must have

$$\lambda_1{}^2/\lambda_2{}^2 \geq \sigma_0{}^2 \qquad \text{or} \qquad \lambda_1{}^2\lambda_3 \geq \sigma_0 \simeq 0.2956$$

In other words, the initial deformation begins to loose its stability whenever $\lambda_1{}^2\lambda_3 \simeq 0.2956$.

[1] This equation is obtained by adding to both sides of the third boundary condition the identity $2\lambda_2{}^2[\alpha - (\lambda_1{}^2 + \lambda_3{}^2)\beta]u_{k,k} = 0$.

Another special case which leads to the direct evaluation of relevant quantities corresponds to the neo-Hookean material in which $\gamma = 0$. One then obtains

(4.4.37)

$$\begin{aligned}
\bar{p} &= -2ikA\alpha\lambda_2{}^2(1-\sigma_0{}^2)\exp k[-x_2 + i(\mathbf{n}\cdot\mathbf{x} - ct)] \\
u_1 &= n_1 A[\exp(-kx_2) - \tfrac{1}{2}(1+\sigma_0{}^2)\exp(-k\sigma_0 x_2)]\exp ik(\mathbf{n}\cdot\mathbf{x} - ct) \\
u_2 &= iA[\exp(-kx_2) - \tfrac{1}{2}\sigma_0^{-1}(1+\sigma_0{}^2)\exp(-k\sigma_0 x_2)]\exp ik(\mathbf{n}\cdot\mathbf{x} - ct) \\
u_3 &= (n_3/n_1)u_1 \\
c^2 &= (2\alpha/\rho_0)[\lambda_3{}^2 + (\lambda_1{}^2 - \lambda_3{}^2)n_1{}^2]\{1 - (\sigma_0{}^2/\lambda_1{}^2\lambda_3{}^2)[\lambda_3{}^2 + (\lambda_1{}^2 - \lambda_3{}^2)n_1{}^2]^{-1}\}
\end{aligned}$$

where σ_0 is again the real root of (4.4.36). Obviously this wave is polarized in the $(x_2, \mathbf{n})$-plane, since $n_1 u_3 - n_3 u_1 = 0$.

The solution (4.4.37) may constitute a basis for a perturbative solution for waves in a Mooney material propagating in an arbitrary direction parallel to the boundary. For the treatment of this proposition and also the case corresponding to uniaxial extension, i.e., $\lambda_2 = \lambda_3$, we refer the reader to Flavin [1963].

4.5 TORSIONAL AND LONGITUDINAL OSCILLATIONS OF AN INITIALLY STRETCHED CIRCULAR CYLINDER

Consider a circular cylinder with an initial radius a_0 subject to a uniform extension in the axial direction. If the lateral surface of the cylinder is free from tractions, we find that

(4.5.1) $$t_{zz}^0 = (\lambda^2 - \mu^2)(\Phi + \mu^2\Psi)$$

and all the other components of the initial stress tensor vanish. Here λ and μ denote the initial axial and lateral stretches, respectively. The quantities Φ and Ψ defined by (4.2.41) are functions of the invariants:

(4.5.2) $$I = \lambda^2 + 2\mu^2, \qquad II = \mu^2(2\lambda^2 + \mu^2), \qquad III = \lambda^2\mu^4$$

hence they are constant. λ and μ are related by

(4.5.3) $$p + \mu^2\Phi + \mu^2(\lambda^2 + \mu^2)\Psi = 0$$

where p, Φ, and Ψ are defined by (4.2.41).

We now superimpose on the deformed cylinder a time-dependent displacement field $\mathbf{u} = \mathbf{u}(r, z, t)$, which is independent of the angular cylindrical coordinate θ. The components of this displacement vector, in cylindrical coordinates, will be denoted by

$$u_r = u(r, z, t), \qquad u_\theta = v(r, z, t), \qquad u_z = w(r, z, t) \tag{4.5.4}$$

The physical components of the incremental stress tensor $\bar{\mathbf{t}}$ are obtained from (4.2.37) by using (4.5.1) and (4.5.2):

(4.5.5)

$$\bar{t}_{rr} = \alpha_1\left(\frac{\partial u}{\partial r} + \frac{u}{r}\right) - 2p\frac{\partial u}{\partial r} + 2\mu^4\Psi\frac{u}{r} + \alpha_2\frac{\partial w}{\partial z}, \qquad \bar{t}_{rz} = \alpha_5\left(\frac{\partial u}{\partial z} + \frac{\partial w}{\partial r}\right) + t^0_{zz}\frac{\partial u}{\partial z}$$

$$\bar{t}_{\theta\theta} = \alpha_1\left(\frac{\partial u}{\partial r} + \frac{u}{r}\right) + 2\mu^4\Psi\frac{\partial u}{\partial r} - 2p\frac{u}{r} + \alpha_2\frac{\partial w}{\partial z}, \qquad \bar{t}_{r\theta} = \alpha_6\left(\frac{\partial v}{\partial r} - \frac{v}{r}\right)$$

$$\bar{t}_{zz} = \alpha_3\left(\frac{\partial u}{\partial r} + \frac{u}{r}\right) + (\alpha_4 + 2t^0_{zz})\frac{\partial w}{\partial z}, \qquad t_{\theta z} = \alpha_7\frac{\partial v}{\partial z}$$

In (4.5.5) we defined

(4.5.6)

$$\begin{aligned}
\alpha_1 &= 2\mu^4[A + (\lambda^2 + \mu^2)^2 B + \lambda^4\mu^4 C + 2\lambda^2\mu^2(\lambda^2 + \mu^2)D + 2\lambda^2\mu^2 E \\
&\quad + 2(\lambda^2 + \mu^2)F] - 2\mu^2[\Phi + (\lambda^2 + \mu^2)\Psi] \\
\alpha_2 &= 2\lambda^2\mu^2[A + 2\mu^2(\lambda^2 + \mu^2)B + \lambda^2\mu^6 C + \mu^4(3\lambda^2 + \mu^2)D + \mu^2(\lambda^2 + \mu^2)E \\
&\quad + (\lambda^2 + 3\mu^2)F] - 2\mu^2(\Phi + \mu^2\Psi) \\
\alpha_3 &= \alpha_2 - (\lambda^2 - \mu^2)(\Phi + \mu^2\Psi) \\
\alpha_4 &= 2\lambda^4(A + 4\mu^4 B + \mu^8 C + 4\mu^6 D + 2\mu^4 E + 4\mu^2 F) - (\lambda^2 - \mu^2)(\Phi + \mu^2\Psi) \\
\alpha_5 &= \mu^2(\Phi + \mu^2\Psi), \qquad \alpha_6 = \mu^2(\Phi + \lambda^2\Psi), \qquad \alpha_7 = \lambda^2(\Phi + \mu^2\Psi)
\end{aligned}$$

The quantities $A, \ldots, F$ in these expressions were defined in (4.2.41), and they are constant, since they must be evaluated for the values of invariants given by (4.5.2). The equations of motion (4.2.19) in terms of physical components with vanishing body forces are, therefore,

$$\begin{aligned}
&\frac{\partial \bar{t}_{rr}}{\partial r} + \frac{\partial \bar{t}_{rz}}{\partial z} + \frac{1}{r}(\bar{t}_{rr} - \bar{t}_{\theta\theta}) = \rho\frac{\partial^2 u}{\partial t^2} \\
&\frac{\partial \bar{t}_{r\theta}}{\partial r} + \frac{\partial \bar{t}_{\theta z}}{\partial z} + \frac{2}{r}\bar{t}_{r\theta} = \rho\frac{\partial^2 v}{\partial t^2} \\
&\frac{\partial \bar{t}_{rz}}{\partial r} + \frac{\partial \bar{t}_{zz}}{\partial z} + \frac{1}{r}\bar{t}_{rz} = \rho\frac{\partial^2 w}{\partial t^2}
\end{aligned} \tag{4.5.7}$$

If we substitute (4.5.5) into (4.5.7) and arrange the resulting expressions we arrive at the field equations in displacement components:

(4.5.8)

$$\frac{\partial^2 u}{\partial r^2} + \frac{1}{r}\frac{\partial u}{\partial r} - \frac{u}{r^2} + \frac{\alpha_7}{\alpha_1 - 2p}\frac{\partial^2 u}{\partial z^2} + \frac{\alpha_2 + \alpha_5}{\alpha_1 - 2p}\frac{\partial^2 w}{\partial r\,\partial z} = \frac{\rho}{\alpha_1 - 2p}\frac{\partial^2 u}{\partial t^2}$$

$$\frac{\partial^2 v}{\partial r^2} + \frac{1}{r}\frac{\partial v}{\partial r} - \frac{v}{r^2} + \frac{\alpha_7}{\alpha_6}\frac{\partial^2 v}{\partial z^2} = \frac{\rho}{\alpha_6}\frac{\partial^2 v}{\partial t^2}$$

$$\frac{\partial^2 w}{\partial r^2} + \frac{1}{r}\frac{\partial w}{\partial r} + \frac{\alpha_4 + 2t_{zz}^0}{\alpha_5}\frac{\partial^2 w}{\partial z^2} + \frac{\alpha_3 + \alpha_5 + t_{zz}^0}{\alpha_5}\left(\frac{\partial^2 u}{\partial r\,\partial z} + \frac{1}{r}\frac{\partial u}{\partial z}\right) = \frac{\rho}{\alpha_5}\frac{\partial^2 w}{\partial t^2}$$

where we assumed that $\alpha_1 - 2p$, α_5, and α_6 are different from zero. The boundary conditions on the lateral surface are found from (4.2.16):

$$\bar{t}_{rr} = \bar{t}_r, \qquad \bar{t}_{r\theta} = \bar{t}_\theta, \qquad \bar{t}_{rz} = \bar{t}_z + \bar{t}_{zz}^0\,(\partial u/\partial z) \tag{4.5.9}$$

while on a plane $z = \text{const}$:

$$\bar{t}_{zr} = \bar{t}_r, \qquad \bar{t}_{\theta z} = \bar{t}_\theta, \qquad \bar{t}_{zz} = \bar{t}_z \tag{4.5.10}$$

One of the interesting features of the system (4.5.8) is that the second equation is not coupled with the first and third equations. This property enables us to investigate the displacement fields $\mathbf{u} = v\mathbf{e}_\theta$ and $\mathbf{u} = u\mathbf{e}_r + w\mathbf{e}_z$ separately. Evidently the former field corresponds to a torsional motion of the cylinder, while the latter to a longitudinal motion together with a lateral contraction. In the following subsections we study these motions.

A. Free Torsional Oscillations[1]

Torsional oscillations of an infinite cylinder with a free lateral surface are governed by the differential equation

$$\frac{\partial^2 v}{\partial r^2} + \frac{1}{r}\frac{\partial v}{\partial r} - \frac{v}{r^2} + \frac{\alpha_7}{\alpha_6}\frac{\partial^2 v}{\partial z^2} = \frac{\rho}{\alpha_6}\frac{\partial^2 v}{\partial t^2} \tag{4.5.11}$$

and boundary condition

$$\bar{t}_{r\theta} = \alpha_6\left(\frac{\partial v}{\partial r} - \frac{v}{r}\right) = 0 \qquad \text{on} \quad r = a = \mu a_0 \tag{4.5.12}$$

Suppose that a harmonic wave is propagating in the axial direction. Introducing

$$v(r, z, t) = V(r)\exp i(kz - \omega t)$$

[1] This problem was investigated by Green [1961].

into (4.5.11), we obtain

$$V'' + \frac{1}{r} V' + \left(\frac{\rho \omega^2}{\alpha_6} - k^2 \frac{\alpha_7}{\alpha_6} - \frac{1}{r^2} \right) V = 0$$

where primes denote differentiation with respect to r. The solution of the foregoing ordinary differential equation, regular at $r = 0$, is

$$V = AJ_1(\kappa r)$$

where A is an arbitrary constant, $J_1(z)$ is the Bessel function, and

$$\kappa^2 = \frac{\rho k^2}{\alpha_6} \left(\frac{\omega^2}{k^2} - \frac{\alpha_7}{\rho} \right)$$

Using this expression for v in the boundary condition (4.5.12), we get the frequency equation

$$\kappa J_2(\kappa a) = \kappa J_2(\mu \kappa a_0) = 0$$

If v_n is the nth root of the Bessel function $J_2(v) = 0$, then the velocity of propagation for the nth mode is

$$c_n^2 = \frac{\omega_n^2}{k^2} = \frac{\lambda^2(\Phi + \mu^2 \Psi)}{\rho} + \frac{\Phi + \lambda^2 \Psi}{\rho k^2 a_0^2} v_n^2$$

or recalling the definitions of Φ and Ψ, we have

$$c_n^2 = \frac{2\lambda^2}{\rho_0} \left(\frac{\partial \Sigma}{\partial I} + \mu^2 \frac{\partial \Sigma}{\partial II} \right) + \frac{2 v_n^2}{\rho_0 k^2 a_0^2} \left(\frac{\partial \Sigma}{\partial I} + \lambda^2 \frac{\partial \Sigma}{\partial II} \right) \tag{4.5.13}$$

where ρ_0 is the density of the material in its undistorted state. If $\lambda = \mu = 1$, i.e., if the material is initially unstrained, (4.5.13) reduces to

$$c_{n0}^2 = c_2^2[1 + (v_n^2/k^2 a_0^2)]$$

where c_2 is the velocity of shear waves in linear, isotropic elastic solids. The corresponding formula for incompressible materials may simply be obtained by taking $\mu^2 = 1/\lambda$. In particular, we have in Mooney materials

$$c_n^2 = \frac{2\lambda^2}{\rho_0} \left[\alpha + \frac{\beta}{\lambda} + \frac{v_n^2}{k^2 a_0^2} \left(\frac{\alpha}{\lambda^2} + \beta \right) \right]$$

Since $\alpha, \beta > 0$ it is easy to see that the velocity of propagation, in a deformed cylinder, is greater than the velocity in the undeformed cylinder if $\lambda > 1$, and less if $\lambda < 1$. Hence the initial tension causes the velocity of propagation to increase, whereas an initial compression causes it to decrease.

Since the equation of motion and boundary conditions in the torsional motions of an initially stretched circular cylinder differ very little from their classical counterparts, we do not pursue this subject any further. We only mention, however, that the free oscillations of a finite cylinder with length l are of the form

$$v = \sum_{n=0}^{\infty} \sum_{m=1}^{\infty} A_{mn} J_1(\nu_m r/a) \cos(n\pi z/l) \exp(i\omega_{mn} t)$$

where ν_m is the mth root of $J_2(\kappa) = 0$ and

$$\omega_{mn}^2 = \frac{2}{\rho_0 {a_0}^2}\left(\frac{\partial\Sigma}{\partial I} + \lambda^2 \frac{\partial\Sigma}{\partial II}\right){\nu_m}^2 + \frac{2\lambda^2}{\rho_0}\left(\frac{\partial\Sigma}{\partial I} + \mu^2 \frac{\partial\Sigma}{\partial II}\right)\frac{n^2\pi^2}{l^2}$$

and A_{mn} are determined by some given initial conditions.

B. Longitudinal Oscillations[1]

The longitudinal oscillations of an infinite cylinder are governed by the differential equations

(4.5.14)

$$\frac{\partial^2 u}{\partial r^2} + \frac{1}{r}\frac{\partial u}{\partial r} - \frac{u}{r^2} + \frac{\alpha_7}{\alpha_1 - 2p}\frac{\partial^2 u}{\partial z^2} + \frac{\alpha_2 + \alpha_5}{\alpha_1 - 2p}\frac{\partial^2 w}{\partial r\,\partial z} = \frac{\rho}{\alpha_1 - 2p}\frac{\partial^2 u}{\partial t^2}$$

$$\frac{\alpha_3 + \alpha_5 + t_{zz}^0}{\alpha_5}\left(\frac{\partial^2 u}{\partial r\,\partial z} + \frac{1}{r}\frac{\partial u}{\partial z}\right) + \frac{\partial^2 w}{\partial r^2} + \frac{1}{r}\frac{\partial w}{\partial r} + \frac{\alpha_4 + 2t_{zz}^0}{\alpha_5}\frac{\partial^2 w}{\partial z^2} = \frac{\rho}{\alpha_5}\frac{\partial^2 w}{\partial t^2}$$

and boundary conditions

(4.5.15)

$$\left.\begin{aligned}(\alpha_1 - 2p)(\partial u/\partial r) + (\alpha_1 + 2\mu^4\Psi)(u/r) + \alpha_2(\partial w/\partial z) &= 0\\ (\partial u/\partial z) + (\partial w/\partial r) &= 0\end{aligned}\right\} \cdot \text{ on } \quad \begin{aligned} r &= a = \mu a_0 \\ \alpha_5 &\neq 0\end{aligned}$$

If we consider harmonic waves propagating in the axial direction we can write

$$\begin{bmatrix} u(r, z, t) \\ w(r, z, t)\end{bmatrix} = \begin{bmatrix} U(r) \\ W(r)\end{bmatrix} \exp i(kz - \omega t)$$

Introducing the foregoing forms into (4.5.14) and solving the resulting ordinary differential equations, we get the following expressions, regular at $r = 0$:

$$\text{(4.5.16)} \quad \begin{aligned} U(r) &= C_1 J_1(\lambda_1 r) + C_2 J_1(\lambda_2 r) \\ W(r) &= [({\kappa_1}^2 - {\lambda_1}^2)/\beta_1\lambda_1] C_1 J_0(\lambda_1 r) + [({\kappa_1}^2 - {\lambda_2}^2)/\beta_1\lambda_2] C_2 J_0(\lambda_2 r)\end{aligned}$$

[1] See Şuhubi [1965]; also Zorski [1964].

where C_1, C_2 are arbitrary constants and

(4.5.17)

$$\lambda_{1,2}^2 = \tfrac{1}{2}\{\kappa_1{}^2 + \kappa_2{}^2 + \beta_1\beta_2 \pm [(\kappa_1{}^2 - \kappa_2{}^2)^2 + \beta_1{}^2\beta_2{}^2 + 2\beta_1\beta_2(\kappa_1{}^2 + \kappa_1{}^2)]^{1/2}\}$$

$$\kappa_1{}^2 = \frac{\rho\omega^2}{\alpha_1 - 2p} - \lambda^2 \frac{\Phi + \mu^2\Psi}{\alpha_1 - 2p} k^2, \qquad \kappa_2{}^2 = \frac{\rho\omega^2}{\alpha_5} - \frac{\alpha_4 + 2t_{zz}^0}{\alpha_5} k^2$$

$$\beta_1 = \frac{\alpha_2 + \alpha_5}{\alpha_1 - 2p} k, \qquad \beta_2 = \frac{\alpha_3 + \alpha_5 + t_{zz}^0}{\alpha_5} k$$

Inserting (4.5.16) into the boundary conditions (4.5.15) we obtain the following rather complicated frequency equation:

$$\text{(4.5.18)} \quad (k\beta_1 - \lambda_1{}^2 + \kappa_1{}^2)\left[(\alpha_1 - 2p)\lambda_2 + k\alpha_2 \frac{\kappa_1{}^2 - \lambda_2{}^2}{\beta_1\lambda_2}\right] J_1(\lambda_1 a)J_0(\lambda_2 a)$$
$$-(k\beta_1 - \lambda_2{}^2 + \kappa_1{}^2)\left[(\alpha_1 - 2p)\lambda_1 + k\alpha_2 \frac{\kappa_1{}^2 - \lambda_1{}^2}{\beta_1\lambda_1}\right] J_0(\lambda_1 a)J_1(\lambda_2 a)$$
$$-\frac{2(p + \mu^4\Psi)(\lambda_1{}^2 - \lambda_2{}^2)}{a} J_1(\lambda_1 a)J_1(\lambda_2 a) = 0$$

the roots of which determine frequencies corresponding to a given wave number k or vice versa. We now study the same problem for the incompressible materials.

(i) *Incompressible materials.* In this case the constraint tr $\tilde{\mathbf{e}} = 0$ yields

$$\text{(4.5.19)} \qquad (\partial u/\partial r) + (u/r) + (\partial w/\partial z) = 0$$

Using this in (4.5.5) and (4.5.6), recalling that Σ is now independent of III, and an unknown pressure $\bar{p}(r, z, t)$ should be added to the right-hand side of (4.5.5), we find that

$$\bar{t}_{rr} = -\bar{p} + \bar{\alpha}_1 \frac{\partial w}{\partial z} + \bar{\alpha}_2 \frac{u}{r}, \qquad \bar{t}_{\theta\theta} = -\bar{p} + \bar{\alpha}_1 \frac{\partial w}{\partial z} + \bar{\alpha}_2 \frac{\partial u}{\partial r}$$

$$\text{(4.5.20)} \qquad \bar{t}_{zz} = -\bar{p} + (\bar{\alpha}_3 + 2t_{zz}^0)\frac{\partial w}{\partial z}, \qquad \bar{t}_{rz} = \alpha_5\left(\frac{\partial u}{\partial z} + \frac{\partial w}{\partial r}\right) + t_{zz}^0 \frac{\partial u}{\partial z}$$

$$\bar{t}_{r\theta} = \bar{t}_{\theta z} = 0$$

where

(4.5.21)

$$\bar{\alpha}_1 = 2\mu^2(\lambda^2 - \mu^2)[A + \mu^2(\lambda^2 + \mu^2)B + (\lambda^2 + 2\mu^2)F] - 2\mu^2(\Phi + \mu^2\Psi)$$

$$\bar{\alpha}_2 = -2\mu^2(\Phi + \lambda^2\Psi), \qquad \bar{\alpha}_3 = 2(\lambda^2 - \mu^2)(\lambda^2 A + 2B + 3\lambda F) + 2\mu^2(\Phi + \mu^2\Psi)$$

with $\lambda\mu^2 = 1$, and α_5 defined by (4.5.6). Substituting (4.5.20) into the equations of motion, we obtain

$$\frac{\alpha_5 + t_{zz}^0}{\bar{\alpha}_1 + \alpha_5}\frac{\partial^2 u}{\partial z^2} + \frac{\partial^2 w}{\partial r\,\partial z} - \frac{1}{\bar{\alpha}_1 + \alpha_5}\frac{\partial \bar{p}}{\partial r} = \frac{\rho}{\bar{\alpha}_1 + \alpha_5}\frac{\partial^2 u}{\partial t^2}$$

(4.5.22)

$$\frac{\partial^2 w}{\partial r^2} + \frac{1}{r}\frac{\partial w}{\partial r} + \frac{\bar{\alpha}_3 - \alpha_5 + t_{zz}^0}{\alpha_5}\frac{\partial^2 w}{\partial z^2} - \frac{1}{\alpha_5}\frac{\partial \bar{p}}{\partial z} = \frac{\rho}{\alpha_5}\frac{\partial^2 w}{\partial t^2}$$

Considering the same type of harmonic solutions as were previously assumed for u and w, and in addition setting

$$\bar{p}(r, z, t) = P(r)\exp i(kz - \omega t)$$

we find the solution of (4.5.22) and (4.5.19) as

$$P = -[C_1 J_0(\lambda_1 r) + C_2 J_0(\lambda_2 r)]$$

(4.5.23)

$$U = \frac{k^2}{\alpha_5}\left[\frac{C_1}{\lambda_1(\kappa_2{}^2 - \lambda_1{}^2)}J_1(\lambda_1 r) + \frac{C_2}{\lambda_2(\kappa_2{}^2 - \lambda_2{}^2)}J_1(\lambda_2 r)\right]$$

$$W = \frac{ik}{\alpha_5}\left[\frac{C_1}{\kappa_2{}^2 - \lambda_1{}^2}J_0(\lambda_1 r) + \frac{C_2}{\kappa_2{}^2 - \lambda_2{}^2}J_0(\lambda_2 r)\right]$$

where

$$\lambda_{1,2}^2 = \tfrac{1}{2}\{\kappa_2{}^2 + q^2 \pm [(\kappa_2{}^2 + q^2)^2 + 4\kappa_1{}^2 q^2]^{1/2}\}$$

(4.5.24)

$$\kappa_1{}^2 = \frac{\rho\omega^2}{\bar{\alpha}_1 + \alpha_5} - \frac{\alpha_5 + t_{zz}^0}{\bar{\alpha}_1 + \alpha_5}k^2, \qquad \kappa_2{}^2 = \frac{\rho\omega^2}{\alpha_5} - \frac{\bar{\alpha}_3 - \alpha_5 + t_{zz}^0}{\alpha_5}k^2$$

$$q^2 = \frac{\bar{\alpha}_1 + \alpha_5}{\alpha_5}k^2$$

Employing (4.5.23) in the boundary conditions, we arrive at the frequency equation:

(4.5.25)

$$\left[1 + \frac{\bar{\alpha}_1}{\alpha_5}\frac{k^2}{\kappa_2{}^2 - \lambda_2{}^2}\right]J_1(\lambda_1 a)J_0(\lambda_2 a) - \frac{\lambda_1}{\lambda_2}\frac{(\lambda_2{}^2 - k^2)(\kappa_2{}^2 - \lambda_1{}^2)}{(\lambda_1{}^2 - k^2)(\kappa_2{}^2 - \lambda_2{}^2)}$$

$$\cdot\left[1 + \frac{\bar{\alpha}_1}{\alpha_5}\frac{k^2}{\kappa_2{}^2 - \lambda_1{}^2}\right]J_0(\lambda_1 a)J_1(\lambda_2 a) - \frac{\bar{\alpha}_2}{\alpha_5}\frac{k^2(\lambda_1{}^2 - \lambda_2{}^2)}{\lambda_2 a(\kappa_2{}^2 - \lambda_2{}^2)(\lambda_1{}^2 - k^2)}$$

$$\cdot J_1(\lambda_1 a)J_1(\lambda_2 a) = 0$$

This equation is not much simpler than (4.5.18), corresponding to compressible materials for computational purposes. We shall, however, see that in the case of Mooney materials some approximate results may be drawn from (4.5.25).

(ii) *Mooney materials.* In this case (4.5.21) reduces to

$$\bar{\alpha}_3 = -\bar{\alpha}_1 = 2\alpha_5 = 4\mu^2(\alpha + \mu^2\beta), \qquad \bar{\alpha}_2 = -4\mu^2(\alpha + \lambda^2\beta)$$

and (4.5.24) reduces to

$$-\kappa_1{}^2 = \kappa_2{}^2 = \frac{\rho\omega^2}{2\mu^2(\alpha + \mu^2\beta)} - \frac{\lambda^2}{\mu^2}k^2, \qquad \lambda_1{}^2 = \kappa_2{}^2, \qquad \lambda_2{}^2 = -k^2$$

Now (4.5.25) degenerates into

(4.5.26)

$$J_0(\kappa_2 a)I_1(ka) - \eta\frac{k^2 + \kappa_2{}^2}{2k^2\kappa_2 a}J_1(\kappa_2 a)I_1(ka) - \frac{(k^2 - \kappa_2{}^2)^2}{4k^3\kappa_2}J_1(\kappa_2 a)I_0(ka) = 0$$

where I_0 and I_1 are the modified Bessel functions, and

$$\eta = (\alpha + \lambda^2\beta)/(\alpha + \mu^2\beta) = (1 + \lambda^2\gamma)/(1 + \mu^2\gamma), \qquad \gamma = \beta/\alpha$$

When the deformed radius a of the cylinder is small, we may expand the ordinary and modified Bessel functions appearing in (4.5.26) into power series and retain only terms up to and including the second order in a. Thus

$$(4.5.27) \quad 4 - \eta(1 + \xi) - (1 - \xi)^2 - \tfrac{1}{8}k^2a^2\,[8\xi - \eta\xi(1 + \xi) + (1 + \xi)^3] = 0$$

where $\xi = \kappa_2{}^2/k^2$. The second-order approximation to the wave velocity $c = \omega/k$, after some manipulation, is obtained as

$$(4.5.28) \quad c^2 = 2\frac{(\lambda^2 + 2\mu^2)\alpha + 3\mu^4\beta}{\rho}\left[1 - \frac{4 - \eta}{8[3 + (\lambda^2/\mu^2) - \eta]}k^2a^2\right]$$

The case $\lambda = 1$ corresponds to the cylinder without an initial stretch. For this case, the classical formula

$$c_0{}^2 = [6(\alpha + \beta)/\rho](1 - \tfrac{1}{8}k^2a^2)$$

is valid. Hence we may rewrite (4.5.28) as

$$(4.5.29) \quad \frac{c^2}{c_0{}^2} = \frac{1}{\lambda^2}\frac{\lambda(2 + \lambda^3) + 3\gamma}{3(1 + \gamma)}\left(1 + \frac{\lambda^3 - 1}{3 - \eta + \lambda^3}\frac{k^2a^2}{8}\right)$$

The square root of the predominant factor in (4.5.29), which is the ratio of wave velocities c/c_0, with a^2 in the first approximation neglected, is plotted versus λ in Fig. 4.5.1 for $\gamma = 0.8$. The minimum value $c/c_0 \simeq 0.924$ occurs at $\lambda = 1.396$.

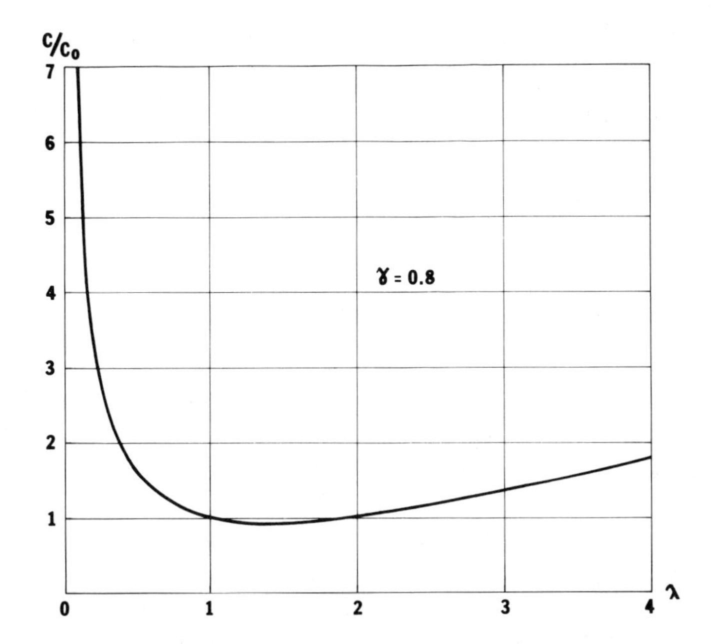

Fig. 4.5.1 Wave velocity as a function of stretch (*after* Şuhubi [1965]).

4.6 OSCILLATIONS OF AN INITIALLY TWISTED CIRCULAR CYLINDER[1]

To illustrate the anticipated inherent difficulties when dealing with problems in which the initial stress field is inhomogeneous, in this section we consider the oscillations of an infinite circular cylinder subject to an initial pure torsion. As is well known, the problem of finite torsion of an *incompressible* isotropic circular cylinder can be solved exactly without taking a special elastic material constitution into account,[2] or in other words, without specifying the form of the strain energy function. If the constant twist per unit length is denoted by K, then one can show that the components of the deformation tensor $\overset{-1}{\mathbf{c}}$ are given by

$$
\begin{aligned}
&\overset{-1}{c}_{rr} = 1, \qquad \overset{-1}{c}_{r\theta} = \overset{-1}{c}_{rz} = 0 \\
&\overset{-1}{c}_{\theta\theta} = 1 + K^2 r^2, \qquad \overset{-1}{c}_{\theta z} = K r^2, \qquad \overset{-1}{c}_{zz} = 1
\end{aligned}
\tag{4.6.1}
$$

[1] See Demiray and Şuhubi [1970].

[2] See, for instance, Eringen [1962, Section 53].

and the components of the initial stress tensor are found to be

$$t^0_{rr} = -2K^2 \int_r^a r \frac{\partial \Sigma}{\partial I}\, dr, \qquad t^0_{\theta\theta} = 2K^2\left(r^2 \frac{\partial \Sigma}{\partial I} - \int_r^a r \frac{\partial \Sigma}{\partial I}\, dr\right)$$

(4.6.2)
$$t^0_{zz} = -2K^2\left(r^2 \frac{\partial \Sigma}{\partial II} + \int_r^a r \frac{\partial \Sigma}{\partial I}\, dr\right), \qquad t^0_{\theta z} = 2Kr\left(\frac{\partial \Sigma}{\partial I} + \frac{\partial \Sigma}{\partial II}\right)$$

$$t^0_{\theta r} = t^0_{rz} = 0$$

The lateral surface of the cylinder is free from tractions, i.e.,

(4.6.3)
$$t^0_{rr} = t^0_{r\theta} = t^0_{rz} = 0 \qquad \text{on} \quad r = a$$

where a is the radius of the cylinder, which is, incidentally, the same as its radius in the undeformed state. The strain energy Σ is a function of invariants

(4.6.4)
$$I = II = 3 + K^2 r^2$$

We now superimpose an infinitesimal displacement field $\mathbf{u}(r, z, t)$ on the initially twisted cylinder. The increments in the existing stress field induced by such displacements have been computed (Green and Spencer [1958]). The resulting expressions are quite complicated, and the integration of the differential field equations can only be carried out after the specification of Σ as functions of the invariants I and II.

Here we consider only a very simple material, namely, Mooney material. Even in this case the equations of motion cannot be integrated exactly. To derive the governing differential equations, we first obtain the components of the incremental stress tensor $\bar{\mathbf{t}}$. After some manipulations, and having made use of the incompressibility condition (4.5.19), it follows from (4.2.50) that

(4.6.5)

$$\bar{t}_{rr} = -\bar{p} + 2t^0_{rr} \frac{\partial u}{\partial r} + 2[2\alpha + 2\beta + 2\beta K^2 r^2 + \alpha K^2(a^2 - r^2)] \frac{\partial u}{\partial r}$$
$$+ 4\beta K^2 r u + 4\beta K r \frac{\partial v}{\partial z}$$

$$\bar{t}_{\theta\theta} = -\bar{p} + 2t^0_{\theta\theta} \frac{u}{r} + 2t^0_{\theta z} \frac{\partial v}{\partial z} + 2[2\alpha + 2\beta + 2\beta K^2 r^2 + \alpha K^2(a^2 - r^2)] \frac{u}{r}$$
$$+ 4\beta K^2 r^2 \frac{\partial u}{\partial r}$$

$$\bar{t}_{zz} = -\bar{p} + 2t^0_{zz} \frac{\partial w}{\partial z} + 2[2\alpha + 2\beta + 2\beta K^2 r^2 + \alpha K^2(a^2 - r^2)] \frac{\partial w}{\partial z}$$

$$\bar{t}_{r\theta} = t_{rr}^0 \frac{\partial v}{\partial r} - t_{\theta\theta}^0 \frac{v}{r} + t_{\theta z}^0 \frac{\partial u}{\partial z} - [2\alpha + 2\beta + \alpha K^2(a^2 - r^2)]\left(\frac{\partial v}{\partial r} - \frac{v}{r}\right)$$

$$- 2\beta K r\left(\frac{\partial u}{\partial z} + \frac{\partial w}{\partial r}\right)$$

$$\bar{t}_{rz} = t_{rr}^0 \frac{\partial w}{\partial r} - t_{\theta z}^0 \frac{v}{r} + t_{zz}^0 \frac{\partial u}{\partial z} + [2\alpha + 2\beta + 2\beta K^2 r^2 + \alpha K^2(a^2 - r^2)]\left(\frac{\partial u}{\partial z} + \frac{\partial w}{\partial r}\right)$$

$$- 2\beta K r \frac{\partial v}{\partial r} + 2\beta K v$$

$$\bar{t}_{\theta z} = - t_{\theta z}^0 \frac{\partial u}{\partial r} + t_{zz}^0 \frac{\partial v}{\partial z} + [2\alpha + 2\beta + 2\beta K^2 r^2 + \alpha K^2(a^2 - r^2)] \frac{\partial v}{\partial z}$$

$$+ 4\beta K r \frac{\partial u}{\partial r}$$

where

$$t_{rr}^0 = -\alpha K^2(a^2 - r^2), \qquad t_{\theta\theta}^0 = \alpha K^2(3r^2 - a^2)$$

$$t_{\theta z}^0 = 2(\alpha + \beta)Kr, \qquad t_{zz}^0 = -2K^2[\alpha(a^2 - r^2) + \beta r^2]$$

If we substitute (4.6.5) into the equations of motion (4.2.19) with $u_{k,k} = 0$, we get, after some lengthy manipulations,

$$\frac{\partial^2 u}{\partial z^2} + \psi K r \frac{\partial^2 v}{\partial r\,\partial z} + K(5\psi - 2)\frac{\partial v}{\partial z} - (1 + \psi K^2 r^2)\frac{\partial^2 w}{\partial r\,\partial z}$$

$$- (5\psi - 1)K^2 r \frac{\partial w}{\partial z} - \frac{1}{2(\alpha + \beta)}\frac{\partial \bar{p}}{\partial r} = \frac{\rho}{2(\alpha + \beta)}\frac{\partial^2 u}{\partial t^2}$$

(4.6.6)
$$\frac{\partial^2 v}{\partial r^2} + \frac{1}{r}\frac{\partial v}{\partial r} - \frac{v}{r^2} + \frac{\partial^2 v}{\partial z^2} - \psi K r\left(\frac{\partial^2 w}{\partial r^2} + \frac{3}{r}\frac{\partial w}{\partial r} + \frac{\partial^2 w}{\partial z^2}\right)$$

$$- 2K(2\psi - 1)\frac{\partial u}{\partial z} = \frac{\rho}{2(\alpha + \beta)}\frac{\partial^2 v}{\partial t^2}$$

$$\frac{\partial^2 w}{\partial r^2} + \frac{1}{r}\frac{\partial w}{\partial r} + \frac{\partial^2 w}{\partial z^2} + \psi K^2 r^2\left(\frac{\partial^2 w}{\partial r^2} + \frac{3}{r}\frac{\partial w}{\partial r}\right) + (3\psi - 1)K^2 r \frac{\partial u}{\partial z}$$

$$- \psi K r\left(\frac{\partial^2 v}{\partial r^2} + \frac{1}{r}\frac{\partial v}{\partial r} - \frac{v}{r^2}\right) - \frac{1}{2(\alpha + \beta)}\frac{\partial \bar{p}}{\partial z} = \frac{\rho}{2(\alpha + \beta)}\frac{\partial^2 w}{\partial t^2}$$

where

(4.6.7)
$$\psi = \beta/(\alpha + \beta)$$

On using (4.6.3), the boundary conditions become

$$\left.\begin{aligned} \bar{t}_{rr} &= 0 \\ \bar{t}_{r\theta} + t^0_{\theta\theta}\frac{v}{r} - t^0_{\theta z}\frac{\partial u}{\partial z} &= 0 \\ \bar{t}_{rz} + t^0_{\theta z}\frac{v}{r} - t^0_{zz}\frac{\partial u}{\partial z} &= 0 \end{aligned}\right\} \quad \text{on} \quad r = a \tag{4.6.8}$$

Equations (4.6.6) and (4.6.8) together with (4.5.19) administer the free oscillatory motion of an initially twisted cylinder made of Mooney material.

Since the second equation of (4.6.6) is not uncoupled with the other two, we see that it is impossible to excite pure torsional or pure longitudinal oscillations once the cylinder has an initial twist.

The integration of system (4.6.6) still presents a formidable task. For a neo-Hookean material, a closed-form solution can be found. In the following (Section 4.6. B) a perturbative solution is described for comparatively small twists in the case of Mooney materials.

A. Neo-Hookean Material

In this case $\beta = 0$ and consequently $\psi = 0$. Even now the angular displacement v is still coupled with u and w. However, we can obtain a closed-form solution. Again considering harmonic waves propagating in the axial direction we write

$$\begin{bmatrix} u \\ v \\ w \\ p \end{bmatrix} = \begin{bmatrix} aU(\xi) \\ aV(\xi) \\ aW(\xi) \\ -2(\alpha + \beta)P(\xi) \end{bmatrix} \exp i(kz - \omega t)$$

where the nondimensional radial variable is denoted by $\xi = r/a$. Also defining a nondimensional wave number q, frequency Ω, and twist ε by

$$q = ka, \qquad \Omega^2 = \rho a^2\omega^2/2(\alpha + \beta), \qquad \varepsilon = Ka \tag{4.6.9}$$

we can express the equations of motion (4.6.6) and (4.5.19) as

$$\begin{aligned} P' + (\Omega^2 - q^2)U - 2iq\varepsilon V - iqW' + iq\varepsilon^2\xi W &= 0 \\ 2iq\varepsilon U + V'' + \frac{1}{\xi}V' + \left(\Omega^2 - q^2 - \frac{1}{\xi^2}\right)V &= 0 \\ iqP - iq\varepsilon^2\xi U + W'' + \frac{1}{\xi}W' + (\Omega^2 - q^2)W &= 0 \\ U' + \frac{1}{\xi}U + iqW &= 0 \end{aligned} \tag{4.6.10}$$

where primes denote differentiation with respect to ξ. The boundary conditions (4.6.8) take the form

$$\left.\begin{aligned} P + 2U' &= 0 \\ V' - V &= 0 \\ iqU + W' &= 0 \end{aligned}\right\} \quad \text{for} \quad \xi = 1 \tag{4.6.11}$$

The longitudinal and torsional oscillations become uncoupled, as they should be, in (4.6.10) whenever $\varepsilon = 0$, i.e., whenever there is no initial twist. The solution to (4.6.10) can now be written as

$$\begin{aligned} U &= i \sum_{j=1}^{3} \frac{A_j}{(\lambda_j^2 + q^2)^{1/2}} J_1(\lambda_j \xi), \qquad V = \sum_{j=1}^{3} A_j J_1(\lambda_j \xi) \\ W &= -\sum_{j=1}^{3} \frac{\lambda_j}{q(\lambda_j^2 + q^2)^{1/2}} A_j J_0(\lambda_j \xi) \\ P &= i\varepsilon \sum_{j=1}^{3} \frac{A_j}{(\lambda_j^2 + q^2)^{1/2}} \left[\varepsilon \xi J_1(\lambda_j \xi) - \frac{2\lambda_j}{q(\lambda_j^2 + q^2)^{1/2}} J_0(\lambda_j \xi) \right] \end{aligned} \tag{4.6.12}$$

where A_1, A_2, and A_3 are arbitrary constants, and λ_1^2, λ_2^2, and λ_3^2 are three roots of the following bicubic equation:

$$(\lambda^2 + q^2 - \Omega^2)^2 (\lambda^2 + q^2) - 4\varepsilon^2 q^2 = 0 \tag{4.6.13}$$

If we use (4.6.12) in the boundary conditions (4.6.11), we obtain the frequency equation for free oscillations in the form

$$\begin{vmatrix} Q_1 & Q_2 & Q_3 \\ R_1 & R_2 & R_3 \\ S_1 & S_2 & S_3 \end{vmatrix} = 0$$

where

$$Q_j = \lambda_j J_2(\lambda_j), \qquad R_j = \frac{\lambda_j^2 - q^2}{q(\lambda_j^2 + q^2)^{1/2}} J_1(\lambda_j)$$

$$S_j = \left[\frac{2 + \varepsilon^2}{(\lambda_j^2 + q^2)^{1/2}} - \frac{4\varepsilon}{q(\lambda_j^2 + q^2)} \right] J_1(\lambda_j) - \left[\frac{2\lambda_j}{(\lambda_j^2 + q^2)^{1/2}} - \frac{2\lambda_j \varepsilon}{q(\lambda_j^2 + q^2)} \right] J_2(\lambda_j)$$

Since neo-Hookean materials are only of academic interest, it seems not worthwhile to pursue the calculations any further.

B. Mooney Material

The harmonic waves propagating in the axial direction satisfy the following set of ordinary differential equations which follow from (4.6.6) and (4.5.19):

$$
\begin{aligned}
&P' + (\Omega^2 - q^2)U + iq\varepsilon\psi\xi V' + iq\varepsilon(5\psi - 2)V \\
&\qquad - iq(1 + \varepsilon^2\psi\xi^2)W' - iq\varepsilon^2(5\psi - 1)\xi W = 0 \\
&-2iq\varepsilon(2\psi - 1)U + V'' + \frac{1}{\xi}V' \\
&\qquad + \left(\Omega^2 - q^2 - \frac{1}{\xi^2}\right)V - \varepsilon\psi\xi\left(W'' + \frac{3}{\xi}W' - q^2 W\right) = 0 \\
&iqP + iq\varepsilon^2(3\psi - 1)\xi U - \psi\varepsilon\xi\left(V'' + \frac{1}{\xi}V' - \frac{1}{\xi^2}V\right) \\
&\qquad + W'' + \frac{1}{\xi}W' + (\Omega^2 - q^2)W + \psi\varepsilon^2\xi^2\left(W'' + \frac{3}{\xi}W'\right) = 0 \\
&U' + \frac{1}{\xi}U + iqW = 0
\end{aligned}
\tag{4.6.14}
$$

The boundary conditions are

$$
\left.
\begin{aligned}
P + 2(1 + \varepsilon^2\psi)U' + 2\varepsilon^2\psi U + 2iq\varepsilon\psi V &= 0 \\
-iq\varepsilon\psi U + V' - V - \varepsilon\psi W' &= 0 \\
i(1 + \varepsilon^2\psi)qU - \varepsilon\psi(V' - V) + (1 + \varepsilon^2\psi)W' &= 0
\end{aligned}
\right\} \quad \text{for} \quad \xi = 1
\tag{4.6.15}
$$

Next we seek a perturbative solution to the preceding equations with the perturbation parameter $\varepsilon = Ka$ assumed to be small. The solution is assumed to have a power series expansion in ε:

$$
U = \sum_{n=0}^{\infty} \varepsilon^n U_n(\xi), \ldots, \qquad \Omega = \sum_{n=0}^{\infty} \varepsilon^n \Omega_n, \qquad \Omega^2 = \sum_{n=0}^{\infty} \sum_{m=0}^{n} \varepsilon^n \Omega_{n-m}\Omega_m
$$

This scheme is intended to develop a recurrence process by which various order terms in these expansions may be determined.

Inserting the above expressions into (4.6.14) and (4.6.15) and equating the coefficients of like powers of ε to zero, we find the following recurrence relations:

(4.6.16)

$$P_n' + (\Omega_0^2 - q^2)U_n - iqW_n' = -\sum_{m=1}^{n}\sum_{k=0}^{m}\Omega_{m-k}\Omega_k U_{n-m} - iq[\psi\xi V'_{n-1} + (5\psi - 2)V_{n-1}] + iq\psi\xi^2 W'_{n-2} + iq(5\psi - 1)\xi W_{n-2}$$

$$V''_n + \frac{1}{\xi}V_n' + \left(\Omega_0^2 - q^2 - \frac{1}{\xi^2}\right)V_n = 2iq(2\psi - 1)U_{n-1} - \sum_{m=1}^{n}\sum_{k=0}^{m}\Omega_{m-k}\Omega_k V_{n-m} + \psi\xi\left(W''_{n-1} + \frac{3}{\xi}W'_{n-1} - q^2 W_{n-1}\right)$$

$$iqP_n + W''_n + \frac{1}{\xi}W_n' + (\Omega_0^2 - q^2)W_n = -iq(3\psi - 1)\xi U_{n-2} + \psi\xi\left(V''_{n-1} + \frac{1}{\xi}V'_{n-1} - \frac{1}{\xi^2}V_{n-1}\right) - \sum_{m=1}^{n}\sum_{k=0}^{m}\Omega_{m-k}\Omega_k W_{n-m} - \psi\xi^2\left(W''_{n-2} + \frac{3}{\xi}W'_{n-2}\right)$$

$$U'_n + \frac{1}{\xi}U_n + iqW_n = 0, \qquad \text{for} \quad n = 0, 1, 2, \ldots$$

and

(4.6.17)

$$\left.\begin{aligned} P_n + 2U_n' &= -2\psi(U'_{n-2} + U_{n-2}) - 2iq\psi V_{n-1} \\ V'_n - V_n &= iq\psi U_{n-1} + \psi W'_{n-1} \\ W'_n + iqU_n &= -iq\psi U_{n-2} + \psi(V'_{n-1} - V_{n-1}) - \psi W'_{n-2} \end{aligned}\right\} \begin{array}{c} \xi = 1 \\ \text{for } n = 0, 1, 2, \ldots \end{array}$$

Starting with $n = 0$, for which the right-hand side of (4.6.16) and (4.6.17) vanishes, we can build step by step a solution to any desired degree by using the foregoing recurrence relations, provided that we remain within the domain of convergence of the perturbation series. Here we carry out computations up to the second degree only.

We are mainly concerned with oscillations which are predominantly torsional in character. A similar analysis can, of course, be performed for oscillations with predominant longitudinal components. We begin with $n = 0$ by supposing that

$$U_0 = W_0 = P_0 \equiv 0$$

That this assumption is compatible with the field equations is quite obvious from the fact that it corresponds to the solution of the classical theory, i.e., to $\varepsilon = 0$. Then V_0 is easily found to be

$$V_0 = J_1(\kappa\xi) \tag{4.6.18}$$

where we wrote 1 for an arbitrary constant amplitude factor, and

$$\kappa^2 = \Omega_0{}^2 - q^2 \tag{4.6.19}$$

The boundary condition on V_0 gives

$$J_2(\kappa) = 0 \tag{4.6.20}$$

Therefore, if ν_m is the mth root of this transcendental equation, then the zeroth-order nondimensional frequency for the mth mode of oscillation is

$$\Omega_0{}^2 = q^2 + \nu_m{}^2 \tag{4.6.21}$$

We can now calculate the first-order terms. It follows from (4.6.16) and (4.6.17) that

$$\begin{gathered} P_1' + \kappa^2 U_1 - iqW_1' = -iq[\psi\xi V_0' + (5\psi - 2)V_0] \\ V_1'' + \frac{1}{\xi}V_1' + \left(\kappa^2 - \frac{1}{\xi^2}\right)V_1 = -2\Omega_0\Omega_1 V_0 \\ iqP_1 + W_1'' + \frac{1}{\xi}W_1' + \kappa^2 W_1 = \psi\xi\left(V_0'' + \frac{1}{\xi}V_0' - \frac{1}{\xi^2}V_0\right) \\ U_1' + \frac{1}{\xi}U_1 + iqW_1 = 0 \end{gathered} \tag{4.6.22}$$

and

$$\left.\begin{aligned} P_1 + 2U_1' &= -2iq\psi V_0 \\ V_1' - V_1 &= 0 \\ iqU_1 + W_1' &= \psi(V_0' - V_0) = 0 \end{aligned}\right\} \quad \text{for} \quad \xi = 1 \tag{4.6.23}$$

where κ was defined in (4.6.19) and determined by (4.6.21). The solution of $(4.6.22)_2$ is readily obtained as

$$V_1 = (\Omega_0\Omega_1/\kappa)\xi J_0(\kappa\xi)$$

in which the homogeneous part of the solution is absorbed into V_0. Then $(4.6.23)_2$ yields

$$\Omega_1\Omega_0 J_1(\kappa)/\kappa = 0$$

Since $J_1(\kappa)/\kappa \neq 0$ for κ satisfying (4.6.20), we must have

$$\Omega_1 = 0 \tag{4.6.24}$$

which means that there is no first-order correction to the frequency of waves. The solution of the remaining equations can be written after somewhat tedious manipulations as

(4.6.25)

$$P_1 = AI_0(q\xi) + \frac{2iq\kappa(3\psi - 1)}{\kappa^2 + q^2} J_0(\kappa\xi)$$

$$U_1 = BJ_1(\kappa\xi) - \frac{q}{\kappa^2 + q^2} AI_1(q\xi) - \frac{iq\psi}{4} \xi^2 J_1(\kappa\xi) + iq\frac{(2\psi - 1)q^2 - \psi\kappa^2}{\kappa(\kappa^2 + q^2)} \xi J_0(\kappa\xi)$$

$$W_1 = \frac{i}{q}\left\{B\kappa J_0(\kappa\xi) - \frac{q^2}{\kappa^2 + q^2} AI_0(q\xi) - \frac{iq\psi}{4}\xi[2J_1(\kappa\xi) + \kappa\xi J_0(\kappa\xi)] + iq\frac{2(\psi - 1)q^2 - \psi\kappa^2}{\kappa(\kappa^2 + q^2)}[2J_0(\kappa\xi) - \kappa\xi J_1(\kappa\xi)]\right\}$$

The constants A and B in these expressions are determined from the boundary conditions $(4.6.23)_{1,3}$ as

$$\bar{A} = A/iq(1 - \psi) = \{(\kappa^4 - q^4)[(\gamma - 2)q^2 - 3\gamma\kappa^2 - 4(2\gamma - 1)] + 4q^2[(2\gamma - 1)q^2 + (3\gamma - 1)\kappa^2]\}\{(\kappa^2 + q^2)[(\kappa^2 - q^2)^2 K_0 + 2q(\kappa^2 - 3q^2)K_1]\}^{-1}$$

$$\bar{B} = B/iq(1 - \psi) = \tfrac{1}{4}\{(\kappa^2 - q^2)\{\kappa^2(\kappa^4 - q^4)\gamma + 8[\gamma\kappa^4 - 2(2\gamma - 1)\kappa^2 q^2 + (\gamma - 1)q^4]\}K_0 + 2qK_1\{\kappa^2[\gamma\kappa^4 + 10\gamma\kappa^2 q^2 - (7\gamma - 8)q^4] + 8[\gamma\kappa^4 - 2\gamma\kappa^2 q^2 + 3(\gamma - 1)q^4]\}\}\{(\kappa^2 + q^2)[(\kappa^2 - q^2)^2 K_0 + 2q(\kappa^2 - 3q^2)K_1]\}^{-1}$$

where

$$K_0 = I_0(q)/J_1(\kappa), \qquad K_1 = I_1(q)/J_1(\kappa), \qquad \gamma = \beta/\alpha$$

Obviously $\{\varepsilon U_1, \varepsilon W_1\} \exp[i(kz - \omega t)]$ represent first order longitudinal oscillations that accompany the torsional oscillations.

Next we try to find a correction to the frequency of waves. In order to accomplish this we need only to consider the second-order equations in $(4.6.16)_2$ and $(4.6.17)_2$. These are written as

$$V_2'' + \frac{1}{\xi} V_2' + \left(\kappa^2 - \frac{1}{\xi^2}\right) V_2 = -2\Omega_0 \Omega_2 V_0 + 2iq(2\psi - 1)U_1$$

$$+\psi\xi \left(W_1'' + \frac{3}{\xi} W_1' - q^2 W_1\right) \tag{4.6.26}$$

$$V_2' - V_2 = \psi(iqU_1 + W_1'), \qquad \text{for} \quad \xi = 1$$

where we used (4.6.19) and (4.6.24). The solution of these nonhomogeneous ordinary differential equations may be represented as follows:

$$V_2 = C_1\xi^4 J_1(\kappa\xi) + C_2\xi^3 J_0(\kappa\xi) + C_3\xi^2 J_1(\kappa\xi) + C_4\xi J_0(\kappa\xi) + C_5 I_1(q\xi)$$

where the homogeneous solution is absorbed into V_0 and

$$C_1 = -\frac{\kappa^2 + q^2}{32}\psi^2, \qquad C_2 = \frac{(1-\psi)\psi}{24\kappa}[3\gamma\kappa^2 + (3\gamma - 2)q^2]$$

$$C_3 = \frac{(1-\psi)\psi}{12\kappa^2}\left[5\gamma\kappa^2 - (2-\gamma)q^2 + \frac{14\gamma\kappa^2 - 2(3-2\gamma)q^2}{\kappa^2 + q^2} + 3\kappa^2(\kappa^2 + q^2)\bar{B}\right]$$

$$C_4 = \frac{\Omega_0\Omega_2}{\kappa} - \frac{(1-\psi)^2}{\kappa}\left\{-\frac{2\gamma[(1-\gamma)q^2 + \kappa^2]}{\kappa^2 + q^2} + [\gamma\kappa^2 + (1-\gamma)q^2]\bar{B}\right\}$$

$$C_5 = \frac{2q^3(1-\psi)^2(2\gamma - 1)}{(\kappa^2 + q^2)^2}\bar{A}$$

Introducing V_2 and (4.6.25) into $(4.6.26)_2$ we obtain a second-order correction to the nondimensional frequency as

$$\Omega_2 = [q^2(1-\psi)^2/\Omega_0]\,\delta(\kappa, q, \gamma) \tag{4.6.27}$$

where

(4.6.28)

$$\begin{aligned}
\delta(\kappa, q, \gamma) = {} & \{(\kappa^2 + q^2)[(\kappa^2 - q^2)^2 K_0 + 2q(\kappa^2 - 3q^2)K_1]\}^{-1} \\
& \cdot\{(\kappa^2 + q^2)^{-2}[2q^2(2\gamma - 1)K_0 - 2q[2(2\gamma - 1) + q^2 + \kappa^2]K_1] \\
& \cdot\{-(\kappa^4 - q^4)[(2-\gamma)q^2 + 3\gamma\kappa^2 + 4(2\gamma - 1)] + 4q^2[(2\gamma - 1)q^2 \\
& + (3\gamma - 1)\kappa^2]\} + \tfrac{1}{8}q^{-2}[\gamma\kappa^2 + (\gamma + 2)q^2]\{(\kappa^2 - q^2)[\kappa^2(\kappa^4 - q^4)\gamma \\
& + 8[\gamma\kappa^4 - 2(2\gamma - 1)\kappa^2 q^2 - (1-\gamma)q^4]]K_0 \\
& + 2q[\kappa^2[\gamma\kappa^4 + 10\gamma\kappa^2 q^2 + (8 - 7\gamma)q^4] \\
& + 8[\gamma\kappa^4 - 2\gamma\kappa^2 q^2 - 3(1-\gamma)q^4]]K_1\}\} - \tfrac{1}{24}q^{-2}\gamma[\gamma\kappa^2 + 2(4\gamma + 1)q^2] \\
& + \tfrac{1}{6}\kappa^{-2}q^{-2}(\kappa^2 + q^2)^{-1}[\gamma(\kappa^2 + q^2)(\kappa^2 - 6) - 12q^2][(1-\gamma)q^2 + \kappa^2]
\end{aligned}$$

Accordingly, within this approximation, the frequency of oscillations can be written as

$$\omega = \frac{1}{a}\left[\frac{2(\alpha+\beta)}{\rho}\right]^{1/2} \Omega_0\left(1 + K^2a^2\frac{\Omega_2}{\Omega_0}\right)$$

or

$$\omega = \frac{1}{a}\left[\frac{2(\alpha+\beta)}{\rho}\right]^{1/2} \Omega_0\left[1 + \frac{K^2a^4k^2(1-\psi)^2}{\Omega_0{}^2}\,\delta(\kappa, q, \gamma)\right]$$

where Ω_0 is given by (4.6.19) and δ by (4.6.28). The frequency of the mth mode follows after replacing κ in the relevant expressions by the mth root ν_m of $J_2(\nu) = 0$.

4.7 OTHER SOLUTIONS

To conclude this chapter, we give here a brief review of the problems investigated by various authors. The literature is not abundant on this subject despite the linearity of the governing differential equations. This, however, should be expected since the differential equations for the superimposed infinitesimal displacements are too complicated, except for special cases of initial stresses that yield constant coefficients. Yet even those display great computational difficulties. However, the system being linear, it is more amenable to approximate or numerical processes than its nonlinear counterpart.

It is not surprising that, although the derivation of fundamental equations has been available for some time, the application to specific problems is relatively new. In addition to the works which have already been cited in previous sections, we mention a few others.

Guo investigated the stability and vibration of a circular plate subjected to a finite initial in-plane deformation [1962b,c] and then extended his results to plates of arbitrary shape [1962d]. He also studied the nonrotational vibrations and stability of an incompressible circular cylinder of finite length, extended and inflated initially [1962e,f]. Some special infinitesimal motions of an infinite isotropic body—homogeneously deformed initially parallel to a plane and inhomogeneously in the perpendicular direction with the deformation depending only on the distance to the plane, which is parallel to the homogeneous deformation axes—have been discussed by Ramakanth [1965a], who also derived the equations for radial oscillations of a radially prestressed sphere [1965b].

APPENDIX A

Tensor Analysis

A.1 SCOPE OF APPENDIX A

A short account of tensor analysis is provided here to equip the reader with adequate tools to understand the passage from a Cartesian coordinate system to curvilinear coordinates. However, for those who are not concerned with the expressions of various equations in curvilinear coordinates this appendix is not essential, except possibly for the compatibility equations of strain.

The concept of tensor is a natural extension of the concept of vector. At the base of both is the fundamental and useful idea of invariance. Since physical laws are expected to be coordinate invariant, tensors provide a natural and powerful tool in the formulation of these laws in a systematic fashion. The formulation in arbitrary curvilinear coordinates presents no particular difficulty and, in fact, provides new insight and suggests various generalizations.

In Section A.2 we present a discussion on curvilinear coordinates, base vectors, and metric tensors. Definitions of tensors and their algebra are given in Section A3. In Section A.4 we derive the relations between physical and tensor components. Section A.5 is devoted to tensor calculus. Covariant differentiation, differential operators, gradient, divergence, and curl are defined and illustrated. The Green–Gauss and Stokes theorems, essential to the derivation of local balance laws, are presented. Section A.6 contains an account of the concept of curvature, which is important in the derivation of the compatibility conditions of strain. A brief discussion of a two-point tensor field is provided in Section A.7.

There exist large numbers of books and monographs of various sophistication on the subject. We recommend for beginners Michal [1947], Sokolnikoff [1951], McConnell [1957], and Thomas [1961a]; for intermediate level readers, Eisenhart [1926] and Synge and Schild [1949]; and for advanced readers, Schouten [1951], Yano [1957], and Sternberg [1965]. An account particularly suitable to continuum physics can be found in Eringen [1971].

A.2 CURVILINEAR COORDINATES

If between the rectangular coordinates z^1, z^2, z^3, or z^k ($k = 1, 2, 3$) of a geometrical point in three-dimensional Euclidean space E_3, and three variables x^k a correspondence can be established, we say that there exists a *coordinate transformation* between z^k and x^k. This is expressed in the form of three equations:

(A.2.1) $$z^k = z^k(x^1, x^2, x^3), \qquad k = 1, 2, 3$$

If this correspondence is one to one, then there exists a unique inverse of (A.2.1) in the form

(A.2.2) $$x^k = x^k(z^1, z^2, z^3), \qquad k = 1, 2, 3$$

It can be shown that such a unique inverse exists in some neighborhood of z^k if the Jacobian in that neighborhood does not vanish, i.e.,

(A.2.3) $$J \equiv \det\left(\frac{\partial z^k}{\partial x^l}\right) = \begin{vmatrix} \partial z^1/\partial x^1 & \partial z^1/\partial x^2 & \partial z^1/\partial x^3 \\ \partial z^2/\partial x^1 & \partial z^2/\partial x^2 & \partial z^2/\partial x^3 \\ \partial z^3/\partial x^1 & \partial z^3/\partial x^2 & \partial z^3/\partial x^3 \end{vmatrix} \neq 0$$

For a fixed set of values of z^1, z^2, and z^3, each of Eqs. (A.2.1) gives a surface. The three surfaces so obtained are noncoincident and intersect each other at a single point p marked with specific values of x^k. These surfaces are called *curvilinear coordinate surfaces*. The intersection lines of any pair are called *curvilinear coordinate lines*, and the point of intersection of the three coordinate surfaces identify the *curvilinear coordinates* x^k of p (Fig. A.2.1).

The position vector $\mathbf{p}$ of p in rectangular coordinates is given by

(A.2.4) $$\mathbf{p} = z^k \mathbf{i}_k$$

where $\mathbf{i}_k$ ($k = 1, 2, 3$) are the unit rectangular base vectors. Here and throughout the repeated indices, in diagonal position, indicate summation over the range (1, 2, 3), unless they carry an underscore, e.g.,

$$z^k \mathbf{i}_k \equiv z^1 \mathbf{i}_1 + z^2 \mathbf{i}_2 + z^3 \mathbf{i}_3 \qquad z^{\underline{k}} \mathbf{i}_{\underline{k}} \equiv \text{anyone of} \quad (z^1 \mathbf{i}_1, z^2 \mathbf{i}_2, z^3 \mathbf{i}_3)$$

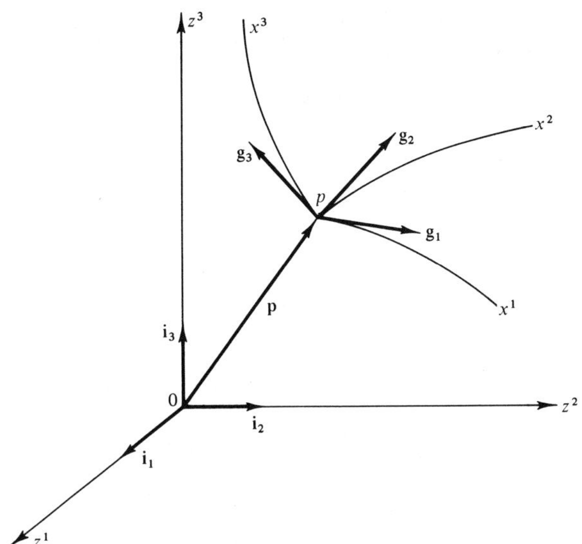

Fig. A.2.1 Curvilinear coordinates.

The *base vectors* $\mathbf{g}_k(x^1, x^2, x^3)$ *are defined by*

(A.2.5) $$\mathbf{g}_k(\mathbf{x}) \equiv \partial\mathbf{p}/\partial x^k = (\partial z^m/\partial x^k)\mathbf{i}_m$$

Multiplying (A.2.5) by $\partial x^k/\partial z^n$ we also obtain

(A.2.6) $$\mathbf{i}_n = (\partial x^k/\partial z^n)\mathbf{g}_k$$

The base vectors $\mathbf{g}_k$ are tangential to the coordinate curves.

An infinitesimal vector at p can be expressed as

(A.2.7) $$d\mathbf{p} = (\partial\mathbf{p}/\partial x^k)\, dx^k = \mathbf{g}_k\, dx^k$$

The scalar product of this with itself gives the square of the arc length

(A.2.8) $$ds^2 = g_{kl}(\mathbf{x})\, dx^k\, dx^l$$

where $g_{kl}(\mathbf{x})$, called the metric tensor, is defined by

(A.2.9) $$g_{kl}(\mathbf{x}) = \mathbf{g}_k \cdot \mathbf{g}_l = \delta_{mn}(\partial z^m/\partial x^k)\, \partial z^n/\partial x^l$$

Here δ_{mn} is the *Kronecker delta*, which is 1 when any two indices are the same and zero otherwise. The name metric tensor for g_{kl} is justified, since when they are known we can calculate the length of any vector and the angle between two vectors. Note that, in general, $g_{kl} \neq 0$ for $k \neq l$. Thus $\mathbf{g}_k$ may not be orthogonal to $\mathbf{g}_l$. The curvilinear coordinates are said to be *orthogonal* if and only if $g_{kl} = 0$ everywhere when $k \neq l$. We also note that $g_{\underline{kk}}$ is not necessarily equal to unity. Thus $\mathbf{g}_k$ are *not* necessarily unit vectors.

The *reciprocal base vectors* $\mathbf{g}^k(\mathbf{x})$ are constructed by finding the solution of the nine equations:

$$\mathbf{g}^k \cdot \mathbf{g}_l = \delta^k{}_l \tag{A.2.10}$$

where $\delta^k{}_l$ is the Kronecker delta. It can be verified that the solution of (A.2.10) is

$$\mathbf{g}^k = g^{kl}(\mathbf{x})\mathbf{g}_l \tag{A.2.11}$$

where g^{kl} is the reduced cofactor in the determinant of g_{kl}, i.e.,

$$g^{kl}(\mathbf{x}) = (\text{cofactor } g_{kl})/g, \qquad g \equiv \det g_{kl} \tag{A.2.12}$$

From (A.2.11), by scalar product with $\mathbf{g}_m$ and $\mathbf{g}^m$, we obtain

$$g^{kl}g_{lm} = \delta^k{}_m, \qquad g^{km} = \mathbf{g}^k \cdot \mathbf{g}^m \tag{A.2.13}$$

An arbitrary vector $\mathbf{v}$ at p may be decomposed into its components v^k or v_k by

$$\mathbf{v} = v^k\mathbf{g}_k = v_k\,\mathbf{g}^k \tag{A.2.14}$$

The components v^k, called *contravariant*, and v_k, called *covariant*, are not identical, in general, since sets $\mathbf{g}_k$ and $\mathbf{g}^k$ are neither orthogonal nor of unit magnitudes. By taking the scalar products of (A.2.14) with $\mathbf{g}^l$ and $\mathbf{g}_l$ and using (A.2.9), (A.2.10) and (A.2.13), we establish the relationships

$$v^k = g^{kl}v_l, \qquad v_k = g_{kl}v^l \tag{A.2.15}$$

Thus when v_l is known we can determine v^k and vice versa. This process is known as *raising* and *lowering* indices.

Example (*cylindrical coordinates*). The cylindrical coordinates x^k are defined by their relations to the rectangular coordinates z^k by

$$z^1 = x^1 \cos x^2, \qquad z^2 = x^1 \sin x^2, \qquad z^3 = x^3 \tag{A.2.16}$$

The Jacobian J, in this case, is given by

$$J = \begin{vmatrix} \cos x^2 & -x^1 \sin x^2 & 0 \\ \sin x^2 & x^1 \cos x^2 & 0 \\ 0 & 0 & 1 \end{vmatrix} = x^1$$

Hence the unique inverse of (A.2.16) exists everywhere except at $x^1 = 0$, and this is given by

$$x^1 = [(z^1)^2 + (z^2)^2]^{1/2}, \qquad x^2 = \arctan(z^2/z^1), \qquad x^3 = z^3$$

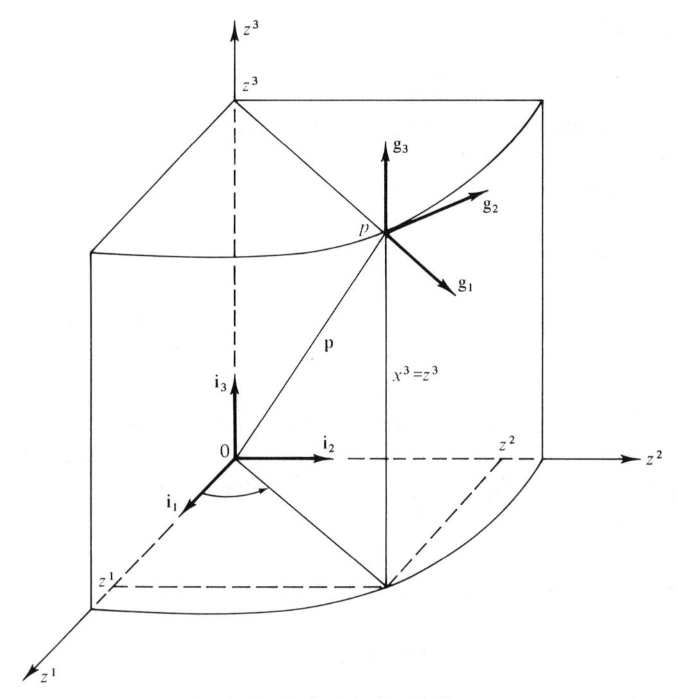

Fig. A.2.2 Cylindrical coordinates.

The coordinate surfaces are circular cylinders with common axis x^3, vertical planes through the x^3 axis, and planes perpendicular to the x^3 axis (Fig. A.2.2). The base vectors are

$$\begin{aligned} \mathbf{g}_1 &= (\cos x^2)\mathbf{i}_1 + (\sin x^2)\mathbf{i}_2 \\ \mathbf{g}_2 &= -(x^1 \sin x^2)\mathbf{i}_1 + (x^1 \cos x^2)\mathbf{i}_2 \\ \mathbf{g}_3 &= \mathbf{i}_3 \end{aligned} \tag{A.2.17}$$

The metric tensors are given by

$$\|g_{kl}\| = \begin{bmatrix} 1 & 0 & 0 \\ 0 & (x^1)^2 & 0 \\ 0 & 0 & 1 \end{bmatrix}, \qquad \|g^{kl}\| = \begin{bmatrix} 1 & 0 & 0 \\ 0 & (x^1)^{-2} & 0 \\ 0 & 0 & 1 \end{bmatrix} \tag{A.2.18}$$

Since $g_{kl} = 0$ for $k \neq l$, the cylindrical coordinates are orthogonal coordinates.

A.3 TENSORS

Definition 1. *A set of quantities* $A^{k_1 k_2 \cdots k_p}{}_{l_1 l_2 \cdots l_q}$ *are called relative tensors of weight* w, *contravariant with respect to* p *indices (superscripts), and covariant with respect to* q *indices (subscripts) if, upon transformation of coordinates they transform according to the rule*

$$\text{(A.3.1)} \quad A^{k_1' k_2' \cdots k_p'}{}_{l_1' l_2' \cdots l_q'}(\mathbf{x}') = \left| \frac{\partial x^r}{\partial x^{s'}} \right| A^{k_1 k_2 \cdots k_p}{}_{l_1 l_2 \cdots l_q}(\mathbf{x}) \cdot \frac{\partial x^{k_1'}}{\partial x^{k_1}} \frac{\partial x^{k_2'}}{\partial x^{k_2}} \cdots \frac{\partial x^{k_p'}}{\partial x^{k_p}} \frac{\partial x^{l_1}}{\partial x^{l_1'}} \frac{\partial x^{l_2}}{\partial x^{l_2'}} \cdots \frac{\partial x^{l_q}}{\partial x^{l_q'}}$$

where $|\partial x^r/\partial x^{s'}| \equiv \det(\partial x^r/\partial x^{s'})$ *and* $x^{k'} = x^{k'}(x^1, x^2, x^3)$ *denote the coordinate transformation*[1] *between* x^k *and* $x^{k'}$. *The total number of indices* $(p + q)$ *is called the rank (or the order) of the tensor. When* $w = 0$ *the tensor is called absolute.*

Example 1 (*relative scalar*). In this case $p = q = 0$, i.e.,

$$\text{(A.3.2)} \qquad A'(\mathbf{x}') = |\partial x^r/\partial x^{s'}|^w A(\mathbf{x})$$

When $w = 0$, A is an *absolute scalar*. Temperature is an example of an absolute scalar.

Example 2 (*relative vectors*)

$$\text{(A.3.3)} \quad A^{k'}(\mathbf{x}') = |\partial x^r/\partial x^{s'}|^w A^k(\mathbf{x})\, \partial x^{k'}/\partial x^k \qquad \text{(relative contravariant vector)}$$

$$\text{(A.3.4)} \quad A_{k'}(\mathbf{x}') = |\partial x^r/\partial x^{s'}|^w A_k(\mathbf{x})\, \partial x^k/\partial x^{k'} \qquad \text{(relative covariant vector)}$$

For example, the differential vector dx^k is an absolute contravariant vector, since according to the chain rule

$$dx^{k'} = (\partial x^{k'}/\partial x^k)\, dx^k$$

which agrees with (A.3.3), with $w = 0$ and $A^k \equiv dx^k$. Similarly, the partial derivative of an absolute scalar is an absolute covariant vector, since

$$\partial \phi/\partial x^{k'} = (\partial \phi/\partial x^k)\, \partial x^k/\partial x^{k'}$$

which agrees with (A.3.4), with $w = 0$ and $A_k \equiv \partial \phi/\partial x^k$.

[1] Note that primed and unprimed indices denote different coordinates.

Example 3 (second-order tensors)

$$A^{k'l'}(\mathbf{x}') = |\partial x^r/\partial x^{s'}|^w A^{kl}(\mathbf{x})(\partial x^{k'}/\partial x^k)\ \partial x^{l'}/\partial x^l \qquad \text{(contravariant)} \tag{A.3.5}$$

$$A_{k'l'}(\mathbf{x}') = |\partial x^r/\partial x^{s'}|^w A_{kl}(\mathbf{x})(\partial x^k/\partial x^{k'})\ \partial x^l/\partial x^{l'} \qquad \text{(covariant)} \tag{A.3.6}$$

$$A^{k'}{}_{l'}(\mathbf{x}') = |\partial x^r/\partial x^{s'}|^w A^k{}_l(\mathbf{x})(\partial x^{k'}/\partial x^k)\ \partial x^l/\partial x^{l'} \qquad \text{(mixed)} \tag{A.3.7}$$

Examples of these tensors are g^{kl}, g_{kl} and $\delta^k{}_l$.

Addition. Two tensors of the same order and type may be added or subtracted to obtain another tensor of the same order and type, e.g.,

$$C^{kl}{}_m = A^{kl}{}_m + B^{kl}{}_m$$

Multiplication. The product of two tensors is obtained simply by multiplying their components, e.g.,

$$C^k{}_{lm} = A_{lm} B^k$$

The product has the order of the sum of the orders of the factors.

Contraction. By equating a superscript to a subscript in a tensor, we obtain a summation reducing the order of the original tensor by 2. This operation is called contraction, e.g.,

$$C^k{}_{lk}, \qquad C^k{}_{km}$$

Symmetrization. A tensor may be symmetrized by changing its r subscripts (and/or superscripts) in all possible ways, adding all such tensors, and dividing the result by $r!$. The symmetrization is indicated by placing the symmetrized indices within parentheses. The indices that are not symmetrized are separated by enclosing them in vertical bars, e.g.,

$$\begin{aligned} T_{(ij)} &= \frac{1}{2!}(T_{ij} + T_{ji}) \\ T^m{}_{(ij|k|l)} &\equiv \frac{1}{3!}(T^m{}_{ijkl} + T^m{}_{likj} + T^m{}_{jlki} + T^m{}_{jikl} + T^m{}_{ljki} + T^m{}_{ilkj}) \end{aligned} \tag{A.3.8}$$

Alternation. Alternation is done by exchanging a number of indices and by using a plus sign for an even permutation and a minus sign for an odd permutation. The sum divided by the factorial of the number of indices affected gives the alternation, which is indicated by placing the indices alternated within a bracket. Thus

$$\begin{aligned} T_{[ij]} &= \frac{1}{2!}(T_{ij} - T_{ji}) \\ T^m{}_{[ij|k|l]} &\equiv \frac{1}{3!}(T^m{}_{ijkl} + T^m{}_{likj} + T^m{}_{jlki} - T^m{}_{jikl} - T^m{}_{ljki} - T^m{}_{ilkj}) \end{aligned} \tag{A.3.9}$$

Raising and lowering indices. By use of the metric tensors g^{kl} and g_{kl} we can raise or lower the indices of a given tensor to obtain *associated tensors*, e.g.,

$$A^{k}_{\cdot l} = g^{km} A_{ml}, \qquad A_{l\cdot}^{k} = g^{km} A_{lm}, \qquad A^{kl} = g^{km} g^{ln} A_{mn}$$

A dot (or open space) is inserted in place of the displaced indices. In general, $A^{k}_{\cdot l} \neq A_{l}^{\cdot k}$. When A_{kl} is a symmetric tensor, then $A^{k}_{\cdot l} = A_{l}^{\cdot k}$, and the relative positions of indices in this case is unimportant.

It is not difficult to see that by first raising the indices of a tensor and then lowering them we obtain the same tensor.

A.4 PHYSICAL COMPONENTS

A vector is expressed in terms of its covariant and contravariant components as

$$\mathbf{u} = u^k \mathbf{g}_k = u_k \mathbf{g}^k \tag{A.4.1}$$

Since $\mathbf{g}_k$ and $\mathbf{g}^k$ are not, in general, of unit magnitudes, the u^k and u_k will not have the same physical dimension as $\mathbf{u}$. However, by decomposing $\mathbf{u}$ into its components along unit vectors $\mathbf{e}_k$ lying along the coordinate curves, we obtain the *physical components*, $u^{(k)}$ of $\mathbf{u}$:

$$\mathbf{u} = u^{(k)} \mathbf{e}_k \tag{A.4.2}$$

where $\mathbf{e}_k$ are the unit vectors defined by

$$\mathbf{e}_k = \mathbf{g}_k / (g_{\underline{kk}})^{1/2}$$

For an arbitrary triad of three vectors $\mathbf{e}_k$, $u^{(k)}$ as defined by (A.4.2) are also known as the *anholonomic* components of $\mathbf{u}$. By taking the scalar product of (A.4.2) by $\mathbf{g}^l$, we obtain the relations between the tensor and its physical components:

$$u^{(k)} = u^k (g_{\underline{kk}})^{1/2}, \qquad u^k = u^{(k)} / (g_{\underline{kk}})^{1/2} \tag{A.4.3}$$

By lowering the index of (A.4.3) we get the covariant components of $\mathbf{u}$:

$$u_k = \sum_l g_{kl} u^{(l)} / (g_{ll})^{1/2} \tag{A.4.4}$$

where a summation sign has been inserted in order to avoid the ambiguity in the order of summations, arising from the repetition of the index l more than twice.

The physical components of higher-order tensors are found by their relations to vectors and scalars. For example the stress tensor $t_l^{\ k}$ is related to tractions t^k by

$$t^k = t_l^{\ k} n^l \tag{A.4.5}$$

where t^k and n^l are contravariant vectors. We now replace t^k and n^l by their expressions in terms of their physical components as expressed by $(\text{A.4.3})_2$. This gives

(A.4.6) $$t^{(k)} = t_{(l)}{}^{(k)} n^{(l)}$$

where

(A.4.7) $$t_{(l)}{}^{(k)} \equiv t_l{}^k (g_{\underline{kk}}/g_{\underline{ll}})^{1/2}$$

are called the physical components of $t_l{}^k$.

When the curvilinear coordinates are orthogonal,

$$g^{\underline{kk}} = 1/g_{\underline{kk}}; \qquad g_{kl} = 0, \qquad k \neq l$$

and (A.4.7) may also be written as

(A.4.8) $$t_{(l)}{}^{(k)} = t_{kl}/(g_{\underline{kk}}\, g_{\underline{ll}})^{1/2} = t^{kl}(g_{\underline{kk}}\, g_{\underline{ll}})^{1/2} = t_l{}^k (g_{\underline{kk}})^{1/2}/(g_{\underline{ll}})^{1/2}$$

For nonorthogonal coordinates and/or nonsymmetrical tensors, other possibilities exist for the definition of physical components. (In this regard, see Eringen [1962, p. 438; 1971, Section 2.4].)

A.5 TENSOR CALCULUS

The partial derivative of $\mathbf{g}_k$ can be calculated by the use of (A.2.5):

$$\partial \mathbf{g}_k/\partial x^l = (\partial^2 z^m/\partial x^k\, \partial x^l)\mathbf{i}_m$$

Upon using (A.2.6) this may be written as

(A.5.1) $$\partial \mathbf{g}_k/\partial x^l = \begin{Bmatrix} n \\ kl \end{Bmatrix} \mathbf{g}_n$$

where $\begin{Bmatrix} n \\ kl \end{Bmatrix}$ are known as the *Christoffel symbols of the second kind* and are defined by

(A.5.2) $$\begin{Bmatrix} n \\ kl \end{Bmatrix} \equiv (\partial^2 z^m/\partial x^k\, \partial x^l)\, \partial x^n/\partial z^m$$

These and the Christoffel symbols of the first kind, $[kl, m]$, are of frequent occurrence in tensor calculus. By using (A.2.9) we can express these in the following forms:

(A.5.3) $$\begin{Bmatrix} n \\ kl \end{Bmatrix} = g^{mn}[kl, m]$$

(A.5.4) $$[kl, m] = \tfrac{1}{2}(g_{km,l} + g_{lm,k} - g_{kl,m})$$

The Christoffel symbols are *not* tensors. They are symmetrical with respect to indices:

(A.5.5) $$\begin{Bmatrix} n \\ kl \end{Bmatrix} = \begin{Bmatrix} n \\ lk \end{Bmatrix}, \qquad [kl, m] = [lk, m]$$

By use of (A.2.11) we also calculate

(A.5.6) $$\partial \mathbf{g}^k / \partial x^l = -\begin{Bmatrix} k \\ lm \end{Bmatrix} \mathbf{g}^m$$

The partial derivative of a vector $\mathbf{u}$ can now be calculated by

$$\frac{\partial \mathbf{u}}{\partial x^k} = \frac{\partial}{\partial x^k}(u^m \mathbf{g}_m) = \frac{\partial u^m}{\partial x^k} \mathbf{g}_m + u^m \frac{\partial \mathbf{g}_m}{\partial x^k} = \left(\frac{\partial u^m}{\partial x^k} + \begin{Bmatrix} m \\ kl \end{Bmatrix} u^l \right) \mathbf{g}_m$$

which may be written as

(A.5.7) $$\partial \mathbf{u} / \partial x^k = u^m{}_{;k} \, \mathbf{g}_m$$

thus defining the *covariant partial derivative* of a contravariant vector by

(A.5.8) $$u^m{}_{;k} \equiv u^m{}_{,k} + \begin{Bmatrix} m \\ kl \end{Bmatrix} u^l$$

Similarly, by differentiating $\mathbf{u} = u_m \mathbf{g}^m$ and using (A.5.6), we obtain

(A.5.9) $$\partial \mathbf{u} / \partial x^k = u_{m;k} \, \mathbf{g}^m$$

(A.5.10) $$u_{m;k} \equiv u_{m,k} - \begin{Bmatrix} l \\ mk \end{Bmatrix} u_l$$

In rectangular coordinates, the Christoffel symbols vanish, so that the covariant derivatives reduce to partial derivatives.

The covariant derivative of an absolute vector is an absolute second-order tensor.

The covariant partial derivatives of higher-order tensors are defined in a similar fashion. For example,

(A.5.11)
$$A^{kl}{}_{;m} \equiv A^{kl}{}_{,m} + \begin{Bmatrix} k \\ mn \end{Bmatrix} A^{nl} + \begin{Bmatrix} l \\ mn \end{Bmatrix} A^{kn}$$
$$A^k{}_{l;m} \equiv A^k{}_{l,m} - \begin{Bmatrix} n \\ lm \end{Bmatrix} A^k{}_n + \begin{Bmatrix} k \\ mn \end{Bmatrix} A^n{}_l$$
$$A_{kl;m} \equiv A_{kl,m} - \begin{Bmatrix} n \\ km \end{Bmatrix} A_{nl} - \begin{Bmatrix} n \\ lm \end{Bmatrix} A_{kn}$$

are third-order absolute tensors.

The covariant derivatives of relative tensors contain an extra term, for example,

$$\phi_{;k} = \phi_{,k} - w\begin{Bmatrix} r \\ kr \end{Bmatrix}\phi \qquad \text{(relative scalar)}$$

$$\text{(A.5.12)} \qquad u^m{}_{;k} = u^m{}_{,k} + \begin{Bmatrix} m \\ kl \end{Bmatrix}u^l - w\begin{Bmatrix} r \\ kr \end{Bmatrix}u^m \qquad \text{(relative vector)}$$

$$A^k{}_{l;m} = A^k{}_{l,m} - \begin{Bmatrix} n \\ lm \end{Bmatrix}A^k{}_n + \begin{Bmatrix} k \\ mn \end{Bmatrix}A^n{}_l - w\begin{Bmatrix} r \\ mr \end{Bmatrix}A^k{}_l \qquad \text{(relative tensor)}$$

If we take the covariant partial derivative of g_{kl}, g^{kl}, $\delta^k{}_l$, and $g \equiv \det g_{kl}$, we find that

$$\text{(A.5.13)} \qquad g_{kl;m} = 0, \qquad g^{kl}{}_{;m} = 0, \qquad \delta^k{}_{l;m} = 0, \qquad g_{;k} = 0$$

This is known as *Ricci's theorem.*

Since $\sqrt{g}$ is a relative scalar of weight 1 from $(\text{A.5.12})_1$ and $(\text{A.5.13})_4$, there follows the useful formula

$$\text{(A.5.14)} \qquad (\sqrt{g})_{,k} = \begin{Bmatrix} r \\ rk \end{Bmatrix}\sqrt{g}$$

The differential operators, the *gradient* of an absolute scalar ϕ and the *divergence* and *curl* of an absolute vector $\mathbf{u}$, are defined by

$$\text{(A.5.15)} \qquad \operatorname{grad}\phi \equiv \phi_{,k}\mathbf{g}^k, \qquad \operatorname{div}\mathbf{u} \equiv u^k{}_{;k}, \qquad \operatorname{curl}\mathbf{u} \equiv \varepsilon^{klm}u_{m;l}\,\mathbf{g}_k$$

where

$$\text{(A.5.16)} \qquad \varepsilon^{klm} \equiv e^{klm}/\sqrt{g}, \qquad \varepsilon_{klm} \equiv \sqrt{g}\,e_{klm}$$
$$e^{123} = e^{312} = e^{231} = -e^{213} = -e^{321} = -e^{132} = 1$$

and all other $e^{klm} = 0$. The alternating symbol e_{klm} is similarly defined. In passing, we note the important relation:

$$\text{(A.5.17)} \qquad \varepsilon_{pkl}\,\varepsilon^{qmn} = \begin{vmatrix} \delta_p{}^q & \delta_k{}^q & \delta_l{}^q \\ \delta_p{}^m & \delta_k{}^m & \delta_l{}^m \\ \delta_p{}^n & \delta_k{}^n & \delta_l{}^n \end{vmatrix}, \qquad \varepsilon_{pkl}\,\varepsilon^{pmn} = \delta_k{}^m\delta_l{}^n - \delta_l{}^m\delta_k{}^n$$
$$\varepsilon_{pkl}\,\varepsilon^{pkn} = 2\delta_l{}^n$$

Sometimes the operator ∇, defined by

$$\text{(A.5.18)} \qquad \nabla \equiv \mathbf{g}^k\,\partial/\partial x^k$$

is conveniently used in expressing (A.5.15). In fact, it can be seen that

$$\operatorname{grad} \phi = \nabla\phi = \mathbf{g}^k \,\partial\phi/\partial x^k$$

(A.5.19) $$\operatorname{div} \mathbf{u} = \nabla \cdot \mathbf{u} = \mathbf{g}^k \cdot \frac{\partial}{\partial x^k}(u^l \mathbf{g}_l) = \mathbf{g}^k \cdot \mathbf{g}_l u^l{}_{;k} = u^k{}_{;k}$$

$$\operatorname{curl} \mathbf{u} = \nabla \times \mathbf{u} = \mathbf{g}^k \times \frac{\partial}{\partial x^k}(u^l \mathbf{g}_l) = \mathbf{g}^k \times \mathbf{g}_l u^l{}_{;k} = \varepsilon^{klm} u_{l,k} \mathbf{g}_m$$

A convenient form of div **u** follows if we use (A.5.14):

(A.5.20) $$\operatorname{div} \mathbf{u} = (1/\sqrt{g})(\sqrt{g}\,u^k)_{,k}$$

The *Laplacian operator* ∇^2, in curvilinear coordinates, takes the form

(A.5.21) $$\nabla^2\phi = \nabla \cdot \nabla\phi = (g^{kl}\phi_{,l})_{;k} = g^{kl}(\phi_{,l})_{;k} = \frac{1}{\sqrt{g}} \frac{\partial}{\partial x^k}\left(\sqrt{g}\, g^{kl} \frac{\partial \phi}{\partial x^l}\right)$$

The *Green–Gauss and Stokes theorems* of vector analysis are expressed as

(A.5.22) $$\int_{\mathscr{V}-\sigma} \operatorname{div} \mathbf{u}\, dv + \int_\sigma [\mathbf{u}] \cdot \mathbf{n}\, da = \oint_{\mathscr{S}-\sigma} \mathbf{u} \cdot \mathbf{n}\, da \qquad \text{(Green–Gauss)}$$

(A.5.23) $$\int_{\mathscr{S}-\gamma} \operatorname{curl} \mathbf{A} \cdot \mathbf{n}\, da + \int_\gamma [\mathbf{A}] \cdot \mathbf{h}\, ds = \oint_{\mathscr{C}-\gamma} \mathbf{A} \cdot d\mathbf{p} \qquad \text{(Stokes)}$$

In (A.5.22) the first integral on the left is over the volume $\mathscr{V}$ excluding a two-sided discontinuity surface σ; the second integral is over σ across which **u** displays the *jump* [**u**]; and the integral on the right is over the surface $\mathscr{S}$ of $\mathscr{V}$ excluding the line of intersection σ with $\mathscr{S}$. The vector **n** is the exterior normal (Fig. A.5.1). Of course, the regularity conditions usual to this theorem are assumed for **u** over $\mathscr{V} - \sigma$ and $\mathscr{S} - \sigma$.

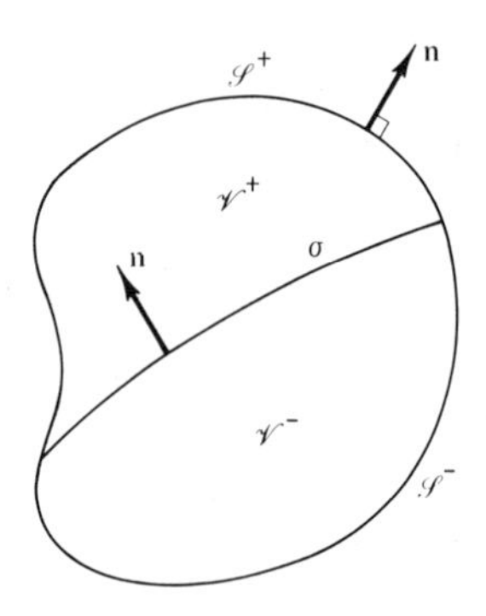

Fig. A.5.1 Region with discontinuity surface.

In (A.5.23) the first integral on the left is over a *two-sided* open surface $\mathscr{S}$ excluding a discontinuity line γ; the second integral is the line integral of the tangential component along γ of the jump $[\mathbf{A}]$; and the line integral on the right is over the boundary curve $\mathscr{C}$ of $\mathscr{S}$ excluding the points of intersection of γ with $\mathscr{C}$ (Fig. A.5.2). The vector $\mathbf{h}$ is the tangent vector to γ. Again,

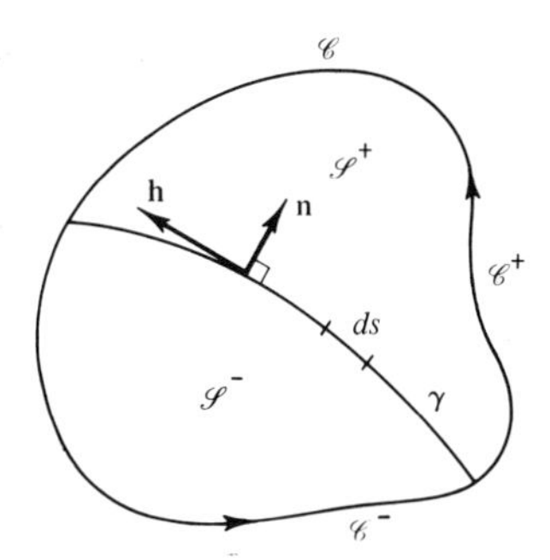

Fig. A.5.2 Surface with discontinuity line.

regularity and smoothness conditions for the validity of (A.5.23) are understood. For proof of these theorems, see Eringen [1971, Section 3.9].

In component forms, the preceding integral theorems read:

$$\text{(A.5.24)} \qquad \int_{\mathscr{V}-\sigma} \frac{1}{\sqrt{g}} \frac{\partial}{\partial x^k} (\sqrt{g}\, u^k)\, dv + \int_\sigma [u^k] n_k\, da = \oint_{\mathscr{S}-\sigma} u^k n_k\, da$$

$$\text{(A.5.25)} \qquad \int_{\mathscr{S}-\gamma} \varepsilon^{klm} A_{m,l}\, n_k\, da + \int_\gamma [A^k] h_k\, ds = \oint_{\mathscr{C}-\gamma} A_k\, dx^k$$

Following, we give the expressions of the differential operators in general orthogonal curvilinear coordinates, in cylindrical, and in spherical coordinates as they are frequently used.

Orthogonal curvilinear coordinates

$$ds^2 = g_{11}(dx^1)^2 + g_{22}(dx^2)^2 + g_{33}(dx^3)^2$$

$$g^{\underline{kk}} = \frac{1}{g_{\underline{kk}}} \qquad \mathbf{g}^k = g^{\underline{kk}} \mathbf{g}_k, \qquad g = g_{11} g_{22} g_{33}$$

$$\text{(A.5.26)} \qquad \begin{Bmatrix} l \\ \underline{kk} \end{Bmatrix} = -\frac{1}{2g_{\underline{ll}}} \frac{\partial g_{\underline{kk}}}{\partial x^l}, \qquad \begin{Bmatrix} \underline{k} \\ kl \end{Bmatrix} = \frac{\partial}{\partial x^l} [\log(g_{\underline{kk}})^{1/2}]$$

$$\begin{Bmatrix} \underline{k} \\ \underline{kk} \end{Bmatrix} = \frac{\partial}{\partial x^k} [\log(g_{\underline{kk}})^{1/2}], \qquad \begin{Bmatrix} k \\ lm \end{Bmatrix} = 0, \qquad k \neq l \neq m$$

(A.5.27)

$$\operatorname{grad}\phi = \frac{1}{(g_{11})^{1/2}}\frac{\partial\phi}{\partial x^1}\mathbf{e}_1 + \frac{1}{(g_{22})^{1/2}}\frac{\partial\phi}{\partial x^2}\mathbf{e}_2 + \frac{1}{(g_{33})^{1/2}}\frac{\partial\phi}{\partial x^3}\mathbf{e}_3$$

$$\operatorname{div}\mathbf{A} = (g_{11}g_{22}g_{33})^{-1/2}\left\{\frac{\partial}{\partial x^1}[(g_{22}g_{33})^{1/2}A^{(1)}] + \frac{\partial}{\partial x^2}[(g_{33}g_{11})^{1/2}A^{(2)}] + \frac{\partial}{\partial x^3}[(g_{11}g_{22})^{1/2}A^{(3)}]\right\}$$

$$\begin{aligned}\operatorname{curl}\mathbf{A} = {} & (g_{22}g_{33})^{-1/2}\left\{\frac{\partial}{\partial x^2}[(g_{33})^{1/2}A^{(3)}] - \frac{\partial}{\partial x^3}[(g_{22})^{1/2}A^{(2)}]\right\}\mathbf{e}_1 \\ & + (g_{33}g_{11})^{-1/2}\left\{\frac{\partial}{\partial x^3}[(g_{11})^{1/2}A^{(1)}] - \frac{\partial}{\partial x^1}[(g_{33})^{1/2}A^{(3)}]\right\}\mathbf{e}_2 \\ & + (g_{11}g_{22})^{-1/2}\left\{\frac{\partial}{\partial x^1}[(g_{22})^{1/2}A^{(2)}] - \frac{\partial}{\partial x^2}[(g_{11})^{1/2}A^{(1)}]\right\}\mathbf{e}_3\end{aligned}$$

$$\nabla^2\phi = (g_{11}g_{22}g_{33})^{-1/2}\left\{\frac{\partial}{\partial x^1}\left[\frac{(g_{22}g_{33})^{1/2}}{(g_{11})^{1/2}}\frac{\partial\phi}{\partial x^1}\right] + \frac{\partial}{\partial x^2}\left[\frac{(g_{33}g_{11})^{1/2}}{(g_{22})^{1/2}}\frac{\partial\phi}{\partial x^2}\right] + \frac{\partial}{\partial x^3}\left[\frac{(g_{11}g_{22})^{1/2}}{(g_{33})^{1/2}}\frac{\partial\phi}{\partial x^3}\right]\right\}$$

Cylindrical coordinates. The cylindrical coordinates (r, θ, z) are defined in terms of rectangular coordinates z^k [we use $x^1 \equiv r$, $x^2 \equiv \theta$, $x^3 \equiv z$, $A^{(1)} \equiv A_r$, $A^{(2)} \equiv A_\theta$, $A^{(3)} \equiv A_z$] by:

$$z^1 = r\cos\theta, \qquad z^2 = r\sin\theta, \qquad z^3 = z$$

$$ds^2 = dr^2 + r^2\,d\theta^2 + dz^2$$

$$g_{11} = g^{11} = g_{33} = g^{33} = 1, \qquad g_{22} = \frac{1}{g^{22}} = r^2$$

$$\begin{Bmatrix}2\\12\end{Bmatrix} = \begin{Bmatrix}2\\21\end{Bmatrix} = \frac{1}{r}, \qquad \begin{Bmatrix}1\\22\end{Bmatrix} = -r, \qquad \text{all other}\begin{Bmatrix}k\\lm\end{Bmatrix} = 0$$

(A.5.28)

$$\operatorname{grad}\phi = \frac{\partial\phi}{\partial r}\mathbf{e}_r + \frac{1}{r}\frac{\partial\phi}{\partial\theta}\mathbf{e}_\theta + \frac{\partial\phi}{\partial z}\mathbf{e}_z$$

$$\operatorname{div}\mathbf{A} = \frac{1}{r}\frac{\partial}{\partial r}(rA_r) + \frac{1}{r}\frac{\partial A_\theta}{\partial\theta} + \frac{\partial A_z}{\partial z}$$

$$\operatorname{curl}\mathbf{A} = \left(\frac{1}{r}\frac{\partial A_z}{\partial\theta} - \frac{\partial A_\theta}{\partial z}\right)\mathbf{e}_r + \left(\frac{\partial A_r}{\partial z} - \frac{\partial A_z}{\partial r}\right)\mathbf{e}_\theta + \left[\frac{1}{r}\frac{\partial}{\partial r}(rA_\theta) - \frac{1}{r}\frac{\partial A_r}{\partial\theta}\right]\mathbf{e}_z$$

$$\nabla^2\phi = \frac{\partial^2\phi}{\partial r^2} + \frac{1}{r}\frac{\partial\phi}{\partial r} + \frac{1}{r^2}\frac{\partial^2\phi}{\partial\theta^2} + \frac{\partial^2\phi}{\partial z^2}$$

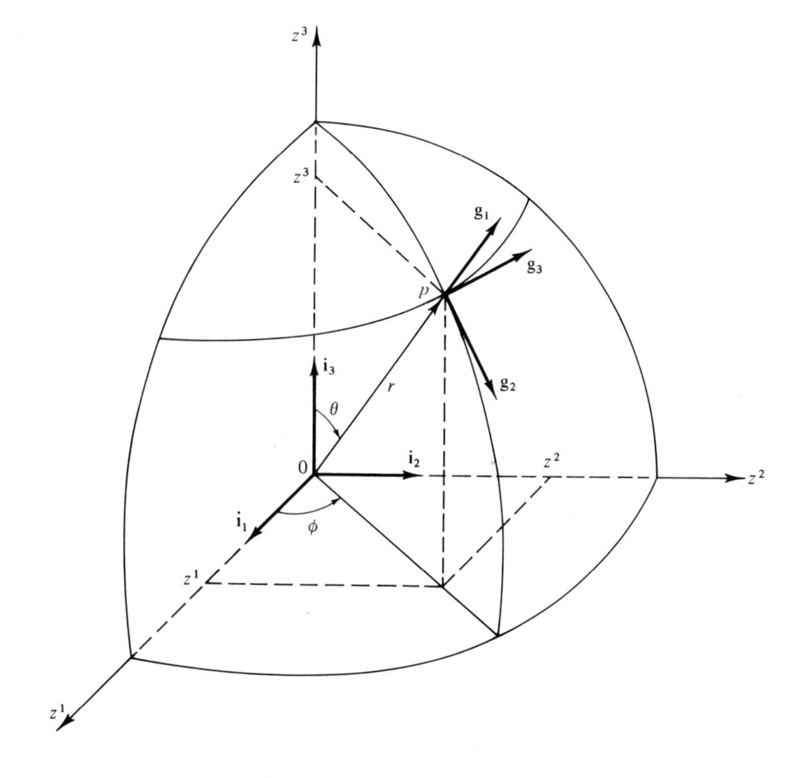

Fig. A.5.3 Spherical coordinates.

Spherical coordinates. The spherical coordinates (r, θ, φ) are defined by (Fig. A.5.3):

$$z^1 = r \sin\theta \cos\varphi, \qquad z^2 = r \sin\theta \sin\varphi, \qquad z^3 = r\cos\theta$$

$$g_{11} = g^{11} = 1, \qquad g_{22} = 1/g^{22} = r^2, \qquad g_{33} = 1/g^{33} = r^2 \sin^2\theta$$

$$g_{kl} = 0, \qquad k \neq l$$

$$\begin{Bmatrix} 1 \\ 22 \end{Bmatrix} = -r, \qquad \begin{Bmatrix} 1 \\ 33 \end{Bmatrix} = -r\sin^2\theta, \qquad \begin{Bmatrix} 2 \\ 12 \end{Bmatrix} = \frac{1}{r}$$

$$\begin{Bmatrix} 2 \\ 33 \end{Bmatrix} = -\sin\theta\cos\theta, \qquad \begin{Bmatrix} 3 \\ 13 \end{Bmatrix} = \frac{1}{r}$$

$$\begin{Bmatrix} 3 \\ 23 \end{Bmatrix} = \cot\theta, \qquad \text{all other} \begin{Bmatrix} k \\ lm \end{Bmatrix} = 0 \tag{A.5.29}$$

$$\operatorname{grad}\phi = \frac{\partial\phi}{\partial r}\mathbf{e}_r + \frac{1}{r}\frac{\partial\phi}{\partial\theta}\mathbf{e}_\theta + \frac{1}{r\sin\theta}\frac{\partial\phi}{\partial\varphi}\mathbf{e}_\varphi$$

$$\operatorname{div} \mathbf{A} = \frac{1}{r^2}\frac{\partial}{\partial r}(r^2 A_r) + \frac{1}{r \sin\theta}\frac{\partial}{\partial\theta}(A_\theta \sin\theta) + \frac{1}{r\sin\theta}\frac{\partial A_\varphi}{\partial\varphi}$$

$$\operatorname{curl} \mathbf{A} = \frac{1}{r\sin\theta}\left[\frac{\partial}{\partial\theta}(A_\varphi \sin\theta) - \frac{\partial A_\theta}{\partial\varphi}\right]\mathbf{e}_r + \left[\frac{1}{r\sin\theta}\frac{\partial A_r}{\partial\varphi} - \frac{1}{r}\frac{\partial}{\partial r}(rA_\varphi)\right]\mathbf{e}_\theta + \frac{1}{r}\left[\frac{\partial}{\partial r}(rA_\theta) - \frac{\partial A_r}{\partial\theta}\right]\mathbf{e}_\varphi$$

$$\nabla^2\phi = \frac{1}{r^2}\frac{\partial}{\partial r}\left(r^2\frac{\partial\phi}{\partial r}\right) + \frac{1}{r^2\sin\theta}\frac{\partial}{\partial\theta}\left(\sin\theta\frac{\partial\phi}{\partial\theta}\right) + \frac{1}{r^2\sin^2\theta}\frac{\partial^2\phi}{\partial\varphi^2}$$

A.6 THE RIEMANN–CHRISTOFFEL CURVATURE TENSOR

The covariant partial derivative of a vector A_k being a second-order tensor, we may take the covariant derivative of this tensor to obtain the second covariant derivative $A_{k;lm}$. The question arises whether the second-order covariant partial derivatives commute, i.e.,

$$A_{k;lm} \overset{?}{=} A_{k;ml}$$

To answer this question we form both sides of this expression and subtract, which gives

$$A_{k;lm} - A_{k;ml} = R^r{}_{klm} A_r \tag{A.6.1}$$

where the fourth-order absolute tensor $R^r{}_{klm}$, known as the *Riemann–Christoffel* tensor, is defined by

$$R^r{}_{klm} \equiv \begin{Bmatrix} r \\ km \end{Bmatrix}_{,l} - \begin{Bmatrix} r \\ kl \end{Bmatrix}_{,m} + \begin{Bmatrix} s \\ km \end{Bmatrix}\begin{Bmatrix} r \\ sl \end{Bmatrix} - \begin{Bmatrix} s \\ kl \end{Bmatrix}\begin{Bmatrix} r \\ sm \end{Bmatrix} \tag{A.6.2}$$

Then the following theorem is clear.

Theorem. *Cross-covariant partial derivatives of any vector are equal if and only if the Riemann–Christoffel tensor vanishes identically.*

By lowering the contravariant index, we obtain a fourth-order covariant tensor called the *curvature tensor*:

$$\begin{aligned} R_{klmn} = g_{kr} R^r{}_{lmn} &= \tfrac{1}{2}(g_{kn,lm} + g_{lm,kn} - g_{km,ln} - g_{ln,km}) \\ &\quad + g^{rs}([lm, s][kn, r] - [ln, s][km, r]) \end{aligned} \tag{A.6.3}$$

Extension of (A.6.1) to higher-order tensors is made in a similar way, for example,

$$A_{kl;mn} - A_{kl;nm} = A_{kr} R^r{}_{lmn} + A_{rl} R^r{}_{kmn} \tag{A.6.4}$$

Both $R^k{}_{lmn}$ and R_{klmn} possess certain symmetry conditions. It can be shown that in an N-dimensional space there are $N^2(N^2-1)/12$ independent components. In three dimensions, the independent components are R_{1212}, R_{1313}, R_{2323}, R_{1213}, R_{2123}, and R_{3132}. In two dimensions, the only component is R_{1212}.

By taking the covariant partial derivative of $R^k{}_{lmn}$ we can show that

$$\text{(A.6.5)} \qquad R^k{}_{lmn;r} + R^k{}_{lnr;m} + R^k{}_{lrm;n} = 0, \qquad R_{klmn;r} + R_{klnr;m} + R_{klrm;n} = 0$$

These are known as *Bianchi's* identities.

The Riemann–Christoffel tensor is a measure of the curvature of the space. When this tensor vanishes, the space is called *flat*. The Euclidean space is a flat space. For other accounts on the subject, we refer the reader to Schouten [1954] and Eringen [1971, Section 87].

A.7 TWO-POINT TENSOR FIELDS

Definition 1. *Quantities $A^k{}_K(\mathbf{x}, \mathbf{X})$ that transform like tensors with respect to indices k and K upon the transformation of each of the two sets of coordinates x^k and X^K are called two-point tensors.*

Thus if

$$\text{(A.7.1)} \qquad x^{k'} = x^{k'}(\mathbf{x}), \qquad X^{K'} = X^{K'}(\mathbf{X})$$

are differentiable coordinate transformations, and if

$$\text{(A.7.2)} \qquad A^{k'}{}_{K'}(\mathbf{x}', \mathbf{X}') = A^m{}_M(\mathbf{x}, \mathbf{X}) \frac{\partial x^{k'}}{\partial x^m} \frac{\partial X^M}{\partial X^{K'}}$$

then $A^k{}_K$ is an absolute two-point tensor field. We notice that the transformation (A.7.2) is like vector transformations with respect to each index. If $\mathbf{g}_k$ and $\mathbf{G}_K$ are the base vectors and $\mathbf{g}^k$ and $\mathbf{G}^K$ their reciprocals in the coordinates x^k and X^K, respectively, then $A^k{}_K$ are the components of the absolute tensor

$$\text{(A.7.3)} \qquad \mathbf{A}(\mathbf{x}, \mathbf{X}) = A^k{}_K(\mathbf{x}, \mathbf{X})\mathbf{g}_k(\mathbf{x})\mathbf{G}^K(\mathbf{X})$$

Examples of two-point tensor fields are *shifters* defined by

$$\text{(A.7.4)} \qquad g^k{}_K(\mathbf{x}, \mathbf{X}) \equiv \mathbf{g}^k(\mathbf{x}) \cdot \mathbf{G}_K(\mathbf{X}), \qquad g^K{}_k(\mathbf{x}, \mathbf{X}) \equiv G^K(\mathbf{X}) \cdot \mathbf{g}_k(\mathbf{x})$$

and deformation gradients

$$\text{(A.7.5)} \qquad x^k{}_{,K} \equiv \partial x^k/\partial X^K, \qquad X^K{}_{,k} \equiv \partial X^K/\partial x^k$$

The two-point tensor character of these quantities is easily shown. For example, by means of the chain rule of differentiation, we have

$$\frac{\partial x^{k'}}{\partial X^{K'}} = \frac{\partial x^{k'}}{\partial x^m} \frac{\partial x^m}{\partial X^M} \frac{\partial X^M}{\partial X^{K'}}$$

which has the form (A.7.2). Higher-order multiple-point tensor fields are similarly defined.

Definition 2. *The total covariant derivative of a two-point tensor field* $A^k{}_K(\mathbf{x}, \mathbf{X})$, *when* $\mathbf{x}$ *is related to* $\mathbf{X}$ *by a mapping* $\mathbf{x} = \mathbf{x}(\mathbf{X})$, *is defined by*

$$A^k{}_{K:L} \equiv A^k{}_{K;L} + A^k{}_{K;l}\, x^l{}_{,L} \tag{A.7.6}$$

where $A^k{}_{K;L}$ denote the covariant partial derivative with the metric G_{KL} used and $\mathbf{x}$ regarded fixed, and $A^k{}_{K;l}$ that with the metric g_{kl} used and $\mathbf{X}$ fixed, i.e.,

$$A^k{}_{K;L} = \frac{\partial A^k{}_K}{\partial X^L} - \begin{Bmatrix} M \\ LK \end{Bmatrix} A^k{}_M, \qquad A^k{}_{K;l} = \frac{\partial A^k{}_K}{\partial x^l} + \begin{Bmatrix} k \\ lm \end{Bmatrix} A^m{}_K$$

Therefore,

$$A^k{}_{K:L} = \frac{\partial A^k{}_K}{\partial X^L} - \begin{Bmatrix} M \\ LK \end{Bmatrix} A^k{}_M + \left(\frac{\partial A^k{}_K}{\partial x^l} + \begin{Bmatrix} k \\ ml \end{Bmatrix} A^m{}_K \right) \frac{\partial x^l}{\partial X^L} \tag{A.7.7}$$

We note that this result follows from differentiating (A.7.3) with respect to X^k and using expressions of the derivatives of $\mathbf{g}_k$ and $\mathbf{G}^K$ in the forms (A.5.1) and (A.5.6). This gives

$$\partial \mathbf{A}/\partial X^K = A^k{}_{L:K}\, \mathbf{g}_k\, \mathbf{G}^L \tag{A.7.8}$$

Applying (A.7.7) to $x^k{}_{,K}(\mathbf{X})$ and noting the absence of $\mathbf{x}$ from the argument of $x^k{}_{,K}$, we obtain

$$(x^k{}_{,K})_{:L} = \frac{\partial^2 x^k}{\partial X^L \partial X^K} - \begin{Bmatrix} M \\ LK \end{Bmatrix} \frac{\partial x^k}{\partial X^M} + \begin{Bmatrix} k \\ lm \end{Bmatrix} \frac{\partial x^m}{\partial X^K} \frac{\partial x^l}{\partial X^L} \tag{A.7.9}$$

We note that (A.7.6) generalizes the total derivative of a scalar function of two variables $\phi(x, X)$ subject to $x = x(X)$, i.e.,

$$\frac{d\phi}{dX} = \frac{\partial \phi}{\partial x} \frac{\partial x}{\partial X} + \frac{\partial \phi}{\partial X}$$

Total covariant differentiation obeys the formal rules of covariant partial differentiation, e.g.,

$$\begin{aligned} g^k{}_{K:M} &= G_{KL:m} = g_{kl:M} = 0 \\ (A^k{}_K B^L{}_l)_{:M} &= A^k{}_{K:M} B^L{}_l + A^k{}_K B^L{}_{l:M} \\ (A^k{}_K + B^k{}_K)_{:M} &= A^k{}_{K:M} + B^k{}_{K:M} \end{aligned} \tag{A.7.10}$$

For other accounts, see Eringen [1971, Part 1].

APPENDIX B

Quasilinear System of Hyperbolic Equations with Two Independent Variables

B.1 CHARACTERISTICS

A quasilinear, homogeneous system of partial differential equations of first order with n dependent variables $(u_1, u_2, \ldots, u_n)$ and two independent variables (x, t) is given as

$$a_{kl} \frac{\partial u_l}{\partial x} + \frac{\partial u_k}{\partial t} = 0, \qquad k, l = 1, 2, \ldots, n \tag{B.1.1}$$

where

$$a_{kl} = a_{kl}(x, t, u_1, u_2, \ldots, u_n)$$

are given functions of their arguments. If we define a column vector $\mathbf{U}$ by

$$\mathbf{U} = \begin{bmatrix} u_1 \\ u_2 \\ \vdots \\ u_n \end{bmatrix}$$

and an $n \times n$ matrix $\mathbf{A}$ by

$$\mathbf{A} = \mathbf{A}(x, t, \mathbf{U}) = \begin{bmatrix} a_{11} & a_{12} & \cdots & a_{1n} \\ a_{21} & a_{22} & \cdots & a_{2n} \\ \vdots & \vdots & & \vdots \\ a_{n1} & a_{n2} & \cdots & a_{nn} \end{bmatrix}$$

then the system (B.1.1) may be conveniently represented in the following matrix form:

(B.1.2)
$$\mathbf{A}\frac{\partial \mathbf{U}}{\partial x} + \frac{\partial \mathbf{U}}{\partial t} = \mathbf{0}$$

In the subsequent analysis we assume that $\mathbf{A} = \mathbf{A}(\mathbf{U})$. The investigation of this case suffices for our investigation of wave propagation in homogeneous elastic solids.

Consider a curve C in the (x, t)-plane and specify the value of $\mathbf{U}$ along this curve by a function $\mathbf{U} = \mathbf{U}(s)$, where s is the arc length along the curve

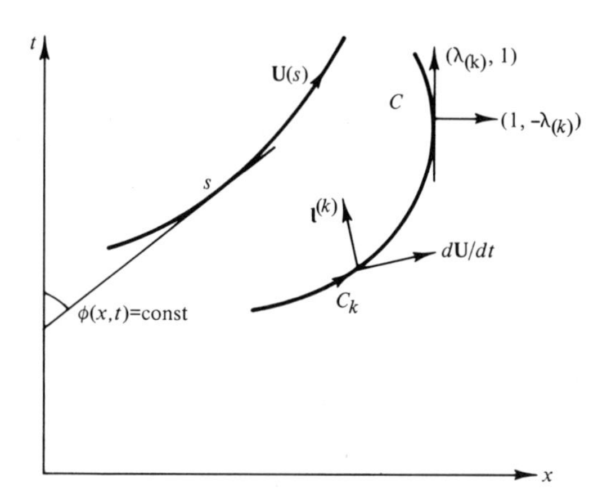

Fig. B.1.1 Variation along characteristics.

measured from a fixed point (Fig. B.1.1). This curve may be represented either in a parametric form

$$x = x(s), \qquad t = t(s); \qquad x = x(t)$$

or as

$$\phi(x, t) = \text{const}$$

The variation of $\mathbf{U}$ along this curve is given by

$$d\mathbf{U} = (\partial \mathbf{U}/\partial x)\, dx + (\partial \mathbf{U}/\partial t)\, dt$$

with

$$(\partial\phi/\partial x)\,dx + (\partial\phi/\partial t)\,dt = 0$$

or

$$dx = \dot{x}\,ds, \qquad dt = \dot{t}\,ds; \qquad \text{or} \qquad dx = (dx/dt)\,dt$$

Denoting the slope of the curve by

$$\lambda = \frac{dx}{dt} = -\frac{\partial\phi/\partial t}{\partial\phi/\partial x} = \frac{\dot{x}}{\dot{t}}$$

we get

$$d\mathbf{U} = \left(\frac{\partial \mathbf{U}}{\partial t} + \lambda\,\frac{\partial \mathbf{U}}{\partial x}\right) dt \qquad \text{on} \quad C \tag{B.1.3}$$

Introducing (B.1.2) into (B.1.3), we obtain

$$(\mathbf{A} - \lambda\mathbf{I})\,\partial\mathbf{U}/\partial x = -d\mathbf{U}/dt \qquad \text{on} \quad C \tag{B.1.4}$$

The right-hand side of (B.1.4) is known from the given data on the curve C. Therefore, if the determinant

$$Q \equiv \det(\mathbf{A} - \lambda\mathbf{I})$$

does not vanish along C, the derivatives $\partial\mathbf{U}/\partial x$ can be determined uniquely from (B.1.4) and then $\partial\mathbf{U}/\partial t$ from (B.1.2). From the given values of $\mathbf{U}$ and its computed first-order derivatives on C, we can determine all derivatives of higher order and obtain a Taylor series expansion for the solution of (B.1.2) in the neighborhood of C. However if $\mathbf{Q}$ vanishes, i.e.,

$$Q = \det(\mathbf{A} - \lambda\mathbf{I}) = 0 \tag{B.1.5}$$

Eq. (B.1.4) cannot have a unique solution along C. Such a curve is called a *characteristic* of the system (B.1.2). Equation (B.1.5) is an nth degree algebraic equation for λ, possessing n roots $\lambda_{(1)}, \lambda_{(2)}, \ldots, \lambda_{(n)}$, which are known as the *eigenvalues* of the matrix $\mathbf{A}$. If all $\lambda_{(k)}$ are real and distinct, the system (B.1.2) is called *totally hyperbolic*. Henceforth we consider only such systems. Therefore, there are n real families of characteristic curves in a totally hyperbolic system with n dependent variables determined upon the integration of equations

$$dx/dt = \lambda_{(k)} \qquad \text{or} \qquad \lambda_{(k)}(\partial\phi/\partial x) + (\partial\phi/\partial t) = 0, \qquad k = 1, 2, \ldots, n \tag{B.1.6}$$

It is clear that the discontinuities in the derivatives of $\mathbf{U}$ can take place only on a characteristic curve, since the derivatives are uniquely determined on any other curve. Thus if the analytic character of the solution to (B.1.2) differs in some regions in the (x, t)-plane, these regions must be separated by a characteristic curve.

For linear equations in which $\mathbf{A} = \mathbf{A}(x, t)$, Eq. (B.1.6) can be integrated immediately. However, in quasilinear systems $\lambda_{(k)}$'s contain unknown functions $u_1, u_2, \ldots, u_n$ so that the characteristic curves cannot be obtained without the solution vector $\mathbf{U}$.

If we identify x as a spatial and t as a temporal independent variable we see that λ's are of the dimension of velocity. Hence a real characteristic curve in the (x,t)-plane represents a wave front $x = \text{const}$, in physical space, propagating with a speed λ in the x-direction.

It is now clear from (B.1.4) that $\mathbf{U}$ cannot be specified arbitrarily on a characteristic curve C_k. Since the determinant $\mathbf{Q}$ vanishes, from a well-known theorem of linear algebra it follows that a solution to (B.1.4) exists, even if not uniquely, if and only if $d\mathbf{U}/dt$ is orthogonal to the homogeneous solution of the adjoint equation. See, for instance, Friedmann [1956, Section 45]. Thus if $\mathbf{l}^{(k)}$ is an eigenvector of $\mathbf{A}^{\mathrm{T}}$ corresponding to $\lambda_{(k)}$, i.e.,[1]

$$(\mathbf{A}^{\mathrm{T}} - \lambda_{(k)}\mathbf{I})\mathbf{l}^{(k)} = \mathbf{0} \tag{B.1.7}$$

where $\mathbf{A}^{\mathrm{T}}$ is the transpose of $\mathbf{A}$, and

$$\mathbf{l}^{(k)} = \begin{bmatrix} l_1^{(k)} \\ \vdots \\ l_n^{(k)} \end{bmatrix}$$

then the following conditions are to be satisfied:

$$\mathbf{l}^{(k)} \cdot d\mathbf{U} = 0 \qquad \text{on} \quad C_k, \qquad k = 1, 2, \ldots, n \tag{B.1.8}$$

This is a differential form for $du_1, \ldots, du_n$ on the characteristic curves C_k corresponding to the eigenvalue, or *characteristic speed*, $\lambda_{(k)}$. The coefficient $\mathbf{l}^{(k)}$ is a function of $\mathbf{U}$ only as a consequence of our assumption on the matrix $\mathbf{A}$. If this form is integrable, then (B.1.8) results in an expression of the form

$$J^{(k)} = J^{(k)}(u_1, \ldots, u_n) = J^{(k)}(\mathbf{U})$$

which is called a *Riemann invariant* of the system (B.1.2) since this function is constant on the characteristic curves C_k belonging to the family associated with $\lambda_{(k)}$. For integrable forms, we have

$$l_i^{(k)} = \partial J^{(k)}/\partial u_i \qquad \text{or} \qquad \mathbf{l}^{(k)} = \partial J^{(k)}/\partial \mathbf{U}$$

Therefore, the components of $\mathbf{l}^{(k)}$ must satisfy the compatibility conditions

$$\partial l_i^{(k)}/\partial u_j = \partial l_j^{(k)}/\partial u_i$$

[1] Note that $\mathbf{l}^{(k)\mathrm{T}}$ is a left eigenvector of $\mathbf{A}$, namely,

$$\mathbf{l}^{(k)\mathrm{T}}(\mathbf{A} - \lambda_{(k)}\mathbf{I}) = \mathbf{0}, \qquad \mathbf{l}^{(k)\mathrm{T}} = [l_1^{(k)}, \ldots, l_n^{(k)}]$$

(cf. Eringen [1971, Section 1.11].)

which impose $n(n-1)/2$ conditions on n numbers $\lambda_i^{(k)}$. Since homogeneous Eqs. (B.1.7) comprise $(n-1)$ conditions, n numbers $\lambda_i^{(k)}$ must satisfy $(n-1)(n+2)/2$ conditions. This shows that, in general, it is impossible to find Riemann invariants, except for the case $n = 2$, which are indeed very handy in the solution of boundary and initial value problems in hyperbolic systems.

The hyperbolic system (B.1.2) may be replaced by a linearly equivalent one by considering derivatives along characteristics. If we multiply (B.1.2) by $\mathbf{l}^{(k)}$ and use (B.1.7), we obtain

$$\text{(B.1.9)} \qquad \mathbf{l}^{(k)} \cdot \left(\frac{\partial \mathbf{U}}{\partial t} + \lambda_{(k)} \frac{\partial \mathbf{U}}{\partial x} \right) = 0, \qquad k = 1, 2, \ldots, n$$

Since $(\lambda_{(k)}, 1)$ is a vector along the characteristic curve C_k [cf. (B.1.6) and Fig. B.1.1], the differentiation in the kth characteristic direction is simply

$$\frac{d}{dk} = \frac{\partial}{\partial t} + \lambda_{(k)} \frac{\partial}{\partial x}$$

Therefore, (B.1.9) becomes

$$\text{(B.1.10)} \qquad \mathbf{l}^{(k)} \cdot d\mathbf{U}/dk = 0$$

which could also be written directly from (B.1.8). This simple formula may lead us to define the *generalized Riemann invariants*, which can be calculated in a class of problems of practical importance. To this end we first consider the right eigenvectors $\mathbf{r}^{(k)}$ of the operator $\mathbf{A}(\mathbf{U})$, which satisfy

$$\text{(B.1.11)} \qquad (\mathbf{A} - \lambda_{(k)} \mathbf{I})\mathbf{r}^{(k)} = \mathbf{0}$$

Following Lax [1957] we define the generalized Riemann invariants $J^{(k)} = J^{(k)}(\mathbf{U})$ as functions satisfying the relations

$$\text{(B.1.12)} \qquad \mathbf{r}^{(k)} \cdot \operatorname{grad} J^{(k)} = 0$$

where

$$\operatorname{grad} J^{(k)} = \begin{bmatrix} \partial J^{(k)}/\partial u_1 \\ \vdots \\ \partial J^{(k)}/\partial u_n \end{bmatrix} = \frac{\partial J^{(k)}}{\partial \mathbf{U}}$$

Since $\mathbf{U}$ constitutes an n dimensional vector space and the vectors grad $J^{(k)}$ are orthogonal to a single vector $\mathbf{r}^{(k)}$, they are in the orthogonal complement of $\mathbf{r}^{(k)}$ with dimensionality $n-1$. Therefore, there are exactly $n-1$ linearly independent gradient vectors, and accordingly $n-1$ independent k-Riemann invariants $J_{(l)}^{(k)}$, $l = 1, 2, \ldots, n-1$, corresponding to the kth eigenvalue $\lambda_{(k)}$.

But the right and left eigenvectors of a matrix form a biorthogonal system, that is,

$$\mathbf{l}^{(i)} \cdot \mathbf{r}^{(k)} = 0, \qquad \text{for} \quad i \neq k$$

Thus the $n-1$ vectors $\mathbf{l}^{(i)}$ $(i \neq k)$ are in the same manifold as the gradients of k-Riemann invariants. Hence we can express $\mathbf{l}^{(i)}$ as a linear combination of grad $J^{(k)}_{(l)}$:

$$\mathbf{l}^{(i)} = \sum_{j=0}^{n-1} b_{ij} \operatorname{grad} J^{(k)}_{(j)}, \qquad i = 1, \ldots, n, \quad i \neq k \tag{B.1.13}$$

and use (B.1.13) in (B.1.10) to get

$$\begin{aligned} \sum_{j=1}^{n-1} b_{ij} \operatorname{grad} J^{(k)}_{(j)} \cdot \frac{d\mathbf{U}}{di} &= \sum_{j=1}^{n-1} b_{ij} \frac{\partial J^{(k)}_{(j)}}{\partial u_l} \frac{du_l}{di} \\ &= \sum_{j=1}^{n-1} b_{ij} \frac{dJ^{(k)}_{(j)}}{di} = 0, \qquad i = 1, \ldots, n, \quad i \neq k \end{aligned} \tag{B.1.14}$$

For a given solution $\mathbf{U}$, the coefficients in (B.1.14) are given functions. The last of (B.1.14) is a linear hyperbolic system of $n-1$ equations for $n-1$, k-Riemann invariants $J^{(k)}_{(1)}, J^{(k)}_{(2)}, \ldots, J^{(k)}_{(n-1)}$, and the directional derivatives are evaluated along characteristic curves C_i with $i \neq k$. Therefore, the curve C_k itself is no longer a characteristic curve for the system (B.1.14). From this result we deduce a very important property of the generalized Riemann invariants: *k-Riemann invariants and their derivatives cannot suffer a discontinuity on the characteristic curve C_k, and hence they can be continued across C_k smoothly.*

In the following section we utilize these results to investigate a special class of solutions, called *simple waves*, of the system (B.1.2). These solutions have important practical applications.

B.2 THE SIMPLE WAVE SOLUTIONS

Suppose that in the (x, t)-plane there exists a region of constant state R_0 in which all components of the vector $\mathbf{U}$ are given constants, viz., $\mathbf{U} = \mathbf{U}_0$. Since the matrix $\mathbf{A}$ and all of its eigenvalues $\lambda_{(k)}$ are then constant in R_0, it follows that each of the n families of characteristics consists of parallel straight lines in R_0. Evidently all of the generalized Riemann invariants $J^{(k)}_{(l)}$ $(k = 1, 2, \ldots, n;\ l = 1, 2, \ldots, n-1)$ are constant by definition in the region of constant state R_0. Consider now a region R in the (x, t)-plane adjacent to R_0, in which $\mathbf{U}$ differs from $\mathbf{U}_0$ in its analytic character. Since these two states can only be separated by a characteristic in the (x, t)-plane, we can imagine only *n adjacent states* to the constant state R_0, the boundary of each one being formed by one member of the n families of straight characteristics. Consider now one of the characteristics C_k corresponding to the eigenvalue $\lambda_{(k)}$ (Fig. B.2.1). The generalized Riemann invariants $J_{(l)}^{(k)}$ associated

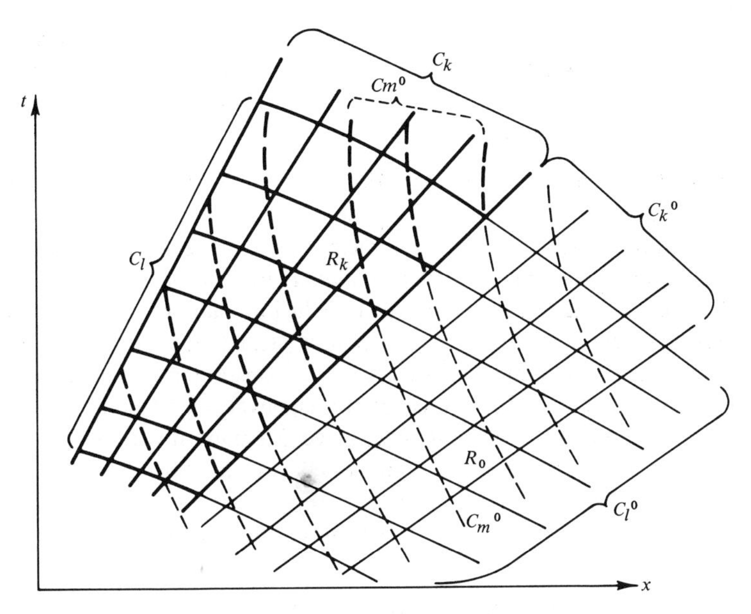

Fig. B.2.1 k-simple waves (wavelets).

with $\lambda_{(k)}$ are constant like all the other dependent variables in R_0. Hence they are naturally constant on the characteristic C_k separating the regions R_k and R_0. However, according to the discussion given in the last paragraph of the preceding section, all k-Riemann invariants associated with the characteristic family C_k can be continued smoothly across the curves C_k, that is, neither they nor their derivatives can suffer a discontinuity across the curves C_k. Therefore, k-Riemann invariants on the other side of the boundary characteristic C_k are also constant, constants being the same in both regions R_0 and R_k. Of course, the generalized Riemann invariants associated with the remaining characteristic families do not enjoy, in general, this important property.

The constancy of k-Riemann invariants in R_k now implies that the solution $u_1, u_2, \ldots, u_n$ to the system (B.1.2) satisfies $n-1$ relations such as

$$J_{(1)}^{(k)}(u_1, u_2, \ldots, u_n) = \text{const}, \qquad J_{(2)}^{(k)}(u_1, u_2, \ldots, u_n) = \text{const}, \ldots$$
$$J_{(n-1)}^{(k)}(u_1, u_2, \ldots, u_n) = \text{const} \tag{B.2.1}$$

whence we deduce that all u_m are expressible as a function of only one variable, say u_1, selected arbitrarily. Here we may remind the reader that if we could determine $J^{(k)}_{(l)}$, $l = 1, 2, \ldots, n-1$, the constants in (B.2.1) are found as the values taken on by the generalized Riemann invariants when $\mathbf{U} = \mathbf{U}_0$.

Instead of expressing all components of $\mathbf{U}$ in terms of any arbitrary one, we may prefer to write all u_m as functions of a single dependent variable $\alpha_{(k)}$ as

$$u_m = u_m(\alpha_{(k)}), \qquad \alpha_{(k)} = \alpha_{(k)}(x, t)$$

This provides freedom in the selection of the most convenient variable depending on the particular nature of the problem under consideration.

A solution $\mathbf{U}$ in which all components are expressed as functions of a single dependent variable is called a *simple wave solution* to the system (B.1.2). Thus, we have the important conclusion that the *solution adjacent to a constant state is a simple wave.* It is now clear that it is possible to construct n simple wave solutions adjacent to a constant state each of which corresponds to a particular choice of the eigenvalue and consequently the characteristic family and relevant generalized Riemann invariants. Now consider one of these simple wave regions, say R_k, and try to construct the k-simple wave solution valid in a region separated from the region of the constant state by a characteristic C_k. To this end let us resort to Eqs. (B.1.10) along characteristics in the region R_k:

$$\mathbf{l}^{(i)} \cdot d\mathbf{U}/di = 0, \qquad i = 1, 2, \ldots, n$$

We have shown that these equations reduce to (B.1.14) for $i \neq k$, and they are satisfied identically in the case of a k-simple wave, since all $J^{(k)}_{(l)}$ are constant everywhere and consequently along characteristics C_i $(i \neq k)$. Therefore, we are left with the equation along the characteristics C_k only:

$$\mathbf{l}^{(k)} \cdot d\mathbf{U}/dk = 0$$

Since all u_l are functions of $\alpha_{(k)}$, then $\mathbf{U} = \mathbf{U}(\alpha_{(k)})$, and we have for *every member* of the k-characteristic family

$$\mathbf{l}^{(k)} \cdot \frac{d\mathbf{U}}{d\alpha_{(k)}} \frac{d\alpha_{(k)}}{dk} = 0 \tag{B.2.2}$$

We deduce that $\alpha_{(k)}$ is constant on the characteristic curves C_k of the k-simple wave. Of course, constants associated with every characteristic will be different in general. Therefore, we reach our first important conclusion: *The solution* $\mathbf{U}$, *and hence all* u_k, *are constant along characteristic curves* C_k *in a k-simple wave.* Since eigenvalues λ are functions of $\mathbf{U}$ only, by (B.1.5) and (B.1.11) we see that $\lambda_{(k)}$ is constant along a characteristic curve C_k. This leads to our second important conclusion: *The characteristic curves C_k in a k-simple wave are all straight lines.* These straight lines are called *wavelets.* In general, they are not parallel to each other. Of course, the other $n-1$ families of characteristics are usually curved. A sketch for a k-simple wave region is shown in Fig. B.2.1.

We now try to obtain the set of k-Riemann invariants $J_{(l)}^{(k)}$ for a k-simple wave. To this end consider (B.1.4) on characteristic curves C_k. Since $\mathbf{U}$ is constant on these curves, the right-hand side of (B.1.4) vanishes, and if we multiply this equation by $d\alpha_{(k)}$ and write

$$(\partial \mathbf{U}/\partial x)\, d\alpha_{(k)} = (d\mathbf{U}/d\alpha_{(k)})\,(\partial \alpha_{(k)}/\partial x)\, d\alpha_{(k)} = d\mathbf{U}\; \partial \alpha_{(k)}/\partial x$$

we get

$$(\mathbf{A} - \lambda_{(k)}\mathbf{I})\, d\mathbf{U}\; \partial \alpha_{(k)}/\partial x = 0 \tag{B.2.3}$$

for the increments $d\mathbf{U}$ in the dependent variable $\mathbf{U}$ on passing from one characteristic to another infinitesimally close one. Clearly, $d\mathbf{U}$ is in the direction of the right eigenvector $\mathbf{r}^{(k)}(\mathbf{U})$ [cf. (B.1.11)]. Hence we may write

$$du_1/r_1^{(k)} = du_2/r_2^{(k)} = \cdots = du_n/r_n^{(k)} \tag{B.2.4}$$

Upon integrating this set of equations we arrive at the expressions of $n-1$ k-Riemann invariants:

$$J_{(l)}^{(k)}(u_1, \ldots, u_n) = a_{(l)}, \qquad l = 1, 2, \ldots, n-1 \tag{B.2.5}$$

where $a_{(l)}$ are $n-1$ arbitrary constants which are determined from the known values $u_{01}, u_{02}, \ldots, u_{0n}$, in the constant state R_0 as

$$a_{(l)} = J_{(l)}^{(k)}(u_{01}, u_{02}, \ldots, u_{0n})$$

That the set $J_{(l)}^{(k)}$ forms k-Riemann invariants follows from their total differential

$$(\partial J_{(l)}^{(k)}/\partial u_m)\, du_m = 0$$

which, in view of (B.2.4), implies that they satisfy (B.1.12). The system (B.2.5) helps to determine $n-1$ dependent variables u_l in terms of one arbitrarily selected u_i, say u_1 (or all u_l in terms of a single parameter $\alpha_{(k)}$). That leaves only one variable to be determined: the values of u_1 (or $\alpha_{(k)}$) which are assigned to each member of the characteristic family C_k. This can be accomplished, for instance, by specifying a boundary condition

$$u_1(0, t) = \bar{u}_1(t) \qquad \text{or} \qquad \alpha_{(k)}(0, t) = \bar{\alpha}_{(k)}(t) \tag{B.2.6}$$

With this, $\lambda_{(k)}$ is determined completely at the points of the time axis. The characteristics passing through these points may therefore be drawn covering the whole k-simple wave region (Fig. B.2.2). The value $\bar{\alpha}_{(k)}(\bar{t})$ of α_k at a point $\bar{t}$

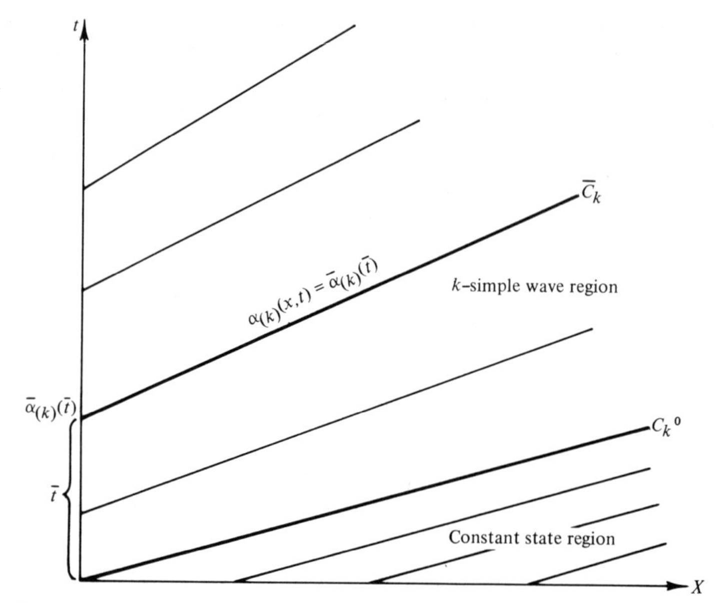

Fig. B.2.2 k-simple wave region near constant state.

is carried away by the characteristic $\bar{C}_k$:

(B.2.7) $$x = \lambda_{(k)}(\bar{\alpha}_{(k)})(t - \bar{t}) = \lambda_{(k)}(\bar{t})(t - \bar{t})$$

passing through this point. Since $\alpha_{(k)}$ is constant along $\bar{C}_k$ it will always be equal to $\bar{\alpha}_{(k)}$ on this particular $\bar{C}_k$, i.e.,

(B.2.8) $$\alpha_{(k)}(x, t) = \bar{\alpha}_{(k)}(\bar{t})$$

If we invert relation (B.2.7) we get

$$\bar{t} = \bar{t}(x, t)$$

so that from (B.2.8) there follows

(B.2.9) $$\alpha_{(k)}(x, t) = \bar{\alpha}_{(k)}[\bar{t}(x, t)]$$

The solution is therefore given by

(B.2.10) $$\mathbf{U} = \mathbf{U}(x, t) = \mathbf{U}(\alpha_{(k)}) = \mathbf{U}[\alpha_{(k)}(x, t)]$$

In Fig. B.2.2 it is assumed that $\bar{t} = 0$ corresponds to the boundary or leading characteristics.

Simple wave solutions are confined to regions that do not contain the points of intersections of corresponding characteristics. For the case of intersecting characteristics, the solution at intersection points may not be unique, since it may acquire different values on different characteristics. Finite discontinuities or jumps in the solution itself may therefore occur at such points, corresponding to a shock formation.

Note also that simple wave solutions do not allow reflections from the boundaries of the region, since the reflected wave violates the underlying requirement that simple waves can only be adjacent to regions of constant state in the (x, t)-plane.

B.3 RIEMANN PROBLEM

One of the fundamental boundary (or initial) value problems for hyperbolic systems is known as the Riemann problem. This problem is formulated in the following way: Find the solution of the system (B.1.2) for $x \geq 0$, under the boundary condition

$$\mathbf{U}(0, t) = \begin{cases} \mathbf{U}_0, & \text{for} \quad t \leq 0 \\ \mathbf{U}_1, & \text{for} \quad t > 0 \end{cases} \tag{B.3.1}$$

where $\mathbf{U}_0$ and $\mathbf{U}_1$ are given constant vectors. The solution to this problem is essentially a *weak solution*, since it is concerned with a boundary discontinuity in the dependent variables. We first prove that the solution, if it is unique, is a function of x/t only. Suppose that $\mathbf{U} = \mathbf{U}(x, t)$ is a solution to the boundary value problem proposed above. Then if K is a positive constant, the function defined by

$$\mathbf{U}_K = \mathbf{U}(Kx, Kt)$$

also satisfies the differential equations and the boundary condition. If we assume that the solution is unique, then we obtain

$$\mathbf{U}(x, t) = \mathbf{U}(Kx, Kt) \tag{B.3.2}$$

This relation is true if and only if

$$\mathbf{U} = \mathbf{U}(x/t) \tag{B.3.3}$$

since (B.3.2) implies that $\mathbf{U}$ is a homogeneous function of x, t of degree zero.

Next we see that the solution $\mathbf{U}$ is constant on the straight lines $x/t = \text{const}$ emerging from the origin of the (x,t)-axes. Therefore, these lines must be characteristics in simple wave regions (Fig. B.3.1). Such a simple wave is called a *centered simple wave*, and it is adjacent to some constant state regions in the (x,t)-plane. If it is known beforehand that a centered k-simple wave connects the given constant states U_l and U_r for $x/t \leq a$ and $x/t \geq b$, respectively, then these constant states must be subject to some restrictions arising from the fact that their k-Riemann invariants are equal and

$$\lambda_{(k)}(\mathbf{U}_l) < \lambda_{(k)}(\mathbf{U}_r) \tag{B.3.4}$$

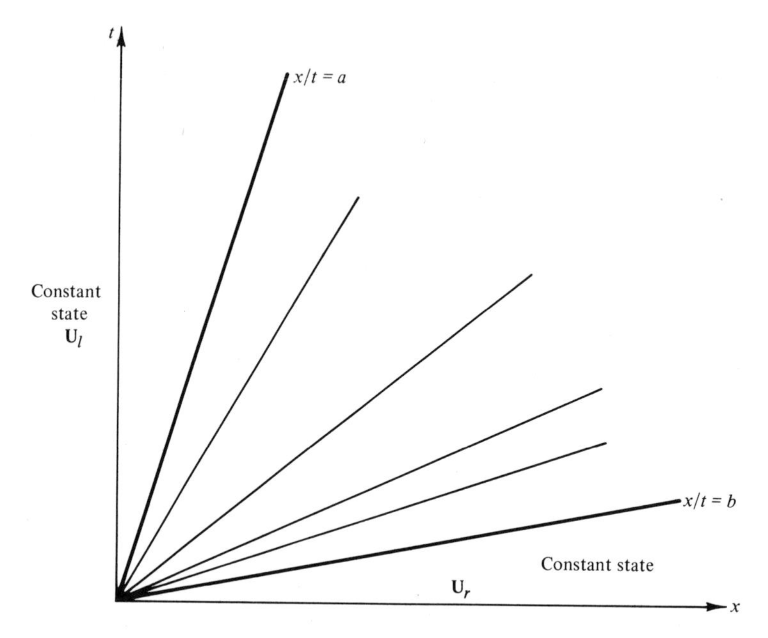

Fig. B.3.1 Centered simple wave.

The simple wave notion has been developed to serve as an effective mathematical tool in treating the problems of compressible fluids. Detailed investigations on hyperbolic systems with two dependent and two independent variables are to be found in Courant and Friedrichs [1948] in connection with discussions on compressible flows. The analysis of Courant and Friedrichs was extended by Lax [1957] to include the hyperbolic systems with n dependent and two independent variables. An account of the theory of simple waves and its application in various branches of physics may be found in a book by Jeffrey and Taniuti [1964].

References

Parentheses following the entries enclose the sections in which text references are made.

Achenbach, J. D. [1967]. Shear waves in finite elastic strain. *Quart. Appl. Math.* **25**, 306–310. (3.11)

Achenbach, J. D., and Reddy, D. P. [1967]. Note on wave propagation in linearly viscoelastic media. *J. Appl. Math. Phys. (ZAMP)* **18**, 141–144. (3.11)

Baker, M., and Ericksen, J. L. [1954]. Inequalities restricting the form of the stress-deformation relations for isotropic elastic solids and Reiner–Rivlin fluids. *J. Wash. Acad. Sci.* **44**, 33–35. (1.18)

Bell, J. F. [1964]. Experiments on large amplitude waves in finite elastic strain. *Proc. Internat. Symp. Second-Order Effects, Haifa 1962*, pp. 173–186. Macmillan, New York. (2.13)

Bell, J. F. [1968]. "The Physics of Large Deformation of Crystalline Solids." Springer-Verlag, Berlin and New York. (2.13, 3.9)

Bell, J. F. [1973]. The experimental foundations of solid mechanics, *in* "Handbuch der Physik," Vol. vi a/1 (C. Truesdell, ed.). Springer-Verlag, Berlin and New York. (2.13)

Benson, R. W., and Raelson, V. J. [1959]. Accoustoelasticity. *Prod. Eng.* **30**, 56–59. (4.3)

Bishop, R. E. D., Gladwell, G. M. L., and Michaelson, S. [1965]. "The Matrix Analysis of Vibration." Cambridge Univ. Press, New York and London. (3.5)

Bland, D. R. [1964a]. On shock waves in hyperelastic media. *Proc. Internat. Symp. Second-Order Effects, Haifa 1962*, pp. 93–108. Macmillan, New York. (2.9, 2.11, 3.7)

Bland, D. R. [1964b]. Dilatational waves and shocks in large displacement isentropic dynamic elasticity. *J. Mech. Phys. Solids* **12**, 245–267. (2.9, 3.5, 3.7, 3.11)

Bland, D. R. [1965a]. On shock structure in a solid. *J. Inst. Math. Appl.* **1**, 56–75. (2.9)

Bland, D. R. [1965b]. Plane isentropic large displacement simple waves in a compressible elastic solid. *J. Appl. Math. Phys. (ZAMP)* **16**, 752–769. (3.5)

Bland, D. R. [1967]. Lagrange Equations for Finite Elasticity, in "Recent Progress in Applied Mechanics," Folke Odqvist Vol., pp. 91–124. Wiley, New York. (3.11)

Bland, D. R. [1969]. "Nonlinear Dynamic Elasticity." Ginn (Blaisdell), Boston, Massachusetts. (3.11)

Carroll, M. M. [1966]. Some results on finite amplitude elastic waves. Report No. AM-66-9, Division of Applied Mechanics, Office of Naval Research, Washington, D.C. (July). (3.10)

Cauchy, A.-L. [1829]. Sur l'équilibre et le mouvement intérieur des corps considérés comme des masses continue. *Ex. de Math.* **4**, 293–319. *Also* "Oeuvres," Vol. 9, pp. 243–369. (4.2)

Çekirge, H. M., and Varley, E. [1973]. Large amplitude waves in bounded media. I. Reflexion and transmission of large amplitude shockless pulses at an interface. *Phil. Trans. Roy. Soc. London, Ser. A* **273**, 261–313. (3.9)

Chen, P. J. [1968a]. Growth of acceleration waves in isotropic elastic materials. *J. Acoust. Soc. Amer.* **43**, 982–987. (2.15)

Chen, P. J. [1968b]. The growth of acceleration waves of arbitrary form in homogeneously deformed elastic materials. *Arch. Rational Mech. Anal.* **30**, 81–89. (2.15)

Chu, B.-T. [1964]. Finite amplitude waves in incompressible perfectly elastic materials. *J. Mech. Phys. Solids* **12**, 45–57. (3.6, 3.11)

Chu, B.-T. [1967]. Transverse shock waves in incompressible elastic solids. *J. Mech. Phys. Solids* **15**, 1–14. (2.9–2.11)

Cole, R. H. [1948]. "Underwater Explosions." Princeton Univ. Press, Princeton, New Jersey. (3.9)

Coleman, B. D., and Noll, W. [1959]. On the thermostatics of continuous media. *Arch. Rational Mech. Anal.* **4**, 97–128. (1.18)

Coleman, B. D., and Noll, W. [1964]. Material symmetry and thermostatic inequalities in finite elastic deformations. *Arch. Rational Mech. Anal.* **15**, 87–111. (1.18)

Coleman, B. D., Gurtin, M. E., and Herrera, I. [1965]. Waves in materials with memory. I. The velocity of one-dimensional shock and acceleration waves. *Arch. Rational Mech. Anal.* **19**, 1–19. (2.9)

Collins, W. D. [1966]. One-dimensional nonlinear wave propagation in incompressible elastic materials. *Quart. J. Mech. Appl. Math.* **19**, 259–328. (3.11)

Collins, W. D. [1967]. The propagation and interaction of one-dimensional non-linear waves in an incompressible isotropic half-space. *Quart. J. Mech. Appl. Math.* **20**, 429–452. (3.11)

Courant, R., and Friedrichs, K. O. [1948]. "Supersonic Flow and Shock Waves." Wiley (Interscience), New York. (3.5, B.3)

Courant, R., and Hilbert, D. [1962]. "Methods of Mathematical Physics," Vol. II. Wiley (Interscience), New York. (3.8)

Cristescu, N. [1967]. "Dynamic Plasticity." Wiley, New York. (3.9)

Currie, P., and Hayes, M. [1969]. Longitudinal and transverse waves in finite elastic strain. Hadamard and Green materials. *J. Inst. Math. Appl.* **5**, 140–161. (3.11)

Davison, L. [1966]. Propagation of plane waves of finite amplitude in elastic solids. *J. Mech. Phys. Solids* **14**, 249–270. (3.7)

Davison, L. [1968]. Perturbation theory of nonlinear elastic wave propagation. *Int. J. Solids Structures* **4**, 301–322. (3.11)

Demiray, H., and Şuhubi, E. S. [1970]. Small torsional oscillations of an initially twisted circular rubber cylinder. *Internat. J. Engrg. Sci.* **8**, 19–30. (4.6)

Dewey, J. [1959]. Strong shocks and stress–strain relations in solids. Aberdeen Proving Ground Ballistic Res. Lab. Rep. No. 1074. (2.9)

Eisenhart, L. P. [1926]. "Differential Geometry." Princeton Univ. Press, Princeton, New Jersey. (A.1)

Ericksen, J. L. [1953]. On the propagation of waves in isotropic incompressible perfectly elastic materials. *J. Rational Mech. Anal.* **2**, 329–337. (2.14)

Ericksen, J. L. [1954]. Deformations possible in every isotropic incompressible perfectly elastic body. *J. Appl. Math. Phys.* (*ZAMP*) **5**, 466–486. (3.2)

Ericksen, J. L., and Toupin, R. A. [1956]. Implications of Hadamard's condition for elastic stability with respect to uniqueness theorems. *Canad. J. Math.* **8**, 432–436. (4.1)

Eringen, A. C. [1954]. The solution of a class of mixed–mixed boundary value problems in plane elasticity. *Proc. 2nd. U. S. Nat. Congr. Appl. Mech.*, 257. (1.17)

Eringen, A. C. [1962]. "Nonlinear Theory of Continuous Media." McGraw-Hill, New York. (1.9, 1.11, 1.14, 1.16, 1.17, 2.13, 3.4, 4.6, A.4)

Eringen, A. C. [1965]. A unified theory of thermomechanical materials. Technical Report, Office of Naval Research, Washington, D.C., *and Internat. J. Engrg. Sci.* **4**, 179 [1966]. (1.13)

Eringen, A. C. [1967]. "Mechanics of Continua." Wiley, New York. (1.5, 1.6, 1.9, 1.11, 1.13, 1.15)

Eringen, A. C. [1971]. Tensor analysis, *in* "Continuum Physics," Vol. I (A.C. Eringen, ed.). Academic Press, New York and London. (2.2, A.1, A.4–A.7, B.1)

Eringen, A. C. [1972a]. Nonlocal polar elastic continua. *Internat. J. Engrg. Sci.* **10**, 1–16. (1.11)

Eringen, A.C. [1972b]. On nonlocal fluid mechanics. *Internat. J. Engrg. Sci.* **10**, 561–575. (1.11)

Eringen, A. C., and Edelen, D. G. B. [1972]. On nonlocal elasticity. *Internat. J. Engrg. Sci.* **10**, 233–248. (1.11)

Eringen, A. C., and Şuhubi, E. S. [1964]. Nonlinear theory of microelastic solids—I & II. *Internat. J. Engrg. Sci.* **2**, 189–204, 389–404. (1.11)

Fine, A. D., and Shild, R. T. [1966]. Second-order effects in the propagation of elastic waves. *Int. J. Solids Structures* **2**, 605–620. (3.11)

Flavin, J. N. [1963]. Surface waves in prestressed Mooney material. *Quart. J. Mech. Appl. Math.* **16**, 441–449. (4.4)

Flory, P. J. [1953]. "Principles of Polymer Chemistry." Cornell Univ. Press, Ithaca, New York. (2.10)

Friedman, B. [1956]. "Principles and Techniques of Applied Mathematics." Wiley, New York. (B.1)

Gradshteyn, I. S., and Ryzhik, I. M. [1965]. "Table of Integrals, Series, and Products," 4th Ed. Academic Press, New York and London. (3.3)

Green, A. E. [1961]. Torsional vibrations of an initially stressed circular cylinder, *in* "Problems of Continuum Mechanics" (Muskhelishvili anniv. vol.), pp. 148–154. Society for Industrial and Applied Mathematics, Philadelphia, Pennsylvania. (4.5)

Green, A. E. [1963]. A note on wave propagation in initially deformed bodies. *J. Mech. Phys. Solids* **11**, 119–126. (2.13, 3.10)

Green, A. E., and Spencer, A. J. M. [1958]. The stability of a circular cylinder under finite extension and torsion. *J. Math. Phys.* **37**, 316–338. (4.6)

Green, A. E., Rivlin, R. S., and Shield, R. T. [1952]. General theory of small elastic deformations superposed on finite elastic deformations. *Proc. Roy. Soc. London Ser. A* **211**, 128–154. (4.1, 4.2)

Green, W. A. [1964]. Growth of plane discontinuities propagating into a homogeneously deformed elastic material. *Arch. Rational Mech. Anal.* **16**, 79–88. Corrections and additional results, *Arch. Rational Mech. Anal.* **19**, 20–23 (1965). (2.15)

Guo, Z.-H. [1962a]. Displacement equations of isentropic motion of a body subject to a finite isothermal initial strain. *Bull. Acad. Polon. Sci. Sér. Sci. Tech.* **10**, 479–483. (4.1)

Guo, Z.-H. [1962b]. The equations of motion of a circular plate subject to initial strain. *Bull. Acad. Polon. Sci. Sér. Sci. Tech.* **10**, 63–70. (4.7)

Guo, Z.-H. [1962c]. The problem of stability and vibration of a circular plate subject to finite initial deformation. *Arch. Mech. Stos.* **14**, 239–252. (4.7)

Guo, Z.-H. [1962d]. The equations of motion of a plate subject to initial homogeneous finite deformation. *Bull. Acad. Polon. Sci. Sér. Sci. Tech.* **10**, 107–113. (4.7)

Guo, Z.-H. [1962e]. Equations of small motion of a cylinder subject to a large deformation. Its natural vibration and stability. *Bull. Acad. Polon. Sci. Sér. Sci. Tech.* **10**, 177–182. (4.7)

Guo, Z.-H. [1962f]. Vibration and stability of a cylinder subject to finite deformation. *Arch. Mech. Stos.* **14**, 757–768. (4.7)

Guo, Z.-H. and Solecki, R. [1963a]. Free and forced finite amplitude oscillations of an elastic thick-walled hollow sphere made of incompressible material. *Arch. Mech. Stos.* **15**, 427–433. (3.4)

Guo, Z.-H., and Solecki, R. [1963b]. Free and forced finite amplitude oscillations of a thick-walled sphere of incompressible material. *Bull. Acad. Polon. Sci. Sér. Sci. Tech.* **11**, 47–52. (3.4)

Guo, Z.-H., and Urbanowski, W. [1963]. Certain stationary conditions in variated states of finite strain. *Bull. Acad. Polon. Sci. Sér. Sci. Tech.* **11**, 27–32. (4.1)

Hadamard, J. [1903]. "Leçons sur la Propagation des Ondes et les Equations de l'Hydrodynamique." Hermann, Paris. (2.3)

Hampton, D., and Huck, P. J. [1968]. Study of wave propagation in confined soils. I.I.T. Research Institute Contract Report S-69-2, Chicago, Illinois. (3.9)

Hayes, M. [1963]. Wave propagation and uniqueness in prestressed elastic solids. *Proc. Roy. Soc. London Ser. A* **274**, 500–506. (4.2)

Hayes, M. [1964]. Uniqueness for the mixed boundary value problem in the theory of small deformations superimposed upon large. *Arch. Rational Mech. Anal.* **16**, 238–242. (4.2)

Hayes, M., and Rivlin, R. S. [1961a]. Propagation of a plane wave in an isotropic elastic material subjected to pure homogeneous deformation. *Arch. Rational Mech. Anal.* **8**, 15–22. (4.3)

Hayes, W., and Rivlin, R. S. [1961b]. Surface waves in deformed elastic materials. *Arch. Rational Mech. Anal.* **8**, 358–380. (4.4)

Hayes, W. [1957]. The vorticity jump across a gasdynamic discontinuity. *J. Fluid Mech.* **2**, 595-600. (2.2)

Hill, R. [1957]. On uniqueness and stability in the theory of finite elastic strain. *J. Mech. Phys. Solids* **5**, 229-241. (4.1)

Hill, R. [1961]. Discontinuity relations in mechanics, *in* "Progress in Solid Mechanics," (I. N. Sneddon and R. Hill, eds.), pp. 246–276. North-Holland Publ., Amsterdam. (2.6)

Hill, R. [1962]. Acceleration waves in solids. *J. Mech. Phys. Solids* **10**, 1–16. (2.12)

Howard, I. C. [1966]. Finite simple waves in a compressible transversely isotropic elastic solid. *Quart. J. Mech. Appl. Math.* **19**, 329–341. (3.11)

Hughes, D. S., and Kelly, J. R. [1953]. Second-order elastic deformation of solids. *Phys. Rev.* (2) **92**, 1145–1149. (4.3)

Huilgol, R. R. [1967]. Finite amplitude oscillations in curvilinearly aelotropic elastic cylinders. *Quart. Appl. Math.* **25**, 293–298. (3.3)

Ishihara, A., Hashitsume, N., and Tatibana, M. [1951]. Statistical theory of rubber-like elasticity, IV. Two dimensional stretching. *J. Chem. Phys.* **19**, 1508–1512. (3.3, 3.11)

Jeffrey, A., and Taniuti, T. [1964]. "Non-Linear Wave Propagation." Academic Press, New York and London. (B.3)

John, F. [1960]. Plane strain problems for a perfectly elastic material of harmonic type. *Comm. Pure Appl. Math.* **13**, 239–296. (3.11)

John, F. [1966]. Plane elastic waves of finite amplitude. Hadamard materials and harmonic materials. *Comm. Pure Appl. Math.* **19**, 309–341. (3.11)
Jouguet, E. [1920a]. Sur les ondes de choc dans les corps solids. *C. R. Acad. Sci. Paris* **171**, 461–464. (2.9)
Jouguet, E. [1920b]. Sur la célérité des ondes dans les corps solides. *C. R. Acad. Sci. Paris* **171**, 512–515. (2.9)
Jouguet, E. [1920c]. Sur la variation d'entropie dans les ondes des choc des solides élastiques. *C. R. Acad. Sci. Paris* **171**, 789–791. (2.9)
Jouguet, E. [1920d]. Applications du principe de Carnot–Clausius aux ondes de choc des solides élastiques. *C. R. Acad. Sci. Paris* **171**, 904–907. (2.9)
Jouguet, E. [1921]. Notes sur la théorie de l'élasticité. *Ann. Toulouse* (3) **12** (1920), 47–92. (2.9)
Kafadar, C. [1972]. On Ericksen's problem. *Arch. Rational Mech. Anal.* **47**, 15–27. (3.2)
Knowles, J. K. [1960]. Large amplitude oscillations of a tube of incompressible elastic material. *Quart. Appl. Math.* **18**, 71–77. (3.3)
Knowles, J. K. [1962]. On a class of oscillations in the finite deformation theory of elasticity. *J. Appl. Mech.* **29**, 283–286. (3.3)
Knowles, J. K., and Jakub, M. T. [1965]. Finite dynamics deformations of an incompressible elastic medium containing a spherical cavity. *Arch. Rat. Mech. Anal.* **18**, 367–378. (3.4)
Ko, W. L. [1963]. "Application of Finite Elasticity Theory to the Behavior of Rubber-Like Materials." PhD Thesis, California Institute of Technology. (2.15, 3.7)
Lax, P. D. [1957]. Hyberbolic systems of conservation laws II. *Comm. Pure Appl. Math.* **10**, 537–566. (3.7, B.1, B.3)
Lichtenstein, L. [1929]. "Grundlagen der Hydromechanik." Springer-Verlag, Berlin and New York. (2.3)
Love, A. E. H. [1944]. "Mathematical Theory of Elasticity." Dover, New York. (1.15)
McConnell, A. J. [1957]. "Applications of Tensor Analysis" (first published as "Applications of the Absolute Differential Calculus" in 1931). Dover, New York. (2.2, A.1)
Michal, A. D. [1947]. "Matrix and Tensor Calculus." Wiley, New York. (A.1)
Müller, I. [1967]. On the entropy inequality. *Arch. Rational Mech. Anal.* **26**, 118–141. (1.13)
Müller, I. [1971]. The coldness, a universal function in thermoelastic bodsie. *Arch. Rational Mech. Anal.* **41**, 319–332. (1.13)
Nowinski, J. L. [1966]. On a dynamic problem in finite elastic shear. *Internat. J. Engrg. Sci.* **4**, 501–510. (3.11)
Nowinski, J. L., and Schultz, A. B. [1964]. Note on a class of finite longitudinal oscillations of thick-walled cylinders. *Proc. Ind. Nat. Congr. Th. Appl. Mech.* [1964], 31–44. (3.11)
Nowinski, J. L., and Wang, S. D. [1966a]. Galerkin's solution to a severely nonlinear problem of finite elastodynamics. *Internat. J. Non-Linear Mech.* **1**, 239–246. (3.3)
Nowinski, J. L., and Wang, S. D. [1966b]. Finite radial oscillations of a spinning thick-walled cylinder. *J. Acoust. Soc. Amer.* **40**, 1548–1553. (3.3)
Pearson, C. E. [1956]. General theory of elastic stability. *Quart. Appl. Math.* **14**, 133–144. (4.1)
Prager, W. [1946]. The general variational principle of the theory of structural stability. *Quart. Appl. Math.* **4**, 378–384. (4.1)
Ramakanth, J. [1965a]. Some problems of propagation of waves in prestressed isotropic bodies. *Proc. Vibration Problems* **6**, 161–172. (4.7)
Ramakanth, J. [1965b]. Radial vibrations of a prestressed sphere. *Bull. Acad. Polon. Sci. Sér. Sci. Tech.* **13**, 401–408. (4.7)
Reddy, D. P., and Achenbach, J. D. [1968]. Simple waves and shock waves in a thin prestressed elastic rod. *J. Appl. Math. Phys.* (*ZAMP*) **19**, 473–485. (3.11)

Sawyers, K. N., and Rivlin, R. S. [1969]. Seismic wave propagation in a self-gravitating anisotropic earth. *Phil. Trans. Roy. Soc. London. Ser. A* **263**, 615–655. (3.11)

Schouten, J. A. [1951]. "Tensor Analysis for Physicists." Oxford Univ. Press, London and New York. (A.1, A.6)

Sedov, L. I. [1970]. "Continuum Mechanics," Vol. I (Russian). (3.5)

Sokolnikoff, I. S. [1951]. "Tensor Analysis: Theory and Applications." Wiley, New York. (A.1)

Sokolnikoff, I. S. [1956]. "Mathematical Theory of Elasticity." McGraw-Hill, New York. (1.15)

Sternberg, S. [1965]. "Lectures on Differential Geometry." Prentice-Hall, Englewood Cliffs, New Jersey. (A.1)

Stoker, J. J. [1950]. "Nonlinear Vibrations in Mechanical and Electrical Systems." Wiley (Interscience), New York. (3.3)

Şuhubi, E. S. [1965]. Small longitudinal vibration of an initially stretched circular cylinder. *Internat. J. Engrg. Sci.* **2**, 509–517. (4.5)

Şuhubi, E. S. [1970]. The growth of acceleration waves of arbitrary form in deformed hyperelastic materials. *Internat. J. Engrg. Sci.* **8**, 699–710. (2.15)

Şuhubi, E. S. [1972]. On the propagation of shock waves in hyperelastic solids. *Bull. Tech. Univ. Istanbul* **25**, 116–125. (2.11)

Synge, J. L., and Schild, A. [1949]. "Tensor Calculus." Univ. of Toronto Press, Toronto. (A.1)

Thomas, T. Y. [1957]. Extended compatibility conditions for the study of surfaces of discontinuity in continuum mechanics. *J. Math. Phys.* **4**, 335–350. (2.2)

Thomas, T. Y. [1961a]. "Concepts from Tensor Analysis and Differential Geometry." Academic Press, New York and London. (2.2, 2.15, A.1)

Thomas, T. Y. [1961b]. "Plastic Flow and Fracture in Solids." Academic Press, New York and London. (2.2, 2.15)

Thomas, T. Y. [1966]. The general theory of compatibility conditions. *Internat. J. Engrg. Sci.* **4**, 207–233. (2.5)

Thurston, R. N. [1965]. Effective elastic coefficients for wave propagation in crystals under stress. *J. Acoust. Soc. Amer.* **37**, 348–356. (4.3)

Thurston, R. N., and Brugger, K. [1964]. Third-order elastic constants and the velocity of small amplitude elastic waves in homogeneously stressed media. *Phys. Rev.* (2) **133** *A*, 1604–1610. (4.3)

Tokuoka, T. [1969]. Ultrasonic wave propagations in nonhomogeneously and dynamically deformed isotropic elastic materials. *Mem. Fac. Engrg. Kyoto Univ.* **31**, 86–94. (4.3)

Tokuoka, T., and Iwashimizu, Y. [1948]. Acoustical birefringence of ultrasonic waves in deformed isotropic elastic materials. *Int. J. Solids Structures* **4**, 383–389. (4.3)

Tokuoka, T., and Saito, M. [1969a]. Elastic wave propagations and acoustical birefringence in stressed crystals. *J. Acoust. Soc. Amer.* **45**, 1241–1246. (4.3)

Tokuoka, T., and Saito, M. [1969b]. Relations between stress and ultrasonic velocity in deformed crystals. *J. Soc. Mater. Sci. Japan* **18**, 727–730 (in Japanese with an abstract in English). (4.3)

Toupin, R. A., and Bernstein, B. [1961]. Sound waves in deformed perfectly elastic materials, acoustoelastic effect. *J. Acoust. Soc. Amer.* **33**, 216–225. (4.1, 4.3)

Truesdell, C. [1961]. General and exact theory of waves in finite elastic strain. *Arch. Rational Mech. Anal.* **8**, 263–296. (2.12, 2.13)

Truesdell, C. [1962]. Solutio generalis et accurata problematum quamplurimorum de motu corporum elasticonum incomprimibilium in deformationibus valde magnis. *Arch. Rational Mech. Anal.* **11**, 106–113; addenda: **12**, 427–428 [1963]; **28**, 387–398 [1968]. (3.2)

Truesdell, C. [1964]. Second-order theory of wave propagation in isotropic elastic materials. *Proc. Internat. Symp. Second-Order Effects, Haifa 1962*, pp. 187–199. Macmillan, New York. (2.13)

Truesdell, C. [1966]. Existence of longitudinal waves. *J. Acoust. Soc. Amer.* **40**, 729–730. (2.12)

Truesdell, C., and Noll, W. [1965]. "The Nonlinear Field Theories of Mechanics," Handbuch der Physik, Vol. III/3. Springer-Verlag, Berlin and New York. (1.18, 2.9, 2.13, 3.2, 4.1, 4.2)

Truesdell, C., and Toupin, R. [1960]. "The Classical Field Theories," Handbuch der Physik, Vol. III/1. Springer-Verlag, Berlin and New York. (2.2, 2.6, 2.8)

Urbanowski, W. [1959]. Small deformations superposed on finite deformations of a curvilinearly orthotropic body. *Arch. Mech. Stos.* **11**, 223–241. (4.1)

Urbanowski, W. [1961]. Deformed body structure. *Arch. Mech. Stos.* **13**, 277–294. (4.1)

Valanis, K. C., and Sun, C. T. [1968]. Propagation of a large amplitude longitudinal wave in a semi-infinite elastic rod. *Internat. J. Engrg. Sci.* **6**, 735–747. (3.11)

Varley, E. [1965a]. Acceleration waves in viscoelastic materials. *Arch. Rational Mech. Anal.* **19**, 215–225. (2.15)

Varley, E. [1965b]. Simple waves in general elastic materials. *Arch. Rational Mech. Anal.* **20**, 309–328. (3.8)

Varley, E., and Cumberbatch, E. [1965]. Nonlinear theory of wave-front propagation. *J. Inst. Math. Appl.* **1**, 101–112. (3.8)

Verma, P. D. S. [1964]. On waves in finite elastic strain. *Indian J. Mech. Math.* **2**, 17–18. (2.9)

Wang, A. S. D. [1969]. On free oscillations of elastic incompressible bodies in finite shear. *Internat. J. Engrg. Sci.* **7**, 1199–1212. (3.11)

Wang, C. C. [1965]. On the radial oscillations of a spherical thin shell in the finite elasticity theory, *Quart. Appl. Math.* **23**, 270–274. (3.4)

Waterstone, R. J. [1968]. One-dimensional evolutionary discontinuities in compressible elastic materials. *J. Inst. Math. Appl.* **4**, 58–77. (3.11)

Wesołowski, Z. [1964]. Problems of radial and axial oscillations of an elastic cylinder of infinitesimal length. *Proc. Vibration Problems* **5**, 19–29. (3.3)

Widder, D. V. [1947]. "Advanced Calculus." Prentice-Hall, Englewood Cliffs, New Jersey. (1.3)

Wright, T. W. [1973]. Acceleration waves in simple elastic materials. *Arch. Rational Mech. Anal.* **50**, 237–330. (2.15)

Yano, K. [1957]. "The Theory of Lie Derivatives and Its Applications." North-Holland, Amsterdam. (A.1)

Zorski, H. [1964]. On the equations describing small deformation superposed on finite deformation. *Proc. Internat. Symp. Second-Order Effects, Haifa 1962*, pp. 109–128. Macmillan, New York. (4.1, 4.5)

Index

A

B

C